Bayesian Networks in Fault Diagnosis

Practice and Application

Bayesian Networks in Fault Diagnosis

Practice and Application

Editors

Baoping Cai
China University of Petroleum (East China), China

Yonghong Liu
China University of Petroleum (East China), China

Jinqiu Hu
China University of Petroleum (Beijing), China

Zengkai Liu
China University of Petroleum (East China), China

Shengnan Wu
China University of Petroleum (Beijing), China

Renjie Ji
China University of Petroleum (East China), China

World Scientific

NEW JERSEY • LONDON • SINGAPORE • BEIJING • SHANGHAI • HONG KONG • TAIPEI • CHENNAI • TOKYO

Published by

World Scientific Publishing Co. Pte. Ltd.
5 Toh Tuck Link, Singapore 596224
USA office: 27 Warren Street, Suite 401-402, Hackensack, NJ 07601
UK office: 57 Shelton Street, Covent Garden, London WC2H 9HE

British Library Cataloguing-in-Publication Data
A catalogue record for this book is available from the British Library.

BAYESIAN NETWORKS IN FAULT DIAGNOSIS
Practice and Application

ISBN 978-981-3271-48-7

For any available supplementary material, please visit
https://www.worldscientific.com/worldscibooks/10.1142/11021#t=suppl

Typeset by Stallion Press
Email: enquiries@stallionpress.com

Printed in Singapore

Preface

With the rapid development of modern industrial systems, systematic complexity increases constantly. Therefore, fault diagnosis must be utilized to obtain high reliability and availability. Fault diagnosis quickly detects process abnormality and component fault and identifies root causes of these failures by using appropriate models, algorithms, and system observations. Therefore, fault diagnosis system is useful in assisting operations staff to detect, isolate, and identify faults and to aid in troubleshooting.

Data-driven fault diagnosis approach is one of the most popular diagnosis approaches. Different from model-based and signal-based methods, data-driven fault diagnosis does not require a known model or signal patterns, but requires a large number of historical data. The data can be obtained based on either statistical or non-statistical approach. Some typical data-driven fault diagnosis approaches, such as principal component analysis, partial least squares, independent component analysis, support vector machine, neural network, and even fuzzy logic have been researched for years.

Bayesian network (BN) is a probabilistic graphical model that represents a set of random variables and their conditional dependencies via a directed acyclic graph, and is considered to be one of the most useful models in the field of probabilistic knowledge representation and reasoning since it was introduced in early 1980s. Recently, BNs are increasingly used in the field of fault detection and diagnosis because it can solve the uncertainty problem. Uncertainty is a major problem in the fault diagnosis of complex systems. It can be caused by various reasons, such as limitations of design sensors and observers, inherent uncertainty of observed complex systems, unpredictable factors during observation and diagnosis, and incompleteness and inaccuracy of methods and models. In addition, complex and uncertain relationships exist between faults and fault symptoms because a fault may cause multiple symptoms, and multiple faults may cause a single symptom.

The book presents a general review on BN-based fault diagnosis methods firstly, and focuses more on practice and application. Some fault diagnosis cases using BNs are presented. Researchers, professionals, academics and graduate students will better understand the theory and application, and benefit those who are keen to develop real BN-based fault diagnosis system.

We wish to acknowledge the support of the National Natural Science Foundation of China (No. 51779267 and No. 51574263), and the Fundamental Research Funds for the Central Universities (No. 17CX05022 and No. 14CX02197A).

Editors

Baoping Cai
Yonghong Liu
Jinqiu Hu
Zengkai Liu
Shengnan Wu
Renjie Ji
Qingdao, China
20 July, 2018

Contents

Chapter 1

Fault Diagnosis

Fault diagnosis is useful in helping technicians detect, isolate, and identify faults, and troubleshoot. Bayesian network (BN) is probabilistic graphical model that effectively deals with various uncertainty problems. This model is increasingly utilized in fault diagnosis. This chapter presents bibliographical review on the use of BNs in fault diagnosis in the last decades with focus on engineering systems. This work also presents the general procedure of fault diagnosis modeling with BNs; processes include BN structure modeling, BN parameter modeling, BN inference, fault identification, validation, and verification. The chapter provides series of classification schemes for BNs for fault diagnosis, BNs combined with other techniques, and domain of fault diagnosis with BN. This study finally explores current gaps and challenges and several directions for future research.

1.1. Introduction

With the rapid development of modern industrial systems, systematic complexity increases constantly. Therefore, fault diagnosis must be utilized to obtain high reliability and availability. Fault diagnosis quickly detects process abnormality and component fault and identifies root causes of these failures by using appropriate models, algorithms, and system observations. Therefore, fault diagnosis system is useful in assisting operations staff to detect, isolate, and identify faults and to aid in troubleshooting.

In general, fault diagnosis approaches can be classified into three categories: model-based [1, 2], signal-based [3–5], and data-driven approaches [6–8]. In model-based approach, the focus is on establishing mathematical models of complex industrial systems.

These models can be constructed by various identification methods, physical principles, etc. Signal-based approach uses detected signals to diagnose possible abnormalities and faults by comparing detected signals with prior information of normal industrial systems [9]. Usually, difficulty occurs in building accurate mathematical models and obtaining accurate signal patterns for complex industrial and process systems. Data-driven fault diagnosis approach requires a large amount of historical data, rather than models or signal patterns [10]. Therefore, data-driven methods are suitable for complex industrial systems.

The data-driven fault diagnosis approach is also called knowledge-based fault diagnosis approach. Knowledge information can be obtained based on either statistical or nonstatistical approach. The data-driven approach can be categorized into statistics-based and nonstatistics-based fault diagnosis approaches [10]. The former includes principal component analysis [11, 12], partial least squares [13, 14], independent component analysis [15, 16], and support vector machine [17, 18], whereas the latter includes neural network [19, 20], fuzzy logic [21, 22], etc.

BN is an important probabilistic graphical model, which can deal effectively with various uncertainty problems based on probabilistic information representation and inference. As a representational tool, BN is quite attractive for three reasons. First, BN is consistent and completely represents and defines unique probability distribution over network variables. Second, the network is modular; its consistency and completeness are ensured using localized tests, which are only applicable to variables and their direct causes. Third, BN is a compact representation, as it allows specification of exponentially sized probability distribution using polynomial of probabilities [38]. Hence, this tool is deeply researched and widely used in many domains ranging from reliability engineering [24–28, 82], risk analysis [29–31], and safety engineering [32, 33] to resilience engineering [34–36].

During the last decades, BNs are studied and utilized in the domain of fault diagnosis, which is typically a data-driven approach. BN-based fault diagnosis models are established using huge amount of historical data. Diagnosis is conducted by backward analysis with various algorithms [37]. That is, we input observed information into evidence nodes, update posterior probabilities of fault nodes based on BN inference, and identify root causes of failure using identification rules. This research attracted considerable attention, solved some important problems of BN-based fault diagnosis, and focused on challenging problems. A large number of literatures, including theory research and project application, were conference proceedings and technical reports. To our knowledge, literature never reviewed use of BNs in fault diagnosis.

This chapter aims to review latest research results of BN-based fault diagnosis approaches to provide reference and to identify future research directions for fault diagnosis researchers and engineers. The remainder of the chapter is organized as follows. Section 1.2 provides an overview of BNs. Section 1.3 presents general procedures of fault diagnosis with BNs. Section 1.4 introduces types of BNs for fault diagnosis; these types include BNs, dynamic Bayesian networks (DBNs) and object-oriented Bayesian networks (OOBNs). Section 1.5 reviews domains of fault diagnosis with BNs. Section 1.6 identifies a few on-going and upcoming research directions. Section 1.7 summarizes this chapter.

1.2. Overview of BNs

BN is a probabilistic graphical model representing a set of random variables and their conditional dependencies via a directed acyclic graph. Such a network consists of qualitative and quantitative parts. The qualitative part is a directed acyclic graph, in which nodes represent system variables, whereas arcs symbolize dependencies or cause-and-effect relationships among variables. The quantitative part consists of the conditional probabilistic table, which represents the relationship between each node and its parents. In BNs, leaf nodes only have parent nodes but no child nodes, and root nodes have child nodes but no parent nodes [23].

Let us consider n random variables $X_1, X_2, \ldots, X_n$, a directed acyclic graph with n numbered nodes, and suppose node

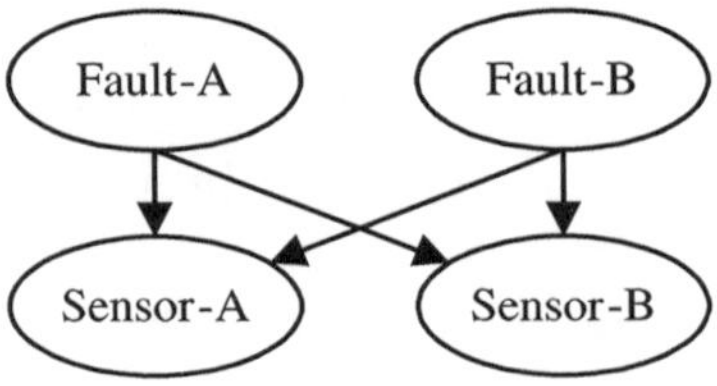

Fig. 1.1. A simple BN-based fault diagnosis model.

j $(1 \leq j \leq n)$ of the graph is associated to Xj variable. The graph is a BN, representing variables $X_1, X_2, \ldots, X_n$ in the following equation:

$$P(X_1, X_2, \ldots, Xn) = \prod_{j=1}^{n} P(X_j|\text{parent}(X_j)), \tag{1.1}$$

where parent (X_j) denotes set of all variables X_i; such an arc connects node i to node j in the graph [131].

Then, let us use an example. Suppose that two faults, Fault-A, Fault-B or both, could affect the sensor's reading, Sensor-A, Sensor-B or both. Suppose that each fault has two states, which are present and absent, and each fault symptom obtained from sensor data has three states, which are higher, lower and normal. The situation can be modeled with BN, as shown in Fig. 1.1.

This model can be used to diagnose the faults, Fault-A and Fault-B, by using the sensor's data, Sensor-A and Sensor-B.

1.3. Procedures of Fault Diagnosis with BNs

Fault diagnosis procedures with BNs consist of BN structure modeling, BN parameter modeling, BN inference, fault identification, and validation and verification. Figure 1.2 provides a detailed flowchart of the process. BN-based fault diagnosis model can be established following the proposed five steps: (1) structure model of BNs for fault diagnosis can be established by using cause-and-effect relationship, mapping algorithms, or structuring learning method; (2) parameter model of BNs can be established by using expert elicitation with noisy models and parameter learning method; (3) exact or approximated inference algorithms can be conducted for BNs inference; (4) system faults can be identified by directly using posterior

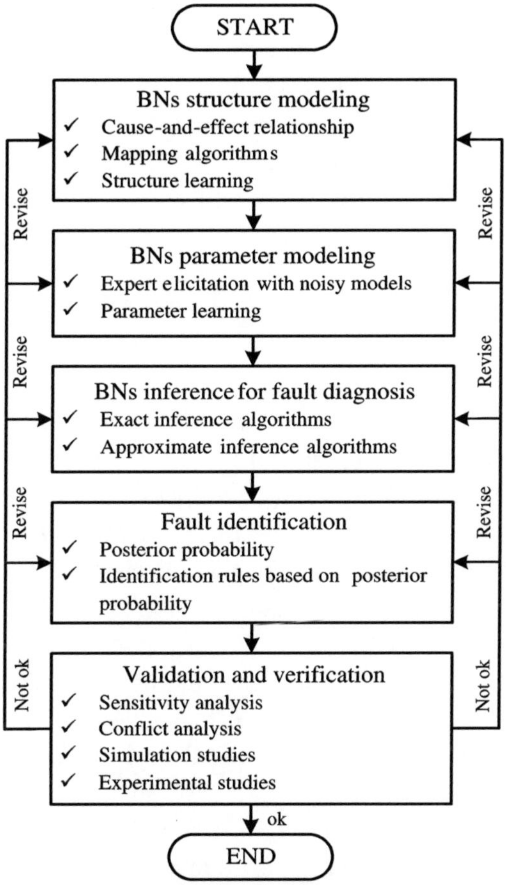

Fig. 1.2. Flowchart of BN-based fault diagnosis.

probability or various fault identification rules; (5) model validation and verification may be conducted through various methods, such as sensitivity analysis, conflict analysis, simulation studies, and experimental studies. When diagnostic results are unsatisfactory, one or several of the first four steps are revised or replaced in the specific method used until satisfactory diagnostic results are obtained.

1.3.1. *BN structure modeling*

Several methods were reported in constructing BN structure models for fault diagnosis. Three main methods include cause-and-effect relationship, mapping algorithms, or structuring learning. Per cause-and-effect relationship method, one reflects on their knowledge and experience about faults and fault symptoms and then captures them into a BN. That is, network structure is established based on causal relationship between faults and fault symptoms.

BN structure is usually composed of two layers of events: fault layer and fault symptom layer. For example, Dey and Stori [39] presented a BN-based process monitor and fault diagnosis method for combining multi-sensor data during machining operation to diagnose root causes of process variations. Root layer and evidence layer consist of structures of proposed BNs for root cause diagnosis. Root layer includes four root nodes, and evidence layer includes 18 evidence nodes. Alaeddini and Dogan [40] developed BNs to model causal relationship among chart patterns, process information, and root cause or assignable cause for root cause diagnosis. Two layers compose the structure. The first represents assignable causes, such as man, machine material, and method; the second layer represents process-specific information, such as performance and operator change and chart patterns, including run and trend.

Aside from two basic layers, other layers representing various information are added and established in BN structure to improve fault diagnostic performances. Zhao *et al.* [41] proposed a three-layer BN for fault detection and diagnosis of chiller. Aside from second and third layer, that is fault layer and fault symptom layer, respectively, researchers added the first layer, that is, additional information layer, to identify possible fault causes. Cai *et al.* [42] developed a BN-based ground-source heat pump fault diagnosis system by fusing multi-sensor data. A new layer, which is observed information layer, is established to connect to the fault layer directly for improving performance and increasing fault diagnosis accuracy. Liu *et al.* [43] constructed an operation procedure layer in BN-based fault diagnosis system of subsea blowout preventers. This fault layer is linked to operation procedure and fault symptom layers. Morgan *et al.* [58] developed a novel BN-based approach for heavy-duty diesel engine fault diagnosis. Aside from typical fault and symptom layers, an input layer is established to represent elemental inputs, and an action layer is established to represent possible actions by operators.

Per mapping algorithms, one automatically synthesizes BN from other types of formal knowledge or models, such as fault tree and bow tie models. Lampis and Andrews [44] proposed a system fault diagnosis method by converting fault trees into BNs. These people built a noncoherent fault tree model for fault diagnosis of systems and mapped it to BNs following a two-step rule. Similarly, Chiremsel *et al.* [58] presented fault detection and diagnosis method for safety-related systems by directly mapping fault tree analysis models to BNs. Lo *et al.* [45] developed a new method for constructing BN structure for fault diagnosis by directly transforming graph models.

According to structure learning, BN structure can be constructed based on learning from data related to faults and fault symptoms when sufficient data are available. Given that learning is an inductive process, the principle of induction guides construction processes, such as maximum likelihood approach and Bayesian approach. For example, Jin *et al.* [46] presented BN-based fault detection and diagnosis method of assembly fixture. These scientists proposed structure learning method by using mutual information test to obtain causal relationships among fixture and sensor nodes. Lin *et al.* [47] developed a method that brought the quality of service management to current internet work. The researchers utilized BNs to conduct service quality diagnosis, in which K2 algorithm was used for BN structure learning. Zhou and Li [101] proposed learning BN structure algorithm for application in fault diagnosis of hydraulic–electrical simulation systems. Scientists used statistical strategies extracted from rule bases to learn BN structures. Rules with strong causality were retained, whereas those with weak causality were discarded.

The former two methods are sometimes known as knowledge representation methods, whereas the third one is known as machine learning method. Networks constructed by knowledge representation methods have a different nature from those constructed by machine learning methods. For example, the former networks had larger sizes and placed harsher computational demands on reasoning algorithms. Moreover, these networks had significant amount of determinism (i.e., probabilities equal to 0 or 1), allowing them to benefit from computational techniques, which may be irrelevant to networks constructed by machine learning method [38]. However, the former two methods, especially the cause-and-effect relationship method, may produce inaccurate models. Although structure learning method is more accurate, difficulty or impossibility sometimes arises from obtaining available and sufficient data for learning. Of course, simulation methods can be used to generate required data to fill this gap.

Some practical applications use jointly two or more BN structure modeling approaches. For example, Bartram [48] developed DBN-based fault diagnosis and prognosis approach for dynamic systems, such as hydraulic actuator systems, in their work, while considering multi-source information. Researchers used expert knowledge and experience to construct primary network structures, and then used DBN structure learning algorithms to determine the remaining structure.

1.3.2. *BN parameter modeling*

BNs parameter model consists of prior probability of root nodes and conditional probability of leaf nodes. Prior probabilities of events are calculated before acquisition of new evidence, observation or information. These data can be achieved from expert knowledge and experience and statistical results of historical, simulated, and experimental data. With increasing prior probability, events have higher probability of happening. In general, in fault diagnosis, we assume that prior probabilities of all fault nodes are identical to emphasize posterior probabilities with new observations [42].

Conditional probabilities expectedly take place when other events are known to occur. These probabilities are usually contained in the conditional probability table (CPT). The various ways for obtaining CPTs generally fall into two main approaches: knowledge elicitation from experts and data-driven parameterization through machine learning methods. Knowledge elicitation from experts has disadvantages, including specificity of exponential growth of parameters. To specify complete CPT for child node m with s_m states and n parent nodes, $(s_m - 1)\prod_{i=1}^{n} s_i$ probabilities should be evaluated; s_i is number of states of parent node i [49]. Suppose that we have variables with n parents, and that each variable has only two values. We then need 2^n independent parameters to completely specify CPT for the variable. Modeling arises as the number of parents, n, becomes larger. Noisy-OR and noisy-MAX models were reported to simplify knowledge elicitation from experts of CPTs.

When a node is Boolean type and has normal and abnormal states, which can be detected by sensors, node CPT can be simplified with noisy-OR model. Supposing that all nodes are Boolean, several causes $X_1, X_2, \ldots, X_n$ lead to an effect Y, and CPT can be generated with only n parameters, and $q_1, q_2, \ldots, q_n$ are obtained using the following:

$$P(Y = 1|X_1, X_2, \ldots, X_n) = 1 - \prod_{i=1}^{n} q_i, \tag{1.2}$$

where q_i is probability for each parent with false Y when X_i is true, and all other parents are false [50].

A node has three states when it is non-Boolean: higher, lower and middle, with which noisy-OR model cannot work. Thus, noisy-MAX model can be used for simplicity. This model is a generalization of the noisy-OR model [50]. Suppose that non-Boolean variables $X_1, X_2, \ldots, X_n$ affect variable Y, CPT can then be established by using the noisy-MAX model through the following:

$$P(Y \leq y|X) = \prod_{\substack{i=1 \\ x_i \neq 0}}^{n} \sum_{y'=0}^{y} q_{i,y'}^{x_i}, \tag{1.3}$$

$$P(Y = y|X) = \begin{cases} P(Y \leq 0|X) & \text{if } y = 0, \\ P(Y \leq y|X) - P(Y \leq y-1|X) & \text{if } y > 0, \end{cases} \tag{1.4}$$

where X represents parent node configurations of Y; $X = x_1, \ldots, x_n$, and $P(Y = 0|X_1 = 0, \ldots, X_n = 0) = 1$.

Recently, Cai *et al.* utilized noisy-OR and noisy-MAX models to determine complete CPTs of BNs for fault diagnosis of complex industrial systems, such as ground-source heat pump [42] and subsea production systems [50]. Kraaijeveld *et al.* [51] proposed a method to simplify and speed up the design in constructing diagnostic BNs, and interaction among variables was modeled by noisy-MAX gates. Similarly, noisy-MAX nodes with LEAK probabilities are used in constructing diagnostic BNs for detecting and diagnosing faults in variable air volume terminals [52] and air handling units [53]. Notably, knowledge elicitation from experts using noisy models may cause errors even in diagnostic results because of independent assumptions of these models.

Noisy models are for local structures only. Each noisy model is based on assumptions about interactions of parents with their common child. When the assumption corresponds to reality, then noisy models are used for local structure. Otherwise, the resulting BN will be an inaccurate model (but it could be a good approximation).

CPTs are also generated through BN parameter learning from historical data. This method is similar to structure learning; its advantage includes more accurate BN parameter model, whereas its disadvantage is the difficult and sometimes impossible acquisition of available and sufficient historical data. For instance, SIAM database was used for learning of CPTs of BN models for railway diagnosis [54]. Parameter learning was utilized to determine CPTs of nodes in static BNs for fault diagnosis models of GMR control systems [55]. Conditional probabilities were learned from a series of training data using Laplace's law of succession in an automated BN-based diagnosis model for universal mobile telecommunications system networks [56]. Barua and Khorasani [57] proposed a novel procedure of CPT generation in BN-based hierarchical fault diagnosis by determining initial probability distribution from performance matrices and producing the remaining distribution with weighted-sum of initial distributions.

1.3.3. *BN inference*

Based on conditional independence assumptions and chain rules, joint probability of variables $U = \{A_1, A_2, \ldots, A_n\}$ can be calculated as follows:

$$P(U) = \prod_{i=1}^{n} P(A_i|\text{Pa}(A_i)), \tag{1.5}$$

where $\text{Pa}(A_i)$ is parent node of A_i in BNs.

BNs can perform backward or diagnostic analyses with various kind of inference algorithms based on *Bayes' theorem*, which is as follows [38]:

$$P(U|E) = \frac{P(E|U)P(U)}{P(E)} = \frac{P(E,U)}{\sum_U P(E,U)}. \tag{1.6}$$

In general, inference algorithms are divided into exact inference and inferences. Exact inference algorithms can compute exact probabilities of variables and include message-passing, conditioning, junction tree, symbolic probabilistic inference, arc reversal/ node reduction, and differential algorithms. Approximate inference computes approximate probabilities of variables with statistical approach. This inference includes algorithms for stochastic sampling, search-based algorithm, and loopy belief propagation algorithm. For complex BNs, inference is NP hard problem.

Exact inference algorithms, for example, junction tree, were used in fault diagnosis of single-tank systems [45] and end-to-end service quality of qualitative diagnosis [47]. Liu *et al.* [59] presented a new concept of node set, that is, maximum quadruple-constrained subset, for improving the exact inference efficiency in BN-based fault diagnosis models. This message-passing algorithm is widely used in inference of fault diagnosis of virtual private networks [60], root cause of process variations [39], and statistical process control [40]. These exact inference algorithms bear exponential complexity in network *treewidth*, a graph-theoretic parameter measuring resemblance of graphs to tree structures. This complexity is handled by exact algorithms for complex BNs.

Complexity in exact inference algorithms leads to a surge of interest in approximate inference algorithms, which are generally independent of *treewidth.* Today, approximate inference algorithms are the only choice for networks having large *treewidth.* Yet, such algorithms lack sufficient local structure. For example, Wiegerinck *et al.* [61] adapted variational methods with tractable structures to develop approximate inference algorithm of BNs for medical diagnosis. Pernestål *et al.* [62] developed an inference algorithm called *updateBN* to update static BNs to account for external interventions in BN-based diagnosis. Zhang and Dong [63] developed inference algorithm using partly missing data to perform fault diagnosis for multi-time-slice DBNs with a mixture of Gaussian output.

Various software are employed in BN inference in fault diagnosis for many applications; these software include Netica [42, 64], Hugin [49, 65], and Matlab/BNT [66].

1.3.4. *Fault identification*

Fault identification is conducted based on posterior probabilities of fault nodes based on provided evidence. Only the probabilities of faults can be given, but definite diagnostic results cannot be drawn per posterior probabilities. Overall, the corresponding fault has higher probability of occurrence with increasing posterior probability.

For some fault diagnoses, the diagnostic result is directly determined by posterior probability. For instance, in BN-based fault diagnosis systems for manufacturing tests of mobile telephone infrastructure, nodes with highest posterior probability of failure state were considered as either tokens to second BN or advice nodes. When nodes were the former, the existing BN was saved as case, and other BNs defined in the token were loaded. In case of the latter, nodes were considered to be advice nodes [67]. In BN-based control loop diagnosis, according to maximum *a posteriori* principle, the mode with the biggest posterior probability is a potential fault mode, and abnormalities related to this fault mode were root causes of failure [68]. In BN-based fault diagnosis system for assembly systems, the status of equipment modules was thought to be hidden or unknown, and the aim is to estimate this status in light of other observed variables within systems, whose statuses, including observed faults, are known or possible to estimate. During diagnosis, the equipment module with the largest probability of being in abnormal state was consequently considered to be the most probable cause of failure, accounting for a specific set of observations [69]. In BN-based fault detection and diagnosis system of complex process operations, researchers utilized sum-product algorithm to calculate posterior probabilities of nodes and utilized max-product algorithm with backtracking to determine the most probable states of nodes. Nodes that were identified to be most probable in failure states represented the most probable faulty process variables, which contributed greatly to failure states of the identified faulty monitored variables [70]. Notably, the used algorithm may cause some errors when posterior probability is used directly for fault identification, because some faults have inherently high prior probability (before BN inference), or simultaneous faults may exist. Therefore, to increase accuracy and robustness of fault diagnosis, a series of fault identification rules were reported for determination of diagnostic results.

In BN-based power system fault diagnosis, a series of fault identification rules were defined as follows: (a) an element must be faulty when its fault belief degree is larger than 70%; (b) an element may be faulty when its fault belief degree is larger than 15% and smaller than 70%; (c) an element is fault-free when its fault belief degree is smaller than 15% [71]. For variable air volume terminals, the BN-based fault diagnosis approach utilized the following rules for fault identification: (a) a fault with largest fault belief is reported when belief is higher than 70%; (b) a fault with largest probability is reported when this belief is 30% higher than the second largest one [52]. In BN-based process parameter fault diagnosis for ceramic shell deformation, posterior probability should have a threshold value between 40% and 100%. Candidate parameters are determined as root nodes when corresponding posterior probabilities are higher than the defined threshold value, whereas other parameters are considered normal nodes [72]. In real-time OOBN-based fault diagnosis approach of industrial systems, two judgment rules were defined per engineering experience: (a) a definite fault is reported when posterior and prior probability for fault nodes have a difference that is greater or equal to 60%; (b) a warning of fault is reported when the difference between posterior and prior probability for fault nodes is between 30% and 60% [49]. In BN-based inverter fault diagnosis, the following identification methods are defined: (a) system reports a single open-circuit of switch with highest posterior probability when it is higher than 70%, or 50% higher than the second highest one; (b) system reports double open-circuit failure of switch with highest and second highest posterior probabilities when both are higher than 70%, or 50% higher than the third highest one [129]. Remarkably, for different fault diagnosis systems, one should develop corresponding different fault identification rules. This goal requires complex work, because rules must be defined, tested, and revised repeatedly until achievement of satisfactory diagnostic performance.

1.3.5. *Verification and validation*

Model verification and validation are significant aspects of fault diagnosis because they provide reasonable confidence to diagnostic results. Verification is defined as the process of determining how accurately a computer program solves equations of mathematical models. It answers the question, "Have I built the system right?" (i.e., Does the system as built meet the performance specifications as stated?) Validation is defined as the process of determining the degree to which a model accurately represents the real world from the perspective of intended users. It answers the question, "Have I built the right system?" (i.e., Is the system model close enough to the physical system, and are the performance specifications and system constraints correct?) [130, 132] Many methods, such as sensitivity analysis, conflict analysis, simulation and experimental research, are used for BN-based fault diagnosis verification and validation. Specifically, sensitivity analysis can be used for model verification, and the four methods can all be used for model validation.

Sensitivity analysis can be conducted by using the mutual information of fault nodes and fault symptom nodes. Let us use an entropy function provided in Eq. (1.6) to represent the uncertainty of a system; mutual information is defined as the uncertainty reducing potential of X given the original uncertainty in T prior to consulting X [73]. Mutual information of T and X can be expressed in Eq. (1.7).

$$H(T) = -\sum_{t} P(t) \log P(t), \tag{1.7}$$

where $P(t)$ is probability distribution of random variable T. The next equation is then used

$$I(T, X) = -\sum_{x}\sum_{t} P(t, x) \log \frac{P(t, x)}{P(t)P(x)}, \tag{1.8}$$

where $P(t)$ is the marginal probability distribution of T, $P(x)$ is the marginal probability distribution of X, and $P(t, x)$ is the joint probability distribution of T and X.

Similarly, Wu *et al.* [64] used sensitivity analysis approach based on three axioms proposed by Jones *et al.* [127] to validate DBN-based risk evaluation models for offshore drilling incident diagnosis. Barua and Khorasani [74] proposed and developed a sensitivity analysis approach to validate the BN-based hierarchical fault diagnosis model by using developed systematic component dependence model methods.

When observations are input to BN-based fault diagnosis model as evidence, conflict may occur, indicating the weak relation of model to evidence. This evidence-driven conflict analysis can be used for detecting possible conflicts in evidence or between the evidence and BN-based fault diagnosis model. This way, we can use conflict analysis approach to validate and verify BN-based fault diagnosis model.

A conflict measure is designed to indicate possible conflicts when joint probability of evidence is less than the product of probabilities of individual pieces of evidence in models. The main assumption is that pieces of evidence are positively correlated such that $P(x) > \prod_{i=1}^{n} P(x_i)$. With this assumption, general conflict measure is defined as follows [75]:

$$\text{conflict}(x) = \text{conflict}(\{x_1, \ldots, x_n\}) = \log \frac{\prod_{i=1}^{n} P(x_i)}{p(x)}. \tag{1.9}$$

This measure implies that positive values of conflict measure (x) indicate possible conflict and therefore an incorrect model. Cai *et al.* [49] utilized this evidence-driven conflict analysis approach to validate an OOBN model for fault detection and diagnosis of complex industrial systems.

In general, sensitivity and conflict analyses are used for partial verification and validation. Simulation and experimental methods can be used for full verification and validation per various simulated and experimental faults. Hu *et al.* [76] used the detected information and file control records to validate intelligent process plant fault diagnosis system developed by hazard and operability study and DBN-integrated methodology. Santos *et al.* [77] used a training data set containing 480 observations to build a BN-based fault diagnosis model of Tennessee Eastman process (TEP) and used a testing data set containing 960 observations to validate modes. Zhao *et al.* [78] used sample data in practical operating conditions to validate a BN-based fault diagnosis model of air handling units. Chien *et al.* [79] conducted experiments to validate the constructed BNs for fault location on distribution feeder and used Pearson correlation coefficients between historical data and derived results for convergent validation. One good and accurate method employs verification and validation of proposed fault diagnosis approach, which is performed by comparing practical and diagnostic faults after development of fault diagnosis approach and corresponding fault diagnosis system; however, this way is not always practical or possible. Many faults cannot be injected and tested. Still, simulation methods can imitate all kinds of faults, which are a good supplement for experimental methods. For instance, Alaeddini and Dogan [40] performed two independent simulation studies to validate performance of BN-based root cause analysis and fault diagnosis approach in control charts while including comparison research with K-nearest neighbor and multi-layer perceptron and evaluation under different conditions. Ling and Mahadevan [80] utilized Bayesian hypothesis testing proposed in [128] to evaluate performance of BN-based structural damage prognosis approach by comparing the predicted and observed data.

1.4. Types of BNs for Fault Diagnosis

1.4.1. *BN for fault diagnosis*

BN, also known as static BN, is most widely used in fault diagnosis. As reviewed in Sec. 1.3, most fault diagnosis applications correspond to BNs. This subsection emphasizes classification but does not repeat literature. When fault diagnosis is involved in temporal system or complex system, inevitable difficulties with static BNs are observed. Therefore, some other types of BNs, such as DBNs and OOBNs, are used to solve these problems, which are described in detail in the next three subsections.

1.4.2. *DBNs for fault diagnosis*

Static BNs are mainly used in modeling and inference for fault detection and diagnosis. Modeling of fault diagnosis does not involve temporal features of faults, fault symptoms, and even the systems themselves. That is, static-BN-based fault diagnosis is performed at a certain time point without considering temporal relations. However, given the same fault symptoms, diagnostic results may be totally different at different time periods because of performance degradation of components [55]. In other words, a new system is more likely to work well than an aged system at a succeeding time point when it works well at the present time. New systems can also increase the accuracy and reliability of fault diagnosis by involving dynamic and temporal features in fault diagnosis models [81]. DBNs are extensional BNs with time-dependent variables and can be used to model temporal evolution of dynamic systems. DBNs include multiple copies of identical nodes, where different copies represent different states of nodes over time. Nodes in the same copies are connected using intra-slice arcs, and nodes in different copies are connected by inter-slice arcs, integrating an entire DBN. DBNs have strong power in modeling, representing, and reasoning of dynamic systems and are therefore increasingly used in dynamic system diagnosis.

Arroyo-Figueroa *et al.* [83–85] introduced a new formalism of BNs, i.e., temporal BN of events, to deal with uncertainty and time for fault detection and diagnosis in dynamic domains. The proposed temporal BN is actually a DBN. Lerner *et al.* [86] proposed a new approach based on the framework of hybrid DBNs for fault diagnosis of complex systems with both discrete and continuous nodes. Kao *et al.* [87] utilized DBN as a knowledge base for a reasoning system to model causal relationships in supply chain, where diagnostic tasks are conducted. Kohda and Cui [88] proposed DBN-based logic diagnosis approach for a safety-related system. Researchers used DBN to model dynamic behaviors of the mentioned system. Gonzalez *et al.* [89] developed a DBN-based real-time online fault detection and diagnosis method for instrument gross error in mass and energy balances. DBN inference was also conducted by combining diagnostic and predictive reasoning. Tobon-Mejia *et al.* [90] proposed wear diagnosis and prognosis approach of computer numerical control tool machine. DBNs were used to model the dynamic behavior of degradation in cutting tools. Duong *et al.* [91] proposed an evaluation approach for confidence level of feedback information for logic diagnosis using DBNs and Markov chain. Yu and Rashid [92] used DBNs to develop networked process monitoring, fault propagation identification, and root cause diagnosis methodology. Zhang and Dong [63] developed multi-time-slice DBN-based process monitoring and fault diagnosis approach for systems with missing data. Zhao *et al.* [93] proposed a DBN-based fault diagnosis and accident progression prediction method for nuclear facilities. Scientists used DBNs to diagnose faults with detected sensor information and to prognose accidents with the detected sensor information and actions of operators. Wu *et al.* [94] proposed the DBN-based decision support method that was used in the study for safety analysis. Posterior probability distributions were used to aid technicians perform online fault diagnosis. Given that the DBN model contains many time slices over a long period, BN inference and fault diagnosis are much slower than BN-based methods.

1.4.3. *OOBNs for fault diagnosis*

The object-oriented approach possesses characteristics, including encapsulation, inheritance, polymorphism, and modularity. This approach is introduced into BNs, forming OOBNs. OOBNs also utilize terminology class and objects from object-oriented approach. Class is a generic network fragment, that is, a BN; object is a fragment generated by instantiating the class [95]. Nodes in class or object can be classified into three categories: input node, output node, and encapsulated node. Input and output nodes are regarded as object interfaces. When an object is encapsulated in other objects, OOBNs provide an approach to achieve hierarchical representation of the model, and each level corresponds to a particular level of abstraction, revealing encapsulated nodes for the current layer of the object.

With comparison of general BNs and DBNs, OOBNs have the following advantages: first, the model supports top-down model construction process. Second, OOBNs are constructed by integrating small and understandable network fragments, benefiting knowledge acquisition and communication between modelers and domain experts. Third, this approach reduces complexity of building BNs and improves reusability of models. Finally, OOBNs have high average rate of convergence and time efficiency because of encapsulation and hierarchy. OOBNs are therefore a powerful and suitable tool for constructing complex models [96].

According to our review, OOBN application is limited when used for fault diagnosis of complex systems. For instance, Cai *et al.* [49] presented a novel OOBN-based real-time fault diagnosis approach. In their study, OOBNs were used to model repetitive structures and components within complex industrial systems. Huang *et al.* [65] used OOBNs to establish a probability-based vehicle fault diagnosis model with four sub-models and three common root causes. Weidl *et al.* [97] developed an OOBN-based methodology for condition monitoring, root cause analysis, and decision support in complex industrial process operations. Jensen *et al.* [98] used OOBNs to model causes of leg disorder in finisher herd, and object-oriented structures were used to reduce specifications of diagnostic networks.

1.4.4. *Other BNs for fault diagnosis*

Aside from DBN and OOBN, other types of BNs were proposed to solve specific problems in fault diagnosis. For example, to improve performance of BNs, Barco *et al.* [99] presented a methodology to model "continuity" in human reasoning; these smooth BNs diagnose faults in mobile communication networks in the presence of imprecise parameters. To achieve better diagnostic performance of modular and distributed industrial systems, Sayed and Lohse [100] utilized a multi-agent Bayesian framework, i.e., multiple-sectioned BNs, to develop condition monitoring and fault diagnosis methodology.

1.5. Domains of Fault Diagnosis with BNs

Although some applications of BNs are present in fault diagnosis in the nonengineering domain, such as biomedicine domain, until now, BNs are mainly used in the engineering domain. Therefore, the current work focuses only on BN-based fault diagnosis in the engineering domain. BNs have distinct advantage over other techniques in handling uncertain information. In fault diagnosis in the engineering domain, uncertainty exists extensively among root cause, fault, fault symptoms, and additional information. For instance, root cause results in multiple faults, and fault generates multiple-fault symptoms. On the contrary, a symptom may be caused by multiple faults, and faults may be caused by multiple root causes. Extra information also has intricate and complex influences on engineering systems, causing various uncertainties in relationships. Publications related to fault diagnosis using BNs are reviewed and summarized according to diagnostic objects following the five main engineering systems, including process, energy, structural, manufacturing, and network systems, as shown in Table 1.1.

1.5.1. *Fault diagnosis for process systems*

Process control is automation control aimed at maintaining controlled variables operating at stable levels with normal variation through adjusting parameters based on various monitored information. Fault diagnosis plays a significant role in process control systems. Qi and Huang [68] proposed a Bayesian-method-based fault node diagnosis approach for control loop. The study used a simulated distillation column process to demonstrate the proposed

Table 1.1. Summary of domains of fault diagnosis with BNs.

Domains	References	Types	Diagnostic objects
Process systems	Qi and Huang [68]	BNs	Control loop
	Jin *et al.* [72]	BNs	Investment casting process
	Liu and Jin [109]	BNs	Assembly process
	Liu *et al.* [110]	BNs	Hot forming process
	Verron *et al.* [66]	BNs	TEP
	Verron *et al.* [112]	BNs	TEP
	Santos *et al.* [77]	BNs	TEP
Energy systems	Xiao *et al.* [52]	BNs	Variable air volume terminals
	Zhao *et al.* [78]	BNs	Air handling units
	Najafi *et al.* [114]	BNs	Air handling units
	Wang *et al.* [115]	BNs	Wind turbine generator
	de Bessa *et al.* [116]	BNs	Wind turbine
	Liu *et al.* [117]	BNs	Solar-assisted hear pump
	Zhu *et al.* [71]	BNs	Transmission power system
	Riascos *et al.* [118]	BNs	Proton exchange membrane fuel cells
	Mengshoel *et al.* [119]	BNs	Electrical power system
Structural systems	Ling and Mahadevan [80]	BNs	Hollow cylindrical component
	Bartram [48]	DBNs	Cantilever beam structure
	Oukhellou *et al.* [54]	BNs	Railway
	Arangio *et al.* [120]	BNs	Suspension bridge
Manufacturing systems	Chan and McNaught [67]	BNs	Manufacturing test
	Sayed and Lohse [69]	BNs	Assembly systems
	Li *et al.* [121]	BNs	Semiconductor manufacturing
Network systems	Khanafer *et al.* [56]	BNs	Universal mobile telecommunications system networks
	Bennacer *et al.* [60]	BNs	Virtual private networks
	Carrera *et al.* [122]	BNs	Internet business service
	Li *et al.* [123]	DBNs	Internet service

fault diagnosis approach. Results showed that diagnostic performance is improved over traditional approaches. Jin *et al.* [72] proposed a BN-based ceramic-shell deformation root-cause analysis and fault diagnosis method to increase dimensional precision in investment-casting process. Similarly, BNs were used to diagnose the variation source of taillight assembly processes with a small and incomplete data set [109]. Liu *et al.* [110] developed a BN-based process monitoring and fault diagnosis approach to study sensor allocation methods for process control application. BNs were used in the study to model cause-and-effect relationship among variables and hot forming processes and cap alignment processes to demonstrate the performance of the developed approach.

TEP simulates actual chemical control process developed by Eastman Chemical Company to evaluate various control strategies and monitoring approaches [111]. TEP was used widely in research and in testing BN-based process system fault diagnosis methodologies. Verron *et al.* [66] presented a novel BN-based fault diagnosis approach for industrial process control. Researchers used the fault database to establish conditional Gaussian networks for fault identification. TEP was used to evaluate the diagnostic performance of this approach. Similarly, diagnostic BNs were constructed to detect and isolate faults in multivariate process [112]. Santos *et al.* [77] used BNs as a classifier to detect and diagnose faults in process systems and assessed the method performance using TEP.

1.5.2. *Fault diagnosis for energy systems*

A review centered on bibliographical survey of BNs in renewable energy systems is given in Ref. [113]; main application fields of BNs are reliability evaluation, complex system analysis, dependability, risk analysis, maintenance, and fault diagnosis. In fault diagnosis of energy systems, BN-based methods are mainly based on cause-and-effect relationship of faults and fault symptoms. Sensors in energy equipment provide information on fault symptoms. Xiao *et al.* [52] presented a diagnostic BN to detect and diagnose faults in variable air volume terminals. A similar BN-based approach was proposed for fault diagnosis in coils and sensors of air handling units [78]. Najafi *et al.* [114] proposed a novel BN-based machine learning approach to diagnose faults for solving problems, such as modeling limitation, measurement constraint, and simultaneous fault identification, existing in air handling units. Wang *et al.* [115] developed a vibration feature recognition system for wind turbine generators based on BN fault diagnosis models. de Bessa *et al.* [116] proposed a time-series- and data-analysis-based fault detection and diagnosis approach for wind turbines. Gibbs sampling method was used to detect faults and employed with fuzzy BN to conduct fault isolation. Liu *et al.* [117] proposed a BN-based fault detection and diagnosis approach for solar-assisted heat pump systems while considering incomplete data and expert knowledge.

For power and cell systems, BN-based fault diagnosis is widely used. Zhu *et al.* [71] used BNs with noisy models to construct fault diagnosis systems of transmission power systems. This proposed model could solve problems of uncertainty and incompleteness of data. Riascos *et al.* [118] used BNs to develop a new fault diagnosis method for proton exchange membrane fuel cells. Mengshoel *et al.* [119] developed a probabilistic method for model-based fault diagnosis for real-world electrical power systems using BNs and arithmetic circuits.

1.5.3. *Fault diagnosis for structural systems*

Structural system is considered a set of interconnected structural components or members that work together to achieve a common function or purpose. These strongly interacting components or members usually cause intrinsic complexity. BNs can deal with the complexity of fault diagnosis for structural systems.

Ling and Mahadevan [80] proposed a Bayesian probabilistic approach to prognose fatigue damage of structural systems and established BNs to model fatigue crack growth. These people used a hollow cylindrical component to demonstrate the proposed prognosis approach. Bartram [48] proposed a DBN-based framework for fault diagnosis and quantification of diagnosis uncertainty when using a particle filter in the presence of multiple faults. Diagnosis of cantilever beam structure was performed in the presence of heterogeneous information. Oukhellou *et al.* [54] proposed a BN-based fault diagnosis approach for structural systems by combining sensor data and structural knowledge. Broken rail diagnosis was used for railway infrastructure to demonstrate the proposed approach. Arangio *et al.* [120] proposed a hierarchical approach to evaluate the integrity of structure systems based on fault information detected and diagnosed by monitoring systems. BNs were used to diagnose possible faults with observed fault symptoms. A 3000 m bridge was used to demonstrate the proposed evaluation approach.

1.5.4. *Fault diagnosis for manufacturing systems*

In various manufacturing and assembly systems, BNs are employed to diagnose faults of the equipment or the process. Chan and McNaught [67] proposed a fault diagnosis system to test manufacturing of mobile-network-based transceiver stations. BNs were used to represent domain knowledge during the test process. A prototype system, which was developed successfully by linking BNs with an intelligent user interface, demonstrated the correctness of the proposed diagnosis method. To eliminate difficulty in establishing BN models with the experience of experts and knowledge restricting applications of BNs in fault diagnosis of manufacturing system, Sayed and Lohse [69] proposed a novel approach to construct BN-based fault diagnosis models from failure mode and results of effect analysis of assembly systems. Li *et al.* [121] proposed a BN-based fault diagnosis method for semiconductor manufacturing equipment. Researchers developed an expert system to identify possible root causes given the posterior probability and generated suggestions and solutions.

1.5.5. *Fault diagnosis for network systems*

In fault diagnosis of network systems, BN applications focused on communication network and Internet network. Khanafer *et al.* [56]

proposed a BN-based automation fault diagnosis method for universal mobile telecommunication system networks. Entropy minimization discretization approach and BNs were integrated to improve diagnostic performance. Bennacer *et al.* [60] developed a BN-based self-diagnostic approach for communication networks. Actual virtual private networks were used to demonstrate the proposed diagnosis approach. Carrera *et al.* [122] developed a multi-agent system for fault diagnosis for management of Internet business services. BNs were used to deal with uncertainty problems in fault diagnosis. Li *et al.* [123] proposed a DBN-based fault diagnosis approach for identification of root causes of Internet service problems. DBNs were used to model dynamic systems and also dealt with complex and uncertain problems in noisy environments.

This section only summarizes five main application domains of BN-based fault diagnosis. Some other applications were also reported but are not popular; such applications include offshore drilling incident diagnosis [64] and supply chain diagnosis [87].

1.6. Discussions and Research Directions

Based on BN-based fault diagnosis presented in this chapter, we identify a few ongoing and upcoming research directions that are of interest to fault diagnosis researchers.

1.6.1. *Integrated big data and BN fault diagnosis methodology*

Big data are a recent data paradigm that describe not only much more voluminous data but also coupling of such data with sophisticated data analytics to acquire new knowledge or insight [124]. A large number of normal and abnormal data are created and collected from the operation of complex industrial systems during long periods. These data can be used for data-driven fault diagnosis. BN-based fault diagnosis approach is typically data-driven. Integrated big data and BN fault diagnosis methodology provide definite orientation for interdisciplinary research, which could be divided into two important parts: The first will study fault feature extraction method from big data, and the second will focus on BN-based fault diagnosis method using these fault features.

1.6.2. *BN-based nonpermanent fault diagnosis*

As reviewed in previous sections, BN-based fault diagnosis methodologies are mainly used to identify various permanent faults. Permanent fault, which is also called hard failure, causes deterioration of system performance and cannot disappear before maintenance or repair. Two types of nonpermanent faults, that is, transient faults and intermittent faults, also exist in various engineering systems. Transient fault occurs with random frequency, is not easily repeatable, and usually cannot destroy systems permanently. Intermittent fault may occur repeatedly in a component and has characteristics of randomness, intermittency, and repeatability. Permanent fault may result from increasing intermittent faults and cause system failure. Transient and intermittent faults are also called soft faults. These faults are very difficult to identify and diagnose, because faults may not occur during system repair. Studies reported about BN/DBN-based fault detection and diagnosis methods for transient or intermittent faults of electronic products [55] and electrical power system [125]. However, significant problems arise regarding the analysis of the nature and causes of these permanent, transient, or intermittent faults and identification of failed components and fault type using BN-based fault diagnosis.

1.6.3. *Fast inference algorithms of BNs for online fault diagnosis*

Real-time online fault diagnosis system is much more useful than off-line one in assisting operations staff to detect, isolate, and identify faults when they occur and to aid troubleshooting. For online fault diagnosis of complex industrial system with huge BNs, especially DBNs, exact and approximate inference algorithms may both be slow. When OOBNs are used, the fault diagnosis model, including thousands of nodes, may be extremely complex. As more evidence are observed, desired hidden variables are estimated to become highly correlated. Traditional inference becomes increasingly expensive, costs lots of time, and hardly performs real-time fault diagnosis. Fast approximate inference algorithms of BNs should be developed for online fault diagnosis. In particular, in OOBN-based fault diagnosis model, structural information encoded in OOBNS, especially encapsulated variables in objects, can improve efficiency and speed of inference, both of which are used to study fast inference algorithms and perform real-time fault diagnosis.

1.6.4. *BNs for closed-loop control system fault diagnosis*

A complex industrial system usually includes numerous closed-loop feedback control subsystems. Previous BN-based methods adopted corresponding relationship of faults and symptoms to build fault diagnosis models, that is, they treated closed-loop feedback control systems as general systems. The studies did not consider effects of closed-loop feedback control algorithms on performance of fault diagnosis. BN is a directed acyclic graph, indicating that networks cannot contain cycles. Challenging problems accompany the establishment of fault diagnosis models of closed-loop feedback control systems with directed acyclic BNs, and investigation of the effects of control on diagnosis need to be performed.

1.6.5. *Fault identification rules*

Although excellent diagnostic performance of the BN-based method in fault diagnosis was demonstrated with various fault identification rules as reviewed above, industrial systems still demonstrate false alarm in fault diagnosis. The false alarm rate is a significant assessment indicator for fault diagnosis, and high false alarm rate cannot be accepted by users of industrial systems. Although some researchers used additionally observed information [42] and probabilistic boundary limit [126] to reduce false alarm rate, false alarm is still a challenging problem. This type of problem can be solved by the development of suitable fault identification rules using posterior probability integrated with prior probability.

1.6.6. *Hybrid fault diagnosis approaches*

To improve diagnostic performance, BNs are utilized with other techniques for fault diagnosis. Combined applications of BNs and other techniques are complementary and can overcome each other's disadvantages. These techniques can be divided into two classes: signal processing techniques and fusion techniques. Signal processing techniques are used to preprocess detected signals; these methods include Haar wavelet [102], ensemble empirical mode decomposition [103], and adaptive statistic test filter [106]. BNs use processed results to perform fault diagnosis. By contrast, fusion techniques are used with BNs together as a hybrid approach to conquer the disadvantages of each other [60, 70, 104, 105, 107, 108]. For example, Bennacer *et al.* [60] used case-based reasoning to reduce the complexity of inference of BN in fault diagnosis for communication networks. Yu *et al.* [70] combined a modified independent component analysis and BNs to overcome limitations in processing variables, which when not monitored, caused inaccuracy of diagnostic results. Agrawal *et al.* [107] used fuzzy logic to identify types and magnitudes of faults and used BNs to analyze root causes of faults. D'Angelo *et al.* [108] utilized fuzzy set theory to solve uncertainty problems in input of BNs.

However, researchers face a series of challenging problems for fault diagnosis with only BNs; these problems include modeling state transition relationship of multi-state components and modeling degradation system for fault diagnosis. Therefore, future research should consider studying hybrid approach with BNs and other fusion techniques. More fusion techniques can be used together with BNs to develop hybrid fault diagnosis approaches.

1.7. Conclusion

Over the past decades, application of BNs on fault diagnosis was well studied by researchers and practitioners. Effort was devoted to formulating BN-based fault diagnosis methodology and developing corresponding fault diagnosis systems. However, related challenges still exist. This chapter provided a literature review of the application of BNs to fault detection and diagnosis.

The chapter presented the general procedure of fault diagnosis modeling with BNs and considered processes included BN structure modeling, BN parameter modeling, BN inference, fault identification, and validation and verification. For each procedure, various methods were reviewed. This work also provided a series of classification schemes of BNs in fault diagnosis. BN types, including static BNs, DBNs, and OOBNs, were reviewed in the application of fault diagnosis. The chapter summarized BNs with other fault diagnosis techniques, such as case-based reasoning and independent component analysis. The domain of fault diagnosis with BNs was classified into two main classes: biomedicine and engineering domains. However, we focused more on the latter. We pointed out the potential problems of applying BNs to fault diagnosis, and upcoming research directions are addressed.

Although we reviewed as much literature as possible, we realize the impossibility of including all publications related to BNs in fault diagnosis, such as publications involving non-English literature. The major contribution of this chapter was presenting a general BN-based fault diagnosis methodology and a series of classification schemes of BNs in fault diagnosis. We hope that this literature review can serve as a useful guide in fault diagnosis with BNs and provide a comprehensive reference for researchers and practitioners.

References

[1] V. Venkatasubramanian, R. Rengaswamy, K. Yin, S. N. Kavuri, "A review of process fault detection and diagnosis: Part I: Quantitative model-based methods," *Computers & Chemical Engineering*, vol. 27, no. 9, pp. 293–311, 2003.

[2] I. Hwang, S. Kim, Y. Kim, C. E. Seah, "A survey of fault detection, isolation, and reconfiguration methods," *IEEE Transactions on Control Systems Technology*, vol. 18, no. 3, pp. 636–653, 2010.

[3] Y. Lei, J. Lin, Z. He, M. J. Zuo, "A review on empirical mode decomposition in fault diagnosis of rotating machinery," *Mechanical Systems and Signal Processing*, vol. 35, no. 1, pp. 108–126, 2013.

[4] P. Henriquez, J. B. Alonso, M. A. Ferrer, C. M. Travieso, "Review of automatic fault diagnosis systems using audio and vibration signals," *IEEE Transactions on Systems, Man, and Cybernetics: Systems*, vol. 44, no. 5, pp. 642–652, 2014.

[5] R. Yan, R. X. Gao, X. Chen, "Wavelets for fault diagnosis of rotary machines: A review with applications," *Signal Processing*, vol. 96, pp. 1–5, 2014.

[6] V. Venkatasubramanian, R. Rengaswamy, S. N. Kavuri, K. Yin, "A review of process fault detection and diagnosis: Part III: Process history based methods," *Computers & Chemical Engineering*, vol. 27, no. 3, pp. 327–346, 2003.

[7] S. Yin, S. X. Ding, X. Xie, H. Luo, "A review on basic data-driven approaches for industrial process monitoring," *IEEE Transactions on Industrial Electronics*, vol. 61, no. 11, pp. 6418–6428, 2014.

[8] S. X. Ding, "Data-driven design of monitoring and diagnosis systems for dynamic processes: A review of subspace technique based schemes and some recent results," *Journal of Process Control*, vol. 24, no. 2, pp. 431–449, 2014.

[9] Z. Gao, C. Cecati, S. X. Ding, "A survey of fault diagnosis and fault-tolerant techniques — Part I: Fault diagnosis with model-based and signal-based approaches," *IEEE Transactions on Industrial Electronics*, vol. 62, no. 6, pp. 3757–3767, 2015.

[10] Z. Gao, C. Cecati, S. X. Ding, "A survey of fault diagnosis and fault-tolerant techniques — Part II: Fault diagnosis with knowledge-based and hybrid/active approaches," *IEEE Transactions on Industrial Electronics*, vol. 62, no. 6, pp. 3768–3774, 2015.

[11] B. Mnassri, M. Ouladsine, "Reconstruction-based contribution approaches for improved fault diagnosis using principal component analysis," *Journal of Process Control*, vol. 33, pp. 60–76, 2015.

[12] B. Jiang, J. Xiang, Y. Wang, "Rolling bearing fault diagnosis approach using probabilistic principal component analysis denoising and cyclic bispectrum," *Journal of Vibration and Control*, DOI: 10.1177/1077546314 547533, 2014.

[13] Y. Zhang, H. Zhou, S. J. Qin, T. Chai, "Decentralized fault diagnosis of large-scale processes using multiblock kernel partial least squares," *IEEE Transactions on Industrial Informatics*, vol. 6, no. 1, pp. 3–10, 2010.

[14] S. X. Ding, S. Yin, K. Peng, H. Hao, B. Shen, "A novel scheme for key performance indicator prediction and diagnosis with application to an industrial hot strip mill," *IEEE Transactions on Industrial Informatics*, vol. 9, no. 4, pp. 2239–2247, 2013.

[15] Z. Wang, J. Chen, G. Dong, Y. Zhou, "Constrained independent component analysis and its application to machine fault diagnosis," *Mechanical Systems and Signal Processing*, vol. 25, no. 7, pp. 2501–2512, 2011.

[16] Y. Guo, J. Na, B. Li, R. F. Fung, "Envelope extraction based dimension reduction for independent component analysis in fault diagnosis of rolling element bearing," *Journal of Sound and Vibration*, vol. 333, no. 13, pp. 2983–2994, 2014.

[17] S. Bansal, S. Sahoo, R. Tiwari, D. J. Bordoloi, "Multiclass fault diagnosis in gears using support vector machine algorithms based on frequency domain data," *Measurement*, vol. 46, no. 9, pp. 3469–3481, 2013.

[18] R. Liu, B. Yang, X. Zhang, S. Wang, X. Chen, "Time-frequency atoms-driven support vector machine method for bearings incipient fault diagnosis," *Mechanical Systems and Signal Processing*, vol. 75, pp. 345–370, 2016.

[19] B. Liang, S. D. Iwnicki, Y. Zhao, "Application of power spectrum, cepstrum, higher order spectrum and neural network analyses for induction motor fault diagnosis," *Mechanical Systems and Signal Processing*, vol. 39, no. 1, pp. 342–360, 2013.

[20] Y. Shatnawi, M. Al-khassaweneh, "Fault diagnosis in internal combustion engines using extension neural network," *IEEE Transactions on Industrial Electronics*, vol. 61, no. 3, pp. 1434–1443, 2014.

[21] J. D. Wu, Y. H. Wang, M. R. Bai, "Development of an expert system for fault diagnosis in scooter engine platform using fuzzy-logic inference," *Expert Systems with Applications*, vol. 33, no. 4, pp. 1063–1075, 2007.

[22] S. A. Khan, M. D. Equbal, T. Islam, "A comprehensive comparative study of DGA based transformer fault diagnosis using fuzzy logic and ANFIS models," *IEEE Transactions on Dielectrics and Electrical Insulation*, vol. 22, no. 1, pp. 590–596, 2015.

[23] J. Pearl, "Bayesian networks: A model of self-activated memory for evidential reasoning," In: *Proceedings of the 7th Conference of the Cognitive Science Society*, University of California, Irvine, 1985.

[24] H. Langseth, L. Portinale, "Bayesian networks in reliability," *Reliability Engineering & System Safety*, vol. 92, no. 1, pp. 92–108, 2007.

[25] P. Weber, G. Medina-Oliva, C. Simon, B. Iung, "Overview on Bayesian networks applications for dependability, risk analysis and maintenance areas," *Engineering Applications of Artificial Intelligence*, vol. 25, no. 4, pp. 671–682, 2012.

[26] B. Cai, Y. Liu, Z. Liu, X. Tian, X. Dong, S. Yu, "Using Bayesian networks in reliability evaluation for subsea blowout preventer control system," *Reliability Engineering & System Safety*, vol. 108, pp. 32–41, 2012.

[27] B. Cai, Y. Liu, Y. Ma, L. Huang, Z. Liu, "A framework for the reliability evaluation of grid-connected photovoltaic systems in the presence of intermittent faults," *Energy*, vol. 93, pp. 1308–1320, 2015.

[28] B. Cai, Y. Liu, Y. Ma, Z. Liu, Y. Zhou, J. Sun, "Real-time reliability evaluation methodology based on dynamic Bayesian networks: A case study of a subsea pipe ram BOP system," *ISA transactions*, vol. 58, pp. 595–604, 2015.

[29] W. S. Wu, C. F. Yang, J. C. Chang, P. A. Château, Y. C. Chang, "Risk assessment by integrating interpretive structural modeling and Bayesian network, case of offshore pipeline project," *Reliability Engineering & System Safety*, vol. 142, pp. 215–224, 2015.

[30] N. Khakzad, "Application of dynamic Bayesian network to risk analysis of domino effects in chemical infrastructures," *Reliability Engineering & System Safety*, vol. 138, pp. 263–272, 2015.

[31] B. Cai, Y. Liu, Z. Liu, X. Tian, Y. Zhang, R. Ji, "Application of Bayesian networks in quantitative risk assessment of subsea blowout preventer operations," *Risk Analysis*, vol. 33, no. 7, pp. 1293–1311, 2013.

[32] M. Hänninen, O. A. Banda, P. Kujala, "Bayesian network model of maritime safety management," *Expert Systems with Applications*, vol. 41, no. 17, pp. 7837–7846, 2014.

[33] A. C. Mbakwe, A. A. Saka, K. Choi, Y. J. Lee, "Alternative method of highway traffic safety analysis for developing countries using Delphi technique and Bayesian network," *Accident Analysis & Prevention*, vol. 93, pp. 135–146, 2016.

[34] N. Yodo, P. Wang, "Resilience modeling and quantification for engineered systems using Bayesian networks," *Journal of Mechanical Design*, vol. 138, no. 3, 031404, 2016.

[35] A. John, Z. Yang, R. Riahi, J. Wang, "A risk assessment approach to improve the resilience of a seaport system using Bayesian networks," *Ocean Engineering*, vol. 11, pp. 136–147, 2016.

[36] S. Hosseini, K. Barker, "Modeling infrastructure resilience using Bayesian networks: A case study of inland waterway ports," *Computers & Industrial Engineering*, vol. 93, pp. 252–266, 2016.

[37] A. Bobbio, L. Portinale, M. Minichino, E. Ciancamerla, "Improving the analysis of dependable systems by mapping fault trees into Bayesian networks," *Reliability Engineering & System Safety*, vol. 71, no. 3, pp. 249–260, 2001.

[38] A. Darwiche, *Modeling and Reasoning with Bayesian Networks*, Cambridge University Press, 2009.

[39] S. Dey, J. A. Stori, "A Bayesian network approach to root cause diagnosis of process variations," *International Journal of Machine Tools and Manufacture*, vol. 45, no. 1, pp. 75–91, 2005.

[40] A. Alaeddini, I. Dogan, "Using Bayesian networks for root cause analysis in statistical process control," *Expert Systems with Applications*, vol. 38, no. 9, pp. 11230–11243, 2011.

[41] Y. Zhao, F. Xiao, S. Wang, "An intelligent chiller fault detection and diagnosis methodology using Bayesian belief network," *Energy and Buildings*, vol. 57, pp. 278–288, 2013.

[42] B. Cai, Y. Liu, Q. Fan, Y. Zhang, Z. Liu, S. Yu, R. Ji, "Multi-source information fusion based fault diagnosis of ground-source heat pump using Bayesian network," *Applied Energy*, vol. 114, pp. 1–9, 2014.

[43] Z. Liu, Y. Liu, B. Cai, C. Zheng, "An approach for developing diagnostic Bayesian network based on operation procedures," *Expert Systems with Applications*, vol. 42, no. 4, pp. 1917–1926, 2015.

[44] M. Lampis, J. D. Andrews, "Bayesian belief networks for system fault diagnostics," *Quality and Reliability Engineering International*, vol. 25, no. 4, pp. 409–426, 2009.

[45] C. H. Lo, Y. K. Wong, A. B. Rad, "Bond graph based Bayesian network for fault diagnosis," *Applied Soft Computing*, vol. 11, no. 1, pp. 1208–1212, 2011.

[46] S. Jin, Y. Liu, Z. Lin, "A Bayesian network approach for fixture fault diagnosis in launch of the assembly process," *International Journal of Production Research*, vol. 50, no. 23, pp. 6655–6666, 2012.

[47] X. Lin, B. Cheng, J. Chen, "Context-aware end-to-end QoS qualitative diagnosis and quantitative guarantee based on Bayesian network," *Computer Communications*, vol. 33, no. 17, pp. 2132–2144, 2010.

[48] G. W. Bartram, "System health diagnosis and prognosis using dynamic Bayesian networks," Ph.D. dissertation, Vanderbilt University, 2013.

[49] B. Cai, H. Liu, M. Xie, "A real-time fault diagnosis methodology of complex systems using object-oriented Bayesian networks," *Mechanical Systems and Signal Processing*, vol. 80, pp. 31–44, 2016.

[50] W. Li, P. Poupart, P. van Beek, "Exploiting structure in weighted model counting approaches to probabilistic inference," *Journal of Artificial Intelligence Research*, vol. 40, pp. 729–765, 2011.

[51] P. Kraaijeveld, M. Druzdzel, A. Onisko, H. Wasyluk, "Genierate: An interactive generator of diagnostic Bayesian network models," In: *Proceeding of the 16th International Workshop Principles Diagnosis*, pp. 175–180, 2005.

[52] F. Xiao, Y. Zhao, J. Wen, S. Wang, "Bayesian network based FDD strategy for variable air volume terminals," *Automation in Construction*, vol. 41, pp. 106–118, 2014.

[53] Y. Zhao, J. Wen, F. Xiao, X. Yang, S. Wang, "Diagnostic Bayesian networks for diagnosing air handling units faults–Part I: Faults in dampers, fans, filters and sensors," *Applied Thermal Engineering*, vol. 111, pp. 1272–1286, 2016.

[54] L. Oukhellou, E. Come, L. Bouillaut, P. Aknin, "Combined use of sensor data and structural knowledge processed by Bayesian network: Application to a railway diagnosis aid scheme," *Transportation Research Part C: Emerging Technologies*, vol. 16, no. 6, pp. 755–767, 2008.

[55] B. Cai, Y. Liu, M. Xie, "A dynamic-Bayesian-network-based fault diagnosis methodology considering transient and intermittent faults," DOI: 10.1109/TASE.2016.2574875, 2017.

[56] R. M. Khanafer, B. Solana, J. Triola, R. Barco, L. Moltsen, Z. Altman, P. Lazaro, "Automated diagnosis for UMTS networks using Bayesian network approach," *IEEE Transactions on Vehicular Technology*, vol. 57, no. 4, pp. 2451–2461, 2008.

[57] A. Barua, K. Khorasani, "Hierarchical fault diagnosis and health monitoring in satellites formation flight," *IEEE Transactions on Systems, Man, and Cybernetics, Part C (Applications and Reviews)*, vol. 41, no. 2, pp. 223–239, 2011.

[58] Z. Chiremsel, R. N. Said, R. Chiremsel, "Probabilistic fault diagnosis of safety instrumented systems based on fault tree analysis and Bayesian network," *Journal of Failure Analysis and Prevention*, vol. 16, no. 5, pp. 747–760, 2016.

[59] D. Liu, Y. Huang, Q. Yu, J. Chen, H. Jia, "A search problem in complex diagnostic Bayesian networks," *Knowledge-Based Systems*, vol. 30, pp. 95–103, 2012.

[60] L. Bennacer, Y. Amirat, A. Chibani, A. Mellouk, L. Ciavaglia, "Self-diagnosis technique for virtual private networks combining Bayesian networks and case-based reasoning," *IEEE Transactions on Automation Science and Engineering*, vol. 12, no. 1, pp. 354–366, 2015.

[61] W. A. Wiegerinck, H. J. Kappen, E. W. ter Braak, W. J. ter Burg, M. J. Nijman, J. P. Neijt, "Approximate inference for medical diagnosis," *Pattern Recognition Letters*, vol. 20, no. 11, pp. 1231–1239, 1999.

[62] A. Pernestål, M. Nyberg, H. Warnquist, "Modeling and inference for troubleshooting with interventions applied to a heavy truck auxiliary braking system," *Engineering Applications of Artificial Intelligence*, vol. 25, no. 4, pp. 705–719, 2012.

[63] Z. Zhang, F. Dong, "Fault detection and diagnosis for missing data systems with a three time-slice dynamic Bayesian network approach," *Chemometrics and Intelligent Laboratory Systems*, vol. 138, pp. 30–40, 2014.

[64] S. Wu, L. Zhang, W. Zheng, Y. Liu, M. A. Lunteigen, "A DBN-based risk assessment model for prediction and diagnosis of offshore drilling incidents," *Journal of Natural Gas Science and Engineering*, vol. 34, pp. 139–158, 2016.

[65] Y. Huang, R. McMurran, G. Dhadyalla, R. P. Jones, "Probability based vehicle fault diagnosis: Bayesian network method," *Journal of Intelligent Manufacturing*, vol. 19, no. 3, pp. 301–311, 2008.

[66] S. Verron, T. Tiplica, A. Kobi, "Fault diagnosis of industrial systems by conditional Gaussian network including a distance rejection criterion," *Engineering Applications of Artificial Intelligence*, vol. 23, no. 7, pp. 1229–1235, 2010.

[67] A. Chan, K. R. McNaught, "Using Bayesian networks to improve fault diagnosis during manufacturing tests of mobile telephone infrastructure," *Journal of the Operational Research Society*, vol. 59, no. 4, pp. 423–430, 2008.

[68] F. Qi, B. Huang, "Bayesian methods for control loop diagnosis in the presence of temporal dependent evidences," *Automatica*, vol. 47, no. 7, pp. 1349–1356, 2011.

[69] M. S. Sayed, N. Lohse, "Ontology-driven generation of Bayesian diagnostic models for assembly systems," *The International Journal of Advanced Manufacturing Technology*, vol. 74, no. 5–8, pp. 1033–1052, 2014.

[70] H. Yu, F. Khan, V. Garaniya, "Modified independent component analysis and Bayesian network-based two-stage fault diagnosis of process operations," *Industrial & Engineering Chemistry Research*, vol. 54, no. 10, pp. 2724–2742, 2015.

[71] Y. Zhu, L. Huo, J. Lu, "Bayesian networks-based approach for power systems fault diagnosis," *IEEE Transactions on Power Delivery*, vol. 21, no. 2, pp. 634–639, 2006.

[72] S. Jin, C. Liu, X. Lai, F. Li, B. He, "Bayesian network approach for ceramic shell deformation fault diagnosis in the investment casting process," *The International Journal of Advanced Manufacturing Technology*, vol. 88, no. 1–4, pp. 1–12, 2016.

[73] J. Pearl, *Probabilistic Reasoning in Intelligent Systems: Networks of Plausible Inference*, Morgan Kaufmann, 2014.

[74] A. Barua, K. Khorasani, "Verification and validation of hierarchical fault diagnosis in satellites formation flight," *IEEE Transactions on Systems, Man, and Cybernetics, Part C (Applications and Reviews)*, vol. 42, no. 6, pp. 1384–1399, 2012.

[75] U. B. Kjaerulff, A. L. Madsen, *Bayesian Networks and Influence Diagrams*, Springer Science+Business Media, 2008.

[76] J. Hu, L. Zhang, Z. Cai, Y. Wang, "An intelligent fault diagnosis system for process plant using a functional HAZOP and DBN integrated methodology," *Engineering Applications of Artificial Intelligence*, vol. 45, pp. 119–135, 2015.

[77] E. B. Santos, N. F. Ebecken, E. R. Hruschka, A. Elkamel, C. M. Madhuranthakam, "Bayesian classifiers applied to the Tennessee Eastman process," *Risk Analysis*, vol. 34, no. 3, pp. 485–497, 2014.

[78] Y. Zhao, J. Wen, S. Wang, "Diagnostic Bayesian networks for diagnosing air handling units faults — Part II: Faults in coils and sensors," *Applied Thermal Engineering*, vol. 90, pp. 145–157, 2015.

[79] C. F. Chien, S. L. Chen, Y. S. Lin, "Using Bayesian network for fault location on distribution feeder," *IEEE Transactions on Power Delivery*, vol. 17, no. 3, pp. 785–793, 2002.

[80] Y. Ling, S. Mahadevan, "Integration of structural health monitoring and fatigue damage prognosis," *Mechanical Systems and Signal Processing*, vol. 28, pp. 89–104, 2012.

[81] B. Huang, "Bayesian methods for control loop monitoring and diagnosis," *Journal of Process Control*, vol. 18, no. 9, pp. 829–838, 2008.

[82] B. Cai, Y. Liu, Y. Zhang, Q. Fan, S. Yu, "Dynamic Bayesian networks based performance evaluation of subsea blowout preventers in presence of imperfect repair," *Expert Systems with Applications*, vol. 40, no. 18, pp. 7544–7554, 2013.

[83] G. Arroyo-Figueroa, L. E. Sucar, A. Villavicencio, "Probabilistic temporal reasoning and its application to fossil power plant operation," *Expert Systems with Applications*, vol. 15, no. 3, pp. 317–324, 1998.

[84] G. Arroyo-Figueroa, Y. Alvarez, L. E. Sucar, "SEDRET — An intelligent system for the diagnosis and prediction of events in power plants," *Expert Systems with Applications*, vol. 18, no. 2, pp. 75–86, 2000.

[85] G. Arroyo-Figueroa, L. E. Sucar, "Temporal Bayesian network of events for diagnosis and prediction in dynamic domains," *Applied Intelligence*, vol. 23, no. 2, pp. 77–86, 2005.

[86] U. Lerner, R. Parr, D. Koller, G. Biswas, "Bayesian fault detection and diagnosis in dynamic systems," In: *AAAI/IAAI*, pp. 531–537, 2000.

[87] H. Y. Kao, C. H. Huang, H. L. Li, "Supply chain diagnostics with dynamic Bayesian networks," *Computers & Industrial Engineering*, vol. 49, no. 2, pp. 339–347, 2005.

[88] T. Kohda, W. Cui, "Risk-based reconfiguration of safety monitoring system using dynamic Bayesian network," *Reliability Engineering & System Safety*, vol. 92, no. 12, pp. 1716–1723, 2007.

[89] R. Gonzalez, B. Huang, F. Xu, A. Espejo, "Dynamic Bayesian approach to gross error detection and compensation with application toward an oil sands process," *Chemical Engineering Science*, vol. 67, no. 1, pp. 44–56, 2012.

[90] D. A. Tobon-Mejia, K. Medjaher, N. Zerhouni, "CNC machine tool's wear diagnostic and prognostic by using dynamic Bayesian networks," *Mechanical Systems and Signal Processing*, vol. 28, pp. 167–182, 2012.

[91] Q. B. Duong, E. Zamai, K. Q. Tran-Dinh, "Confidence estimation of feedback information for logic diagnosis," *Engineering Applications of Artificial Intelligence*, vol. 26, no. 3, pp. 1149–1161, 2013.

[92] J. Yu, M. M. Rashid, "A novel dynamic Bayesian network-based networked process monitoring approach for fault detection, propagation identification, and root cause diagnosis," *AIChE Journal*, vol. 59, no. 7, pp. 2348–2365, 2013.

[93] Y. Zhao, J. Tong, L. Zhang, Q. Zhang, "Pilot study of dynamic Bayesian networks approach for fault diagnostics and accident progression prediction in HTR-PM," *Nuclear Engineering and Design*, vol. 291, pp. 154–162, 2015.

[94] X. Wu, H. Liu, L. Zhang, M. J. Skibniewski, Q. Deng, J. Teng, "A dynamic Bayesian network based approach to safety decision support in tunnel construction," *Reliability Engineering & System Safety*, vol. 134, pp. 157–168, 2015.

[95] T. D. Nielsen, F. V. Jensen, *Bayesian Networks and Decision Graphs*, Springer Science & Business Media, 2009.

[96] D. Koller, A. Pfeffer, "Object-oriented Bayesian networks," In: *Proceedings of the Thirteenth Conference on Uncertainty in Artificial Intelligence*, pp. 302–313, Morgan Kaufmann Publishers Inc, 1997.

[97] G. Weidl, A. L. Madsen, S. Israelson, "Applications of object-oriented Bayesian networks for condition monitoring, root cause analysis and decision support on operation of complex continuous processes," *Computers & Chemical Engineering*, vol. 29, no. 9, pp. 1996–2009, 2005.

[98] T. B. Jensen, A. R. Kristensen, N. Toft, N. P. Baadsgaard, S. Østergaard, H. Houe, "An object-oriented Bayesian network modeling the causes of leg disorders in finisher herds," *Preventive Veterinary Medicine*, vol. 89, no. 3, pp. 237–248, 2009.

[99] R. Barco, L. Díez, V. Wille, P. Lázaro, "Automatic diagnosis of mobile communication networks under imprecise parameters," *Expert Systems with Applications*, vol. 36, no. 1, pp. 489–500, 2009.

[100] M. S. Sayed, N. Lohse, "Distributed Bayesian diagnosis for modular assembly systems — A case study," *Journal of Manufacturing Systems*, vol. 32, no. 3, pp. 480–488, 2013.

[101] Y. Zhou, P. Li, "Application of Bayesian network in failure diagnosis of hydro-electrical simulation system," In: *Intelligent Decision Technologies*, pp. 691–698, Springer, Berlin, 2011.

[102] I. Mandal, "Developing new machine learning ensembles for quality spine diagnosis," *Knowledge-Based Systems*, vol. 73, pp. 298–310, 2015.

[103] Z. Liu, Y. Liu, H. Shan, B. Cai, Q. Huang, "A fault diagnosis methodology for gear pump based on EEMD and Bayesian network," *PLoS One*, vol. 10, no. 5, e0125703, 2015.

[104] H. C. Cho, J. Knowles, M. S. Fadali, K. S. Lee, "Fault detection and isolation of induction motors using recurrent neural networks and dynamic Bayesian modeling," *IEEE Transactions on Control Systems Technology*, vol. 18, no. 2, pp. 430–437, 2010.

[105] F. Sahin, M. Ç. Yavuz, Z. Arnavut, Ö. Uluyol, "Fault diagnosis for airplane engines using Bayesian networks and distributed particle swarm optimization," *Parallel Computing*, vol. 33, no. 2, pp. 124–143, 2007.

[106] K. Li, Q. Zhang, K. Wang, P. Chen, H. Wang, "Intelligent condition diagnosis method based on adaptive statistic test filter and diagnostic bayesian network," *Sensors*, vol. 16, no. 1, pp. 76, 2016.

[107] V. Agrawal, B. K. Panigrahi, P. M. V. Subbarao, "Intelligent decision support system for detection and root cause analysis of faults in coal mills," *IEEE Transactions on Fuzzy Systems*, vol. 25, no. 4, pp. 934–944.

[108] M. F. D'Angelo, R. M. Palhares, L. B. Cosme, L. A. Aguiar, F. S. Fonseca, W. M. Caminhas, "Fault detection in dynamic systems by a fuzzy/Bayesian network formulation," *Applied Soft Computing*, vol. 21, pp. 647–653, 2014.

[109] Y. Liu, S. Jin, "Application of Bayesian networks for diagnostics in the assembly process by considering small measurement data sets," *The International Journal of Advanced Manufacturing Technology*, vol. 65, no. 9–12, pp. 1229–1237, 2013.

[110] K. Liu, X. Zhang, J. Shi, "Adaptive sensor allocation strategy for process monitoring and diagnosis in a Bayesian network," *IEEE Transactions on Automation Science and Engineering*, vol. 11, no. 2, pp. 452–462, 2014.

[111] N. L. Ricker, "Decentralized control of the Tennessee Eastman challenge process," *Journal of Process Control*, vol. 6, no. 4, pp. 205–221, 1996.

[112] S. Verron, J. Li, T. Tiplica, "Fault detection and isolation of faults in a multivariate process with Bayesian network," *Journal of Process Control*, vol. 20, no. 8, pp. 902–911, 2010.

[113] M. Borunda, O. A. Jaramillo, A. Reyes, P. H. Ibargüengoytia, "Bayesian networks in renewable energy systems: A bibliographical survey," *Renewable and Sustainable Energy Reviews*, vol. 62, pp. 32–45, 2016.

[114] M. Najafi, D. M. Auslander, P. L. Bartlett, P. Haves, M. D. Sohn, "Application of machine learning in the fault diagnostics of air handling units," *Applied Energy*, vol. 92, pp. 347–358, 2012.

[115] B. Wang, Y. Wang, X. Chen, "Research on wind turbine generator dynamic reliability test system based on feature recognition," *Research Journal of Applied Sciences, Engineering and Technology*, vol. 6, no. 16, pp. 3065–3071, 2013.

[116] I. V. de Bessa, R. M. Palhares, M. F. D'Angelo, J. E. Chaves Filho, "Data-driven fault detection and isolation scheme for a wind turbine benchmark," *Renewable Energy*, vol. 87, pp. 634–645, 2016.

[117] Z. Liu, Y. Liu, D. Zhang, B. Cai, C. Zheng, "Fault diagnosis for a solar assisted heat pump system under incomplete data and expert knowledge," *Energy*, vol. 87, pp. 41–48, 2015.

[118] L. A. Riascos, M. G. Simoes, P. E. Miyagi, "On-line fault diagnostic system for proton exchange membrane fuel cells," *Journal of Power Sources*, vol. 175, no. 1, pp. 419–429, 2008.

[119] O. J. Mengshoel, M. Chavira, K. Cascio, S. Poll, "Darwiche A, Uckun S. Probabilistic model-based diagnosis: An electrical power system case study," *IEEE Transactions on Systems, Man, and Cybernetics, Part A: Systems and Humans*, vol. 40, no. 5, pp. 874–885, 2010.

[120] S. Arangio, F. Bontempi, M. Ciampoli, "Structural integrity monitoring for dependability," *Structure and Infrastructure Engineering*, vol. 7, no. 1–2, pp. 75–86, 2011.

[121] B. Li, T. Han, F. Kang, "Fault diagnosis expert system of semiconductor manufacturing equipment using a Bayesian network," *International Journal of Computer Integrated Manufacturing*, vol. 26, no. 12, pp. 1161–1171, 2013.

[122] Á. Carrera, C. A. Iglesias, J. García-Algarra, D. Kolařík, "A real-life application of multi-agent systems for fault diagnosis in the provision of an Internet business service," *Journal of Network and Computer Applications*, vol. 37, pp. 146–154, 2014.

[123] Z. Li, L. Cheng, X. S. Qiu, L. Wu, "Fault diagnosis for high-level applications based on dynamic Bayesian network," In: *Asia-Pacific Network Operations and Management Symposium*, pp. 61–70, Springer, Berlin, 2009.

[124] A. McAfee, E. Brynjolfsson, "Big data: the management revolution," *Harvard Business Review*, vol. 90, no. 10, pp. 60–68, 2012.

[125] B. Ricks, O. J. Mengshoel, "Diagnosis for uncertain, dynamic and hybrid domains using Bayesian networks and arithmetic circuits," *International Journal of Approximate Reasoning*, vol. 55, no. 5, pp. 1207–1234, 2014.

[126] S. He, Z. Wang, Z. Wang, X. Gu, Z. Yan, "Fault detection and diagnosis of chiller using Bayesian network classifier with probabilistic boundary," *Applied Thermal Engineering*, vol. 107, pp. 37–47, 2016.

[127] B. Jones, I. Jenkinson, Z. Yang, J. Wang, "The use of Bayesian network modelling for maintenance planning in a manufacturing industry," *Reliability Engineering & System Safety*, vol. 95, no. 3, pp. 267–277, 2010.

[128] R. Rebba, S. Mahadevan, S. Huang, "Validation and error estimation of computational models," *Reliability Engineering & System Safety*, vol. 91, no. 10, pp. 1390–1397, 2006.

[129] B. Cai, Y. Zhao, H. Liu, M. Xie, "A data-driven fault diagnosis methodology in three-phase inverters for PMSM drive systems," *IEEE Transactions on Power Electronics*, DOI: 10.1109/TPEL.2016.2608842, 2016.

[130] National Research Council (US), *Assessing the Reliability of Complex Models: Mathematical and Statistical Foundations of Verification, Validation, and Uncertainty Quantification*, National Academies Press, 2012.

[131] O. Pourret. N. Patrick, M. Bruce, *Bayesian Networks: A Practical Guide to Applications*, John Wiley & Sons, 2008.

[132] G. Vachtsevanos, F. Lewis, M. Roemer, A. Hess, B. Wu. *Intelligent Fault Diagnosis and Prognosis for Engineering Systems*, John Wiley & Sons, 2006.

Chapter 2

Multi-Source Information Fusion-Based Fault Diagnosis of Ground-Source Heat Pump Using Bayesian Network

In order to increase the diagnostic accuracy of ground-source heat pump (GSHP) system, especially for multiple simultaneous faults, this chapter proposes a multi-source information fusion-based fault diagnosis methodology by using Bayesian network (BN) due to the fact that it is considered to be one of the most useful models in the field of probabilistic knowledge representation and reasoning, and can deal with the uncertainty problem of fault diagnosis well. BNs based on sensor data and information observed by human beings are established. Each BN consists of two layers: a fault layer and a fault symptom layer. The BN structure is established according to the cause and effect sequence of faults and symptoms, and the parameters are studied by using noisy-OR and noisy-MAX models. The entire fault diagnosis model is established by combining the two proposed BNs. Six fault diagnosis cases of GSHP system are studied, and the results show that the fault diagnosis model using evidences from only sensor data is accurate for a single fault, but not accurate enough for multiple simultaneous faults. By adding the observed information as evidence, the probability of a fault being present for a single fault of "Refrigerant overcharge" increases to 100% from 99.69%, and the probabilities of fault being present for multiple simultaneous faults of "Non-condensable gas" and "Expansion valve port largen" increase to almost 100% from 61.1% and 52.3%, respectively. In addition, the observed information can correct the wrong fault diagnostic results, such as "Evaporator fouling". Therefore, the multi-source information fusion-based fault diagnosis model using BN can increase the fault diagnostic accuracy greatly.

2.1. Introduction

Ground-source heat pumps (GSHPs) recovering heat from ground, have been widely utilized all over the world. GSHPs result in primary energy consumption reduction of up to 60% compared to conventional heating systems are of great significance in energy saving and environment protection [1–4]. Failure of the heat pump will cause reduction of energy efficiency and increment of environmental pollution. The relevant faults occurring in GSHPs are divided into hard faults and soft faults. Generally, hard faults are easy to detect and estimate, and soft faults are more difficult to discover [5]. The common hard faults include: (a) Compressor hard shutdown; (b) Complete valve choke; (c) Fan stops running, and so on. On the other hand, the common soft faults include: (a) Refrigerant overcharge; (b) Refrigerant leakage; (c) Evaporator fouling, and so on. Various fault diagnosis techniques are developed and used to locate the soft faults exactly in heat pump systems.

Using fault diagnosis techniques, the degradation performance of heat pump systems can be detected early, and the exact reasons for degradation can be diagnosed [6]. Xiao *et al.* [7] presented a fault diagnosis strategy based on a simple regression model and a set of generic rules for centrifugal chillers. Lee *et al.* [8] described a scheme for online fault detection and diagnosis at the subsystem level in an air-handling unit using general regression neural networks, which consisted of process estimation, residual generation, and fault detection and diagnosis. Wang and Cu [9] developed an online strategy to detect, diagnose, and validate sensor faults in centrifugal chillers using principal-component analysis method. Mohanraj *et al.* [10, 11] reviewed the applications of artificial neural networks for refrigeration, air conditioning and heat pumps, and presented the suitability of artificial neural network to predict the performance of a direct expansion solar-assisted heat pump, and the experiments were performed. Li and Braun [12] extended the decoupling-based fault detection and diagnosis method to heat pumps and developed diagnostic features for leakage within check valves and reversing valves Sun *et al.* [13] developed an online sensor fault detection and diagnosis strategy based on data fusion technology to detect faults in the building cooling load direct measurement. Najafi *et al.* [14] developed diagnostic algorithms for air handling units that can address such constraints more effectively, such as modeling limitations, measurement constraints, and the complexity of concurrent faults, by systematically employing machine-learning techniques. Gang and Wang [15] developed artificial neural network models for predicting the temperature of the water exiting the ground heat exchanger. A numerical simulation package of a hybrid GSHP system is adopted for training and testing the model.

Bayesian network (BN) is considered to be one of the most useful models in the field of probabilistic knowledge representation and reasoning, which has been widely used in reliability evaluation and fault diagnosis. Cai *et al.* [16–18] studied the reliability of subsea blowout preventer control system, subsea blowout preventer operations, and human factors on offshore blowouts by using BN or dynamic BN. Langseth and Portinale [19] and Weber *et al.* [20] presented a bibliographical review over the last decade on the application of BN for reliability, dependability, risk analysis, and maintenance. Recently, the application of BN on fault diagnosis has been investigated deeply. Dey and Stori [21] developed and presented a process monitoring and diagnosis approach based on a Bayesian belief network for incorporating multiple process metrics from multiple sensor sources in sequential machining operations to identify the root cause of process variations and provide a probabilistic confidence level of the diagnosis. Sahin *et al.* [22] presented a fault diagnosis system for airplane engines using BNs and distributed particle swarm optimization. Gonzalez *et al.* [23] developed a methodology for the real-time detection and quantification of instrument gross error Zhu *et al.* [24] proposed an active and dynamic method for to achieve rapid and precise diagnosis of crop diseases, using BNs to represent the relationships among the symptoms and crop diseases. However, there are few application of BN in heating, ventilation, and air conditioning systems. Zhao *et al.* [25] proposed a generic intelligent fault detection and diagnosis strategy to simulate the actual diagnostic outlook of chiller experts and developed a three-layer diagnostic BN to diagnose chiller faults based on the BN theory.

In order to increase the diagnostic accuracy, especially for multiple simultaneous faults this work presented a multi-source information fusion-based fault diagnosis methodology for a GSHP system by using the BN method. The proposed BN consists of two layers: a fault layer and a fault symptom layer. The fault symptom layer includes not only sensor data but also observed information, which can increase the fault diagnostic accuracy greatly. The chapter is structured as follows. Section 2 presents the faults and fault symptoms of a GSHP system. In Sec. 3, the fault diagnosis methodology is developed using BN. In Sec. 4, the fault diagnosis results using evidences from sensor data and observed information are researched. Section 5 summarizes the chapter.

2.2. Faults and Fault Symptoms

The schematic diagram of a GSHP system in the heating mode is depicted in Fig. 2.1 [26–28]. The system mainly consists of three major circuits: (a) the ground heat exchanger circuit, (b) the heat pump unit circuit, and (c) the indoor fan coil circuit [29–32]. The ground heat exchanger circuit comprises of a ground heat exchanger

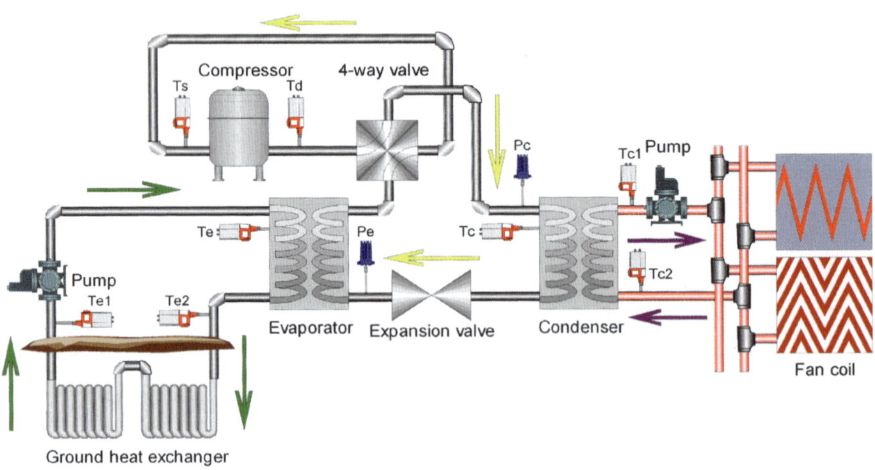

Fig. 2.1. Layout of a GSHP system in the heating mode.

and a water pump; the heat pump unit circuit is composed of a compressor, an evaporator, a condenser, an electronic expansion valve, and a 4-way valve; and the indoor fan coil circuit is made up of several indoor fan coils and a water pump. The Coefficient of Performance (COP) of the GSHP system is 3.5, and the noise can be controlled to less than 65 dB.

As mentioned above, the soft faults of a GSHP system are difficult to detect, and are diagnosed by monitoring the system status. Based on a review of the references and practical experience, eight soft faults are analyzed in this work: (a) Refrigerant overcharge (ReOv); (b) Refrigerant leakage (ReLe); (c) Evaporator fouling (EvFo); (d) Condenser fouling (CoFo); (e) Non-condensable gas (NcGa); (f) Compressor suction or discharge valve leakage (CoVL); (g) Expansion valve port largen (ExPL); and (h) High-pressure pipe line blockage (HPLB) [33–36]. Each fault has two states, i.e., present and absent.

The status of GSHPs is monitored by using temperature sensors and pressure sensors. The fault symptoms therefore include: (a) Evaporating pressure (EvaPr, Pe); (b) Condensing pressure (ConPr, Pc); (c) Evaporating temperature (EvaTe, Te); (d) Condensing temperature (ConTe, Tc); (e) Compressor suction temperature (ComST, Ts); (f) Compressor discharge temperature (ComDT, Td); (g) Evaporator water temperature difference (EvaTD, ΔTe); and (h) Condenser water temperature difference (ConTD, ΔTc) Each fault symptom obtained from the sensor data has three states, higher, lower and normal.

The relationships between faults and symptoms obtained from sensor data are given in Table 2.1 Taking ReLe for example, the refrigerant in the heat pump unit circuit decreases because of refrigerant leakage, making both the evaporating pressure and condensing pressure decrease. The refrigerant discharge superheat temperature therefore increases, making the compressor suction temperature and discharge temperature increase. Due to the fact that the heat pump works in a state of ill health with insufficient refrigerant, the heat absorption capacity and heating capacity decrease; therefore, the evaporating temperature, condensing temperature, evaporator water

Table 2.1. Relationship between faults and symptoms obtained from sensor data.

	Fault symptom							
Fault	EvaPr	ConPr	EvaTe	ConTe	ComST	ComDT	EvaTD	ConTD
ReOv	Higher	Higher	Higher	N/A	Lower	Lower	Normal	Higher
ReLe	Lower	Lower	Lower	Lower	Higher	Higher	Lower	Lower
EvFo	Lower	Lower	Lower	Lower	Lower	Lower	Higher	Higher
CoFo	Lower	Higher	Higher	Higher	Higher	Higher	Lower	Higher
NcGa	Lower	Higher	N/A	Higher	Higher	Higher	Lower	Lower
CoVL	Lower	Lower	Higher	Lower	Higher	Lower	Lower	Lower
ExPL	Higher	Higher	Higher	N/A	Lower	Lower	Lower	Normal
HPLB	Lower	Higher	Lower	Lower	Higher	Higher	Lower	Lower

Table 2.2. Relationship between faults and symptoms obtained from observed information.

Fault	Fault symptom
ReOv	Compressor cannot stop (CoNoS) Compressor surface frost (CoSuF) Compressor vibration (CoVib)
ReLe	Too much foam (ToMuF) Compressor surface frost (CoSuF) Pungent odor (PunOd) Grease stains in wiped joint (GrSWJ)
NcGa	Compressor discharge pressure gauge vibration (CoGaV)
CoVL	Compressor cannot stop (CoNoS)
ExPL	Too much foam (ToMuF) Compressor surface frost (CoSuF)

temperature difference and condenser water temperature difference all decrease.

In addition, several fault symptoms can be observed directly by human beings, such as (a) Compressor cannot stop; (b) Compressor surface frost; and (c) Compressor vibration. These symptoms can help to diagnose the faults of a GSHP system more accurately. The relationship between faults and symptoms obtained from observed information is given in Table 2.2. Similarly, taking ReLe for example, four observed fault symptoms including (a) Too much foam (ToMuF), (b) Compressor surface frost (CoSuF), (c) Pungent odor (PunOd), and (d) Grease stains in wiped joint (GrSWJ) can be caused by refrigerant leakage. Each fault symptom obtained from the observed information has two states, present and absent.

2.3. Fault Diagnosis Methodology

2.3.1. *Fault diagnosis based on sensor data*

The fault diagnosis model of a GSHP system is established by using BN method. Specifically, each BN is constructed in two consecutive steps, which are defining the network structure and then defining the network parameters.

2.3.1.1. *BN structure*

The BN structure is established according to the cause and effect sequence of events. In this work, faults of the GSHP system, such as refrigerant overcharge and evaporator fouling, are the causes, and fault symptoms, such as evaporating pressure is higher, compressor suction temperature is lower, are the consequences. The relationship is denoted by an arc between them. According to the relationship between faults and fault symptoms obtained from sensor data given in Table 2.1, the BNs for fault diagnosis are established as shown in Fig. 2.2. The proposed BN structure consists of two layers: a fault layer and a fault symptom layer. The fault layer consists of eight parent nodes, indicating eight potential faults concerned. The symptom layer consists of eight child nodes, indicating eight fault symptoms obtained from the sensor data. Taking the node EvaPr, for example, it is connected to its eight parent nodes according to eight arcs, which indicates that the fault symptom EvaPr is related to all of the eight faults.

2.3.1.2. *BN parameters*

Prior probabilities and conditional probabilities are required to specify for BNs. The prior probability of an event is the probability of the event computed before the arrival of new evidence or information.

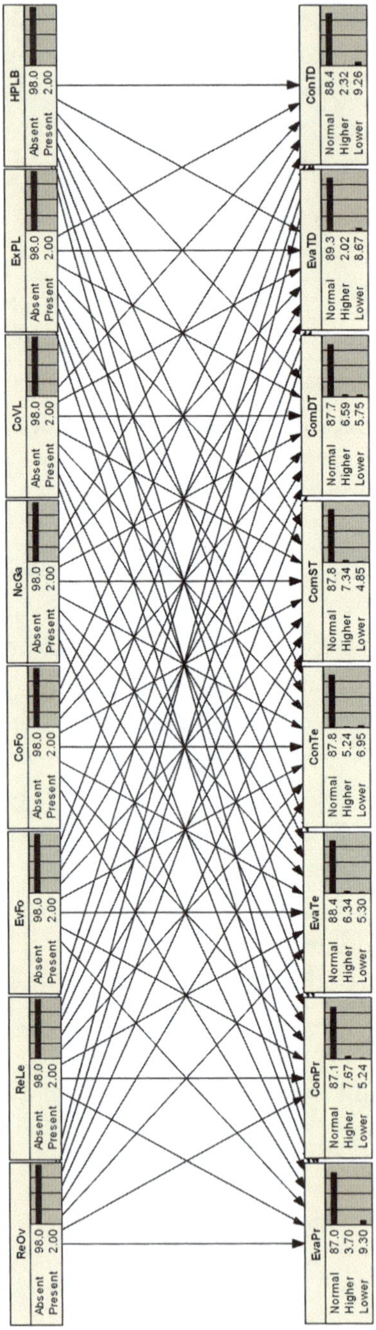

Fig. 2.2. BNs for fault diagnosis using sensor data.

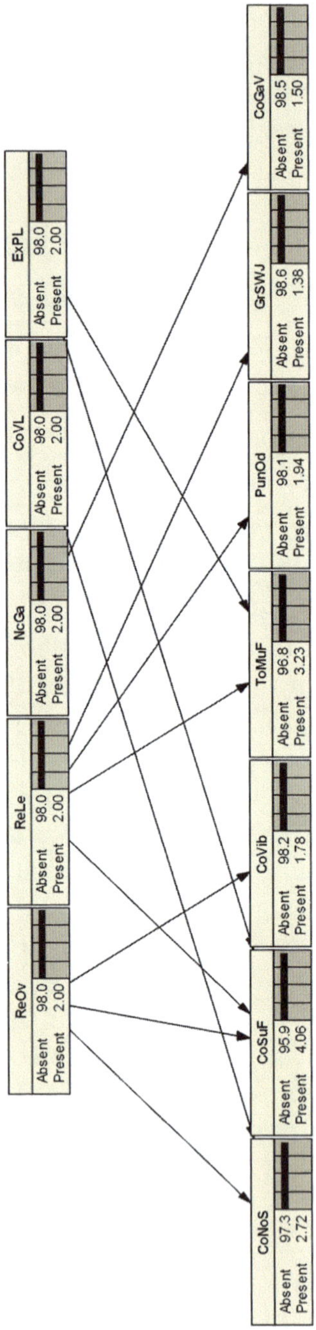

Fig. 2.3. BNs for fault diagnosis using observed information.

It is obtained based on the experiences of the experts and statistical analysis of historical data. The higher the prior probability of an event, the more likely the event is to happen. For the GSHP system, the same prior probabilities of faults are assumed, in order to emphasize the posterior probabilities given new evidences. As shown in Fig. 2.2, the probabilities of faults are all 2%.

A conditional probability is the probability that an event will occur, when another event is known to occur or to have occurred. It is also obtained based on the experiences of experts and statistical analysis of historical data. One of the major issues faced is the exponential growth of the number of parameters in the conditional probability tables. The specification of a complete conditional probability table for a child node m with s_m states and n parent nodes requires the assessment of $(s_m - 1)\prod_{i=1}^{n} s_i$ probabilities, where s_i is the number of states of parent node i [37]. The most common practical solution is the application of noisy-MAX to simplify the conditional probability tables. The noisy gate needs to meet following three assumptions: (a) the child node and all its parents must be variables indicating the degree of presence of an anomaly; (b) each of the parent nodes must represent a cause that can produce the effect (the child node) in the absence of the other causes; and (c) there may be no significant synergies among the causes [38]. Therefore, only $\sum_{1}^{n}(s_m - 1)s_i$ probabilities are required to specify the conditional probability tables, thereby simplifying knowledge acquisition greatly.

Suppose, for example, there are n causes $X_1, X_2, \ldots, X_n$ of Y, by using noisy-MAX, the full conditional probability relationship can be written as [39, 40]

$$P(Y \leq y|X) = \prod_{\substack{i=1 \\ x_i \neq 0}}^{n} \sum_{y'=0}^{y} q_{i,y'}^{x_i}, \tag{2.1}$$

$$P(Y = y|X) = \begin{cases} P(Y \leq 0|X) & \text{if } y = 0, \\ P(Y \leq y|X) - P(Y \leq y-1|X) & \text{if } y > 0. \end{cases} \tag{2.2}$$

where X represents a certain configuration of the parents of Y, $X = x_1, \ldots, x_n$, and $P(Y = 0|X_1 = 0, \ldots, X_n = 0) = 1$.

It can be seen from Table 2.1 that when a fault occurs, the corresponding fault symptom occurs theoretically. For example, the fault ReOv causes the fault symptom EvaPr "higher". However, in practice, the fault symptom is uncertain, for example, the fault ReOv can cause the fault symptoms EvaPr "higher", "lower", or "normal". The existing uncertainty problem is caused due to various reasons, and sensor accuracy and measure uncertainty are the important causes. In the current work, one designer and two repairmen of GSHP systems were invited to determine the relationship between the parent nodes and the child nodes for sensor data, as given in Table 2.3. By using the relationship and Eqs. (2.1) and (2.2), the conditional probability table can be computed.

2.3.2. *Fault diagnosis based on observed information*

2.3.2.1. *BN structure*

It is similar to the BN of fault diagnosis based on sensor data; the BN based on observed information also has two layers, as shown in Fig. 2.3. The fault layer consists of five parent nodes as described above, and the fault symptom layer consist of seven child nodes, as given in Table 2.2. Taking CoSuF for example, the observed information "Compressor surface frost" can be caused by "Refrigerant overcharge", "Refrigerant leakage", and "Expansion valve port largen"; therefore, the arcs from parent nodes ReOv, ReLe and ExPL are connected to the child node CoSuF.

2.3.2.2. *BN parameters*

The prior probabilities of all the parent nodes are also 2% for the present state, as mentioned above. The conditional probability table is simplified by using the noisy-OR model, because the states of the parent and child nodes are both binary. Suppose, for example, there are n causes $X_1, X_2, \ldots, X_n$ of Y, by using noisy-MAX, the full conditional probability relationship can be written as [41]

$$P(Y|X_1, X_2, \ldots, X_n) = 1 - \prod_{1\leq j\leq n} (1 - p_j). \tag{2.3}$$

Table 2.3. Relationship between parent nodes and child nodes for sensor data.

		Parent node (Present)							
Child node	State	ReOv	ReLe	EvFo	CoFo	NcGa	CoVL	ExPL	HPLB
EvaPr	Higher	0.80	0.00	0.11	0.04	0.12	0.11	0.78	0.10
	Lower	0.05	0.95	0.85	0.68	0.59	0.82	0.01	0.89
	Normal	0.15	0.05	0.04	0.28	0.29	0.07	0.21	0.01
ConPr	Higher	0.75	0.00	0.08	0.78	0.99	0.01	0.84	0.73
	Lower	0.05	0.90	0.69	0.05	0.00	0.83	0.05	0.10
	Normal	0.20	0.10	0.23	0.17	0.01	0.16	0.11	0.17
EvaTe	Higher	0.65	0.00	0.20	0.84	0.01	0.82	0.90	0.01
	Lower	0.10	0.92	0.72	0.12	0.01	0.02	0.01	0.80
	Normal	0.25	0.08	0.08	0.04	0.98	0.16	0.09	0.19
ConTe	Higher	0.97	0.05	0.12	0.78	0.89	0.00	0.01	0.05
	Lower	0.00	0.81	0.87	0.10	0.00	0.88	0.02	0.89
	Normal	0.03	0.14	0.01	0.12	0.11	0.12	0.97	0.06
ComST	Higher	0.02	0.69	0.05	0.86	0.59	0.85	0.08	0.84
	Lower	0.86	0.00	0.65	0.14	0.10	0.01	0.69	0.02
	Normal	0.12	0.31	0.30	0.00	0.31	0.14	0.23	0.14
ComDT	Higher	0.00	0.66	0.06	0.89	0.91	0.10	0.10	0.87
	Lower	0.79	0.00	0.88	0.02	0.00	0.59	0.57	0.09
	Normal	0.21	0.34	0.06	0.09	0.09	0.31	0.33	0.04
EvaTD	Higher	0.10	0.10	0.01	0.59	0.03	0.05	0.11	0.12
	Lower	0.02	0.60	0.91	0.05	0.69	0.68	0.68	0.87
	Normal	0.88	0.30	0.08	0.36	0.28	0.27	0.21	0.01
ConTD	Higher	0.79	0.00	0.01	0.12	0.01	0.08	0.08	0.19
	Lower	0.00	0.59	0.92	0.85	0.81	0.85	0.02	0.78
	Normal	0.21	0.41	0.07	0.03	0.18	0.07	0.90	0.03

Similarly, one designer and two repairmen of GSHP systems were invited to determine the relationship between the parent nodes and the child nodes for the observed information, as given in Table 2.4 By using the relationship and Eq. (2.3), the conditional probability table can be computed.

2.3.3. *Multi-source information fusion-based fault diagnosis*

In order to increase the fault diagnostic accuracy of GSHP system, the data and information obtained from sensor and human being

Table 2.4. Relationship between parent nodes and child nodes for observed information.

		Parent node (Present)				
Child node	State	ReOv	ReLe	NcGa	CoVL	ExPL
CoNoS	Present	0.59	—	—	0.78	—
	Absent	0.41	—	—	0.22	—
CoSuF	Present	0.65	0.75	—	—	0.66
	Absent	0.35	0.25	—	—	0.34
CoVib	Present	0.89	—	—	—	—
	Absent	0.11	—	—	—	—
ToMuF	Present	—	0.85	—	—	0.78
	Absent	—	0.15	—	—	0.22
PunOd	Present	—	0.97	—	—	—
	Absent	—	0.03	—	—	—
GrSWJ	Present	—	0.69	—	—	—
	Absent	—	0.31	—	—	—
CoGaV	Present	—	—	0.75	—	—
	Absent	—	—	0.25	—	—

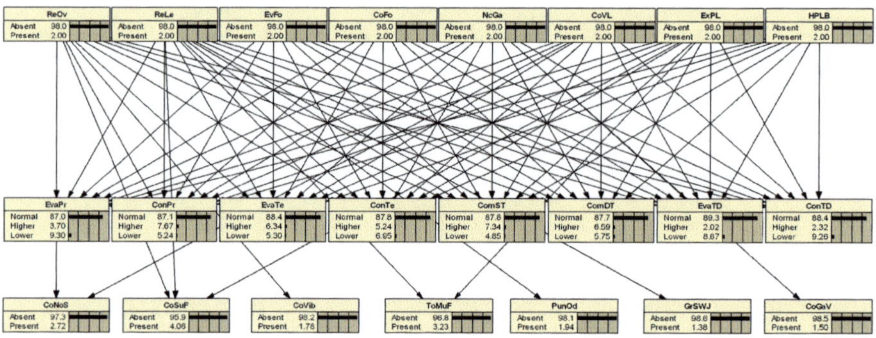

Fig. 2.4. The entire BNs for fault diagnosis.

are fused by using BN method, and the entire fault diagnosis model shown in Fig. 2.4 is established by combining the two sub-models in Figs. 2.2 and 2.3. Taking the fault, "Refrigerant overcharge" for example, it can cause not only a change in the eight sensors data, but also helps to observe information, such as "Compressor cannot stop", "Compressor surface frost", and "Compressor vibration".

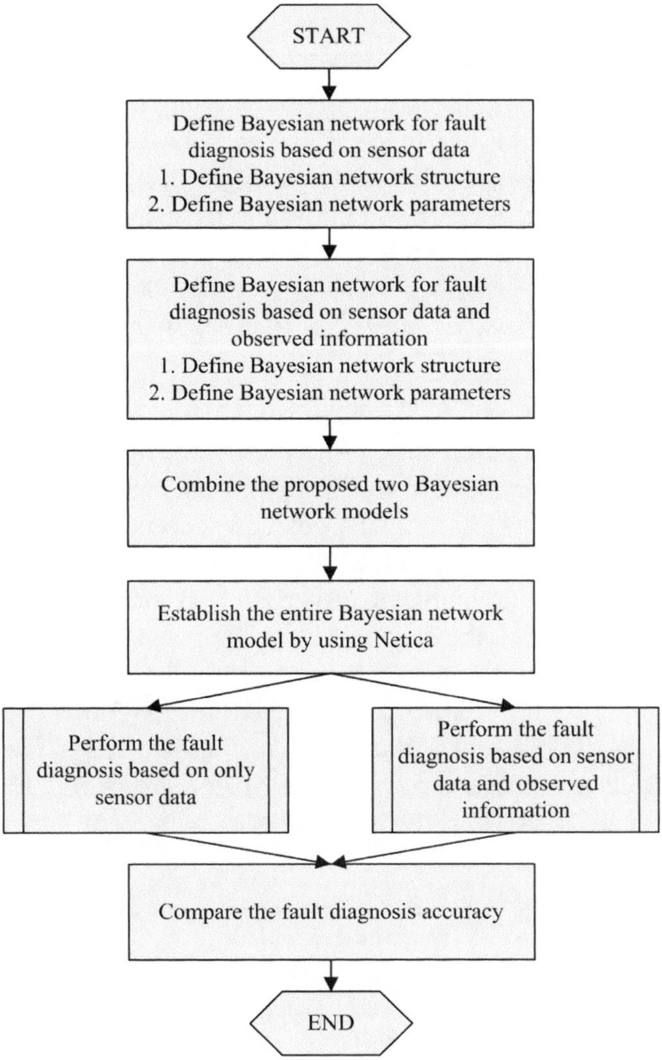

Fig. 2.5. Flow chart for the development of fault diagnosis methodology.

Therefore, the parent node ReOv is connected to 11 child nodes via 11 arcs.

The flow chart for the development of the multi-source information fusion-based fault diagnosis methodology for a GSHP system is given in Fig. 2.5.

2.4. Results and Discussion

In order to research the effects of evidence from sensor data and observed information, the fault diagnosis using evidences from only sensor data and from both sensor data and observed information are performed, and the posterior probability of any given set of a fault is calculated given some fault symptoms by using the approximate reasoning method of the Netica software.

2.4.1. *Fault diagnosis using evidences from only sensor data*

Table 2.5 gives three fault diagnosis cases using evidences from only sensor data, and the fault diagnosis results are shown in Fig. 2.6. For case No. 1, when the data from the eight sensors are set as evidences in the BNs shown in Fig. 2.2, the posterior probabilities of all of the faults are calculated, as shown in Fig. 2.6(a). It can be seen that the probability of a fault being present for "Refrigerant overcharge" is 99.69%, and the probabilities for the other seven fault are almost 0. The diagnostic result is in accordance with the fault found during practical operation. It indicates that the fault diagnosis model using evidences from only sensor data is accurate for a single fault.

The fault diagnostic results of case No. 2 are shown in Fig. 2.6(b), and it can be seen that the probabilities of a fault being present

Table 2.5. Three fault diagnosis cases using evidences from only sensor data.

	Case		
Evidence	No. 1	No. 2	No. 3
EvaPr	Higher	Lower	Lower
ConPr	Higher	Higher	Lower
EvaTe	Higher	Higher	Lower
ConTe	Higher	Higher	Lower
ComST	Lower	Lower	Higher
ComDT	Lower	Lower	Lower
EvaTD	Normal	Lower	Lower
ConTD	Higher	Lower	Lower

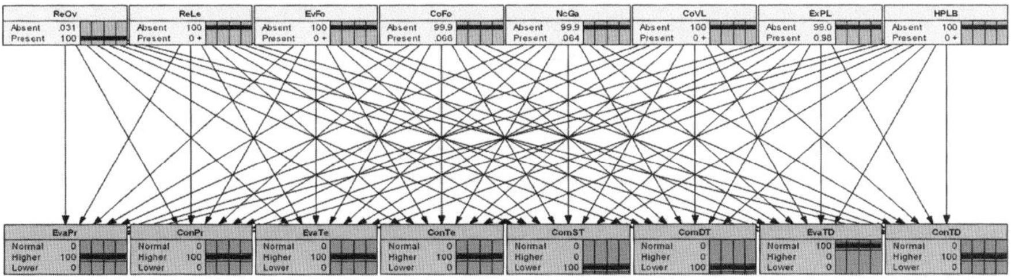

(a)

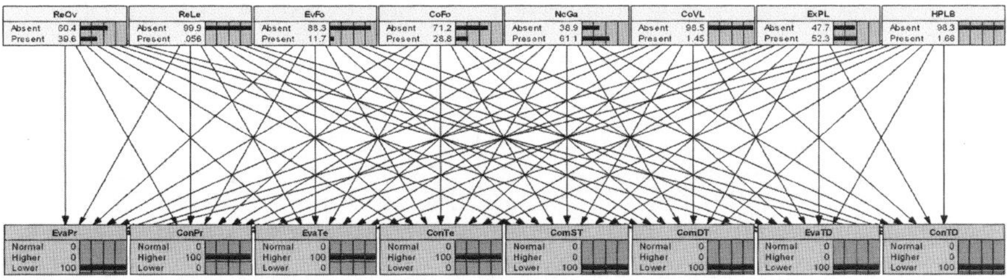

(b)

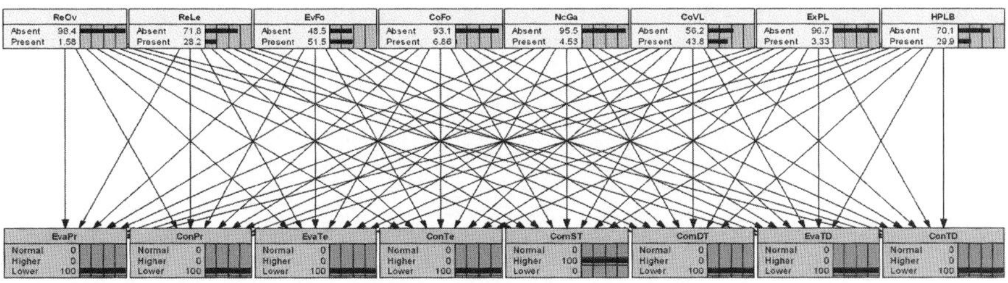

(c)

Fig. 2.6. Three fault diagnosis results using evidences from only sensor data (a) Case No. 1, (b) Case No. 2 and (c) Case No. 3.

for "Non-condensable gas" and "Expansion valve port largen" are more than 50%, indicating that the two faults occur simultaneously. The diagnostic result is in accordance with the faults found during practical operation.

However, for another case, No. 3, with multiple-simultaneous faults as shown in Fig. 2.6(c), the faults "Evaporator fouling" and "Compressor suction or discharge valve leakage" have the maximum posterior probabilities, which are not in accordance with the faults "Refrigerant leakage" and "Compressor suction or discharge valve leakage" found during practical operation. Therefore, the fault diagnosis model using evidences from only sensor data is not accurate enough for multiple-simultaneous faults.

2.4.2. *Fault diagnosis using evidences from sensor data and observed information*

In order to increase the fault diagnostic accuracy, the observed information is also taken as the evidence. Three other fault diagnosis cases using evidence from sensor data and observed information, as shown in Table 2.6, are studied. The fault diagnostic results are shown in Fig. 2.7.

The comparison of fault diagnostic results is given in Fig. 2.8, indicating the effects of the observed information on the fault diagnostic accuracy. It can be seen form Fig. 2.8(a) that when the observed information is set as the evidence for case No. 1+, the probability of a fault present for "Refrigerant overcharge" increases to 100% from 99.69%, while the probabilities of other faults decrease slightly. As shown in Fig. 2.8(b), the probabilities of fault present for "Non-condensable gas" and "Expansion valve port largen" increase to almost 100% from 61.1% and 52.3%, respectively. The two cases

Table 2.6. Three fault diagnosis cases using evidences from sensor data and observed information.

	Case		
Evidence	No. 1+	No. 2+	No. 3+
EvaPr	Higher	Lower	Lower
ConPr	Higher	Higher	Lower
EvaTe	Higher	Higher	Lower
ConTe	Higher	Higher	Lower
ComST	Lower	Lower	Higher
ComDT	Lower	Lower	Lower
EvaTD	Normal	Lower	Lower
ConTD	Higher	Lower	Lower
CoVib	Present	—	—
ToMuF	—	Present	Present
CoGaV	—	Present	—
CoNoS	—	—	Present

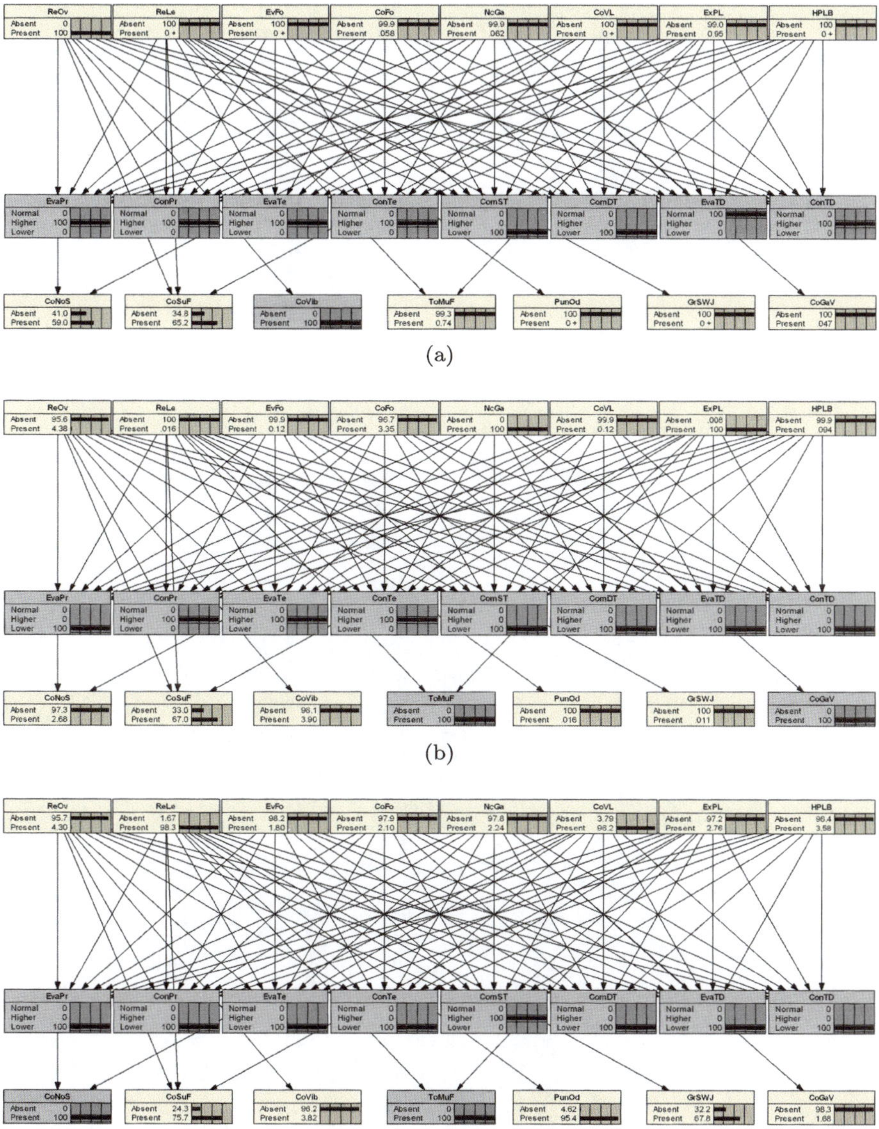

(a)

(b)

(c)

Fig. 2.7. Three fault diagnosis results using evidences from only sensor data (a) Case No. 1+, (b) Case No. 2+, and (c) Case No. 3+.

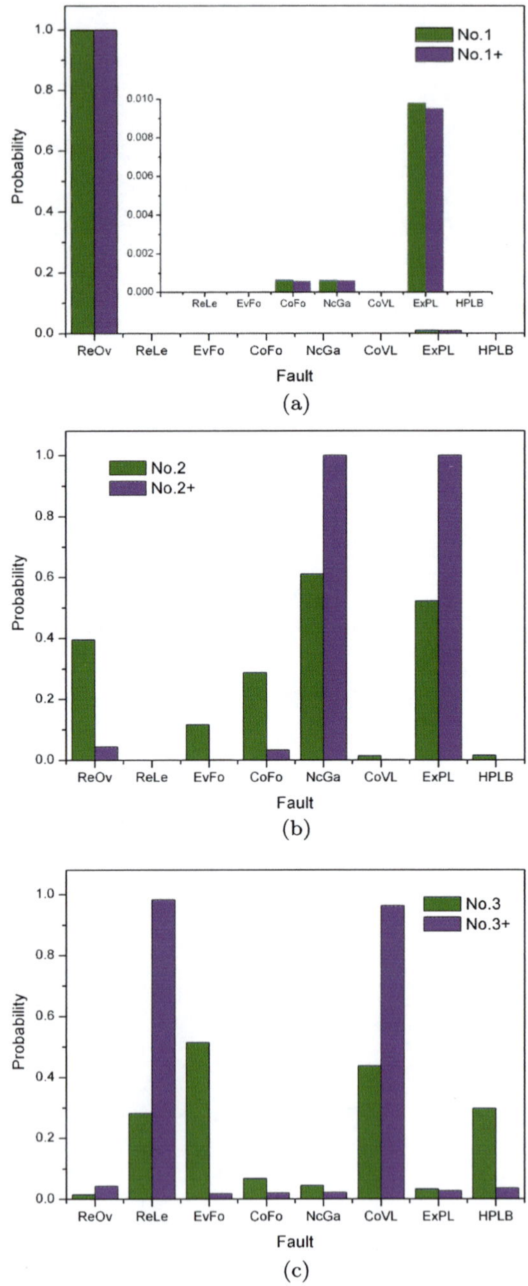

Fig. 2.8. Comparison of fault diagnosis results (a) Case No. 1 and No. 1+, (b) Case No. 2 and No. 2+, and (c) Case No. 3 and No. 3+.

show that the observed information can increase the fault diagnostic accuracy greatly.

As shown in Fig. 2.8(c), the probability of fault present for "Evaporator fouling" decreases to 1.8% from 51.5%, while the probability of "Refrigerant leakage" increases to 98.3% from 28.2%, and the probability of "Compressor suction or discharge valve leakage" increases to 96.2% from 43.8%. The diagnostic result is in accordance with the faults found during practical operation. Therefore, the observed information can correct the wrong fault diagnostic results.

Above all, the multi-source information fusion-based fault diagnosis model can increase the fault diagnostic accuracy greatly.

According to the research described above, it can be seen that the proposed BN-based fault diagnosis methodology is different from other artificial intelligence method-based fault diagnosis, such as that based on artificial neural networks [10] and data fusion technology [13]. The proposed methodology can deal with the uncertainty of fault and fault symptoms well. For example, the present of ReOv can cause EvaPr higher, lower, and normal with three probabilities of 80%, 5%, and 15%, which can be defined in the conditional probability table of BNs. In addition, several new observed information can be added into the fault diagnosis model easily to increase the diagnosis accuracy.

2.5. Conclusion

In order to increase the diagnostic accuracy, especially for multiple-simultaneous faults, the work proposed a multi-source information fusion-based fault diagnosis methodology for a GSHP system.

(1) The entire fault diagnosis model of a GSHP system is established by combining two proposed BNs, which are established according to the cause and effect sequence of faults and fault symptoms, including sensor data and information deserved by human beings.
(2) The fault diagnosis model using evidences from only sensor data is accurate for a single fault, for example, the probability of a fault being present for a single fault of "Refrigerant overcharge" is 99.69%.

(3) The fault diagnosis model using evidences from only sensor data is not accurate enough for multiple-simultaneous faults, for example, the faults "Evaporator fouling" and "Compressor suction or discharge valve leakage" have maximum posterior probabilities of 51.5% and 43.8%, which are not in accordance with the faults found during practical operation.
(4) The observed information can increase the fault diagnostic accuracy greatly for a single fault, for example, the probability of a fault being present for "Refrigerant overcharge" increases to 100% from 99.69%, while the probabilities of other faults decrease slightly.
(5) The observed information can increase the fault diagnostic accuracy greatly as well as correct wrong fault diagnostic results for multiple-simultaneous faults. For example, the probabilities of a fault being present for "Non-condensable gas" and "Expansion valve port largen" increase to almost 100% from 61.1% and 52.3%, respectively.
(6) The cases show that the multi-source information fusion-based fault diagnosis model using BN is effective for a GSHP system.

The work focuses on the BN-based fault diagnosis methodology, and a future scope of this work can be directed toward the development and validation of BN-based GSHP system automatic fault diagnosis software.

References

[1] A. Michopoulos, K. T. Papakostas, N. Kyriakis, "Potential of autonomous ground-coupled heat pump system installations in Greece," *Applied Energy*, vol. 88, pp. 2122–2129, 2011.
[2] Y. H. Bi, X. H. Wang, Y. Liu, H. Zhang, L. G. Chen, "Comprehensive energy analysis of a ground-source heat pump system for both building heating and cooling modes," *Applied Energy*, vol. 86, pp. 2560–2565, 2009.
[3] K. J. Chua, S. K. Chou, W. M. Yang, "Advances in heat pump systems: A review," *Applied Energy*, vol. 87, pp. 3611–3624, 2010.
[4] K. J. Chua, S. K. Chou, W. M. Yang, J. Yan, "Achieving better energy-efficient air conditioning — A review of technologies and strategies," *Applied Energy*, vol. 104, pp. 87–104, 2013.

[5] Y. M. Chen, L. L. Lan, "A fault detection technique for air-source heat pump water chiller heaters," *Energy and Buildings*, vol. 41, pp. 881–887, 2009.

[6] D. Zogg, E. Shafai, H. P. Geering, "Fault diagnosis for heat pumps with parameter identification and clustering," *Control Engineering Practice*, vol. 14, pp. 1435–1444, 2006.

[7] F. Xiao, C. Y. Zheng, S. W. Wang, "A fault detection and diagnosis strategy with enhanced sensitivity for centrifugal chillers," *Applied Thermal Engineering*, vol. 31, pp. 3963–3970, 2011.

[8] W. Y. Lee, J. M. House, N. H. Kyong, "Subsystem level fault diagnosis of a building's air-handling unit using general regression neural networks," *Applied Energy*, vol. 77, pp. 153–170, 2004.

[9] S. W. Wang, J. T. Cui, "Sensor-fault detection, diagnosis and estimation for centrifugal chiller systems using principal-component analysis method," *Applied Energy*, vol. 82, pp. 197–213, 2005.

[10] M. Mohanraj, S. Jayaraj, C. Muraleedharan, "Performance prediction of a direct expansion solar assisted heat pump using artificial neural networks," *Applied Energy*, vol. 86, pp. 1442–1449, 2009.

[11] M. Mohanraj, S. Jayaraj, C. Muraleedharan, "Applications of artificial neural networks for refrigeration, air conditioning and heat pumps — A review," *Renewable and Sustainable Energy Reviews*, vol. 16, pp. 1340–1358, 2012.

[12] H. Li, J. E. Braun, "Decoupling features for diagnosis of reversing and check valve faults in heat pumps," *International Journal of Refrigeration*, vol. 32, pp. 316–326, 2009.

[13] Y. J. Sun, S. W. Wang, G. S. Huang, "Online sensor fault diagnosis for robust chiller sequencing control," *International Journal of Thermal Sciences*, vol. 49, pp. 589–602, 2010.

[14] M. Najafi, D. M. Auslander, P. L. Bartlett, P. Haves, M. D. Sohn, "Application of machine learning in the fault diagnostics of air handling units," *Applied Energy*, vol. 96, pp. 347–358, 2012.

[15] W. J. Gang, J. B. Wang, "Predictive ANN models of ground heat exchanger for the control of hybrid ground source heat pump systems," *Applied Energy* 2013; doi:10.1016/j.rser.2011.10.015.

[16] B. P. Cai, Y. H. Liu, Z. K. Liu, X. J. Tian, X. Dong, S. L. Yu, "Using Bayesian networks in reliability evaluation for subsea blowout preventer control system," *Reliability Engineering & System Safety*, vol. 108, pp. 32–41, 2012.

[17] B. P. Cai, Y. H. Liu, Y. W. Zhang, Q. Fan, Z. K. Liu, X. J. Tian, "A dynamic Bayesian networks modelling of human factors on offshore blowouts," *Journal of Loss Prevention in the Process Industries*, vol. 26, pp. 639–649, 2013.

[18] B. P. Cai, Y. H. Liu, Z. K. Liu, X. J. Tian, Y. Z. Zhang, R. J. Ji, "Application of Bayesian networks in quantitative risk assessment of subsea blowout preventer operations," *Risk Analysis*, vol. 33, pp. 1293–1311, 2013.

[19] H. Langseth, L. Portinale, "Bayesian networks in reliability," *Reliability Engineering & System Safety*, vol. 92, pp. 92–108, 2007.

[20] P. Weber, G. Medina-Oliva, C. Simon, B. Iung, "Overview on Bayesian networks applications for dependability, risk analysis and maintenance areas," *Engineering Applications of Artificial Intelligence*, vol. 25, pp. 671–682, 2012.

[21] S. Dey, J. A. Stori, "A Bayesian network approach to root cause diagnosis of process variations," *International Journal of Machine Tools and Manufacture*, vol. 45, pp. 75–91, 2005.

[22] F. Sahin, M. C. Yavuz, Z. Arnavut, O. Uluyol, "Fault diagnosis for airplane engines using Bayesian networks and distributed particle swarm optimization," *Parallel Computing*, vol. 33, pp. 124–143, 2007.

[23] R. Gonzalez, B. Huang, F. W. Xu, A. Espejo, "Dynamic Bayesian approach to gross error detection and compensation with application toward an oil sands process," *Chemical Engineering Science*, vol. 67, pp. 44–56, 2012.

[24] Y. G. Zhu, D. Y. Liu, G. F. Chen, H. Y. Jia, H. L. Yu, "Mathematical modeling for active and dynamic diagnosis of crop diseases based on Bayesian networks and incremental learning," *Mathematical and Computer Modelling*, vol. 58, pp. 514–523, 2013.

[25] Y. Zhao, F. Xiao, S. W. Wang, "An intelligent chiller fault detection and diagnosis methodology using Bayesian belief network," *Energy and Buildings*, vol. 57, pp. 278–288, 2013.

[26] A. Capozza, M. D. Carli, A. Zarrella, "Investigations on the influence of aquifers on the ground temperature in ground-source heat pump operation," *Applied Energy*, vol. 107, pp. 350–363, 2013.

[27] Z. Sagia, C. Rakopoulos, E. Kakaras, "Cooling dominated hybrid ground source heat pump system application ," *Applied Energy*, vol. 94, pp. 41–47, 2012.

[28] S. J. Self, B. V. Reddy, M. A. Rosen, "Geothermal heat pump systems: Status review and comparison with other heating options," *Applied Energy*, vol. 101, pp. 341–348, 2013.

[29] Y. Man, H. X. Yang, J. G. Wang, Z. H. Fang, "*In situ* operation performance test of ground coupled heat pump system for cooling and heating provision in temperate zone," *Applied Energy*, vol. 97, pp. 913–920, 2012.

[30] C. Montagud, J. M. Corberán, F. Ruiz-Calvo, "Experimental and modeling analysis of a ground source heat pump system," *Applied Energy*, vol. 109, pp. 328–336, 2013.

[31] O. Ozyurt, D. A. Ekinci, "Experimental study of vertical ground-source heat pump performance evaluation for cold climate in Turkey," *Applied Energy*, vol. 88, pp. 1257–1265, 2011.

[32] H. Park, J. S. Lee, W. Kim, Y. Kim, "The cooling seasonal performance factor of a hybrid ground-source heat pump with parallel and serial configurations," *Applied Energy*, vol. 102, pp. 877–884, 2013.

[33] Y. Zhao, S. W. Wang, F. Xiao, "A statistical fault detection and diagnosis method for centrifugal chillers based on exponentially-weighted moving

average control charts and support vector regression," *Applied Thermal Engineering*, vol. 51, pp. 560–572, 2013.

[34] M. Kim, W. V. Payne, P. A. Domanski, S. H. Yoon, C. J. L. Hermes, "Performance of a residential heat pump operating in the cooling mode with single faults imposed," *Applied Thermal Engineering*, vol. 29, pp. 770–778, 2009.

[35] S. H. Yoon, W. V. Payne, P. A. Domanski, "Residential heat pump heating performance with single faults imposed," *Applied Thermal Engineering*, vol. 31, pp. 765–771, 2011.

[36] D. C. Gao, S. W. Wang, Y. J. Sun, F. Xiao, "Diagnosis of the low temperature difference syndrome in the chilled water system of a super high-rise building: A case study," *Applied Energy*, vol. 98, pp. 597–606, 2012.

[37] M. Pradhan, G. Provan, B. Middleton, M. Henrion, "Knowledge engineering for large belief networks," *Proceedings of the Tenth International Conference on Uncertainty in Artificial Intelligence*, San Francisco, 1994.

[38] A. Zagorecki, M. J. Druzdzel, "Knowledge engineering for bayesian networks: How common are noisy-max distribution in practice," *IEEE Transactions on Systems, Man, and Cybernetics, Part A: Systems and Humans*, vol. 43, pp. 186–195, 2013.

[39] F. J. Díez, S. F. Galán, "Efficient computation for the noisy max," *International Journal of Intelligent Systems*, vol. 18, pp. 165–177, 2003.

[40] W. Li, P. Poupart, P. Van Beek, "Exploiting structure in weighted model counting approaches to probabilistic inference." *Journal of Artificial Intelligence Research*, vol. 40, pp. 729–765, 2011.

[41] R. E. Neapolitan, Learning Bayesian Networks, Prentice-Hall, London, 2003.

Chapter 3

A Data-Driven Fault Diagnosis Methodology in Three-Phase Inverters for PMSM Drive Systems

Permanent magnet synchronous motor and power electronics-based three-phase inverter are the major components in the modern industrial electric drive system, such as electrical actuators in an all-electric subsea Christmas tree. Inverters are the weakest components in a drive system, and power switches are the most vulnerable components in inverters. Fault detection and diagnosis of inverters are extremely necessary for improving the drive system reliability. Motivated by solving the uncertainty problem in fault diagnosis of inverters, which is caused by various reasons, such as bias and noise of sensors, this chapter proposes a Bayesian network-based data-driven fault diagnosis methodology of three-phase inverters. Two output line-to-line voltages for different fault modes are measured, the signal features are extracted using the fast Fourier transform, the dimensions of samples are reduced using the principal component analysis, and the faults are detected and diagnosed using Bayesian networks. Simulated and experimental data are used to train the fault diagnosis model as well as validate the proposed fault diagnosis methodology.

3.1. Introduction

Electrical actuators are important components of an all-electric subsea Christmas tree which is used to control the flow of offshore oil and gas out of the well. This type of actuator consists of two permanent magnet synchronous motors (PMSMs), i.e., a "drive" motor and a "clutch" motor, as shown in Fig. 3.1 [1]. The drive motor forces the subsea valve to open against a closing spring. Once the valve is fully open, the drive motor stops and the clutch motor

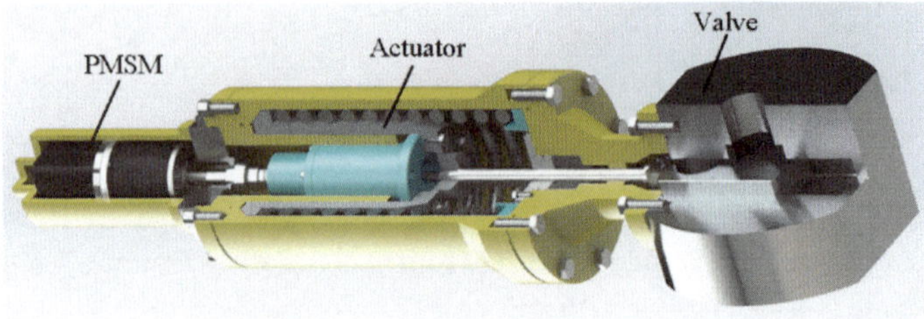

Fig. 3.1. Subsea electric actuator with PMSMs.

ensures that the subsea valve remains open by means of a friction-based mechanism. The speed of the PMSMs in an electric actuator is slow for the sake of safe operation during oil production. These actuators with motors are one of the most important components for the all-electric subsea Christmas tree system. The failure of actuators can lead to devastating consequences, such as oil spill and subsequent ocean pollution [33].

These motors are predominantly fed from sinusoidal pulse width modulation (SPWM) three-phase inverters for various speed operations. The inverters use power switches, such as insulated-gate bipolar transistors (IGBTs) and metal–oxide–semiconductor field-effect transistors (MOSFETs), to control the frequency and shape of the desired AC voltage. These power switches have the advantages of high efficiency, fast switching, and easy control of the gate-signal communications; however, they become faulty because of aging, overloading, or unpredicted operational conditions and are the most vulnerable components in inverters [2]. In this context, fault detection and diagnosis of inverters become extremely necessary.

Most of the faults in inverters are related to power switches in the form of short circuit (SC) and open circuit (OC). An SC fault of power switch is usually caused by overvoltage, overtemperature, or improper gate signal so that both switches in one leg are tuned on. The SC fault will eventually cause a short circuit of source voltage and the flow of high currents. In most cases, the SC fault is detected by the standard protection system, such as fuse and breaker, and the initial SC fault becomes an OC fault. The SC fault is destructive and makes the inverter shut down immediately. Further operation is forbidden and a repair is required. An OC fault of power switch is generally caused by a loss of gate signal or disconnection of wire because of thermal effects. The OC fault does not lead to the drive system shutdown, but can lead to the current imbalance in both the faulty and healthy phases, which degrades the performance of the drive system significantly. Different from SC faults, OC faults do not trigger the standard protection system of inverters and can lead to secondary faults in the other components if no fault detection and diagnosis system is available.

Most fault diagnosis methods for OC and SC faults use output current or voltage signals for diagnostic purposes, namely, current-based methods or voltage-based methods. Current-based methods are independent of inverter parameters and they do not require additional sensors [11]. For example, Park *et al.* [5] proposed a simple fault diagnosis scheme for brushless direct-current motor drives. The algorithm was achieved based on the measured phase current information and operating characteristics of the motor, and it could recover the control performance quickly by fast fault detection and reconfiguration of system topology. Estima *et al.* [9, 10] researched multiple OC fault diagnosis methods in voltage-fed pulse width modulation (PWM) motor drives by using the reference current errors. However, the current signals are load-dependent, and current-based methods have low diagnostic accuracy under no-load and light load operations [17].

Although voltage-based methods require additional voltage sensors to measure the line-to-line voltages, they have some inherent advantages, such as higher immunity to false alarms, loads, and noise. For example, Alavi *et al.* [12] proposed a fault diagnosis method for isolating SC faults in MOSFETs in three-phase voltage source inverters by analyzing the PWM switching signals in voltage spaces. Freire *et al.* [11] proposed a new voltage-based approach — without additional sensors — for OC fault diagnosis in closed-loop controlled PWM AC voltage source converters by using the information contained in the reference voltages available from the control system.

Generally, fault diagnosis methods can be categorized into model-based methods, signal-based methods, and data-driven methods [3, 4]. Over the last several years, model-based and signal-based fault diagnoses of inverters for drive systems are widely researched. An *et al.* [6] proposed a fast-diagnostic method for OC faults in inverters without sensors by the analysis of switching function model of the inverter under both healthy and faulty conditions. Subsequently, they proposed a current residual vector-based fault diagnosis method in order to get rid of the effects of load [7]. Jung *et al.* [8] proposed an OC fault diagnosis method in voltage source inverters for the PMSM drive system by employing the model reference adaptive system techniques. Zhang *et al.* [13, 14] proposed a simple method for single- and double-switch OC fault diagnosis method based on three-phase current distortions in voltage source inverters for vector-controlled induction motor drives. Trabelsi *et al.* [15] proposed a fault detection technique for OC faults of IGBTs in voltage source inverter-fed induction motor drives by analyzing the PWM switching signals and the line-to-line voltage levels during the switching times.

Different from model-based and signal-based methods, data-driven fault diagnosis does not require a known model or signal patterns, but requires a large number of historical data. However, this is not a major drawback because large amounts of data under both healthy and faulty operating conditions can be obtained using reliable simulation tools. Recently, data-driven fault diagnosis methods of inverters are investigated, and some techniques are used for diagnosis. Khomfoi *et al.* [16] proposed fault diagnosis and reconfiguration methods for multilevel inverter using neural network. Output phase voltages of the inverter are used as diagnostic signals to detect faults and their locations. Zidani *et al.* [18] proposed a fault detection and diagnosis method in a PWM voltage source inverter induction motor drive using fuzzy logic. The technique requires the measurement of the output inverter currents to detect intermittent loss of firing pulses in the inverter power switches. Masrur *et al.* [19] proposed a machine learning technique for fault diagnostics in induction motor drives using structured neural network. Most common types of faults including single-switch OC faults, post-SCs,

SCs and the unknown faults can be detected and isolated using the proposed method. Wang *et al.* [20] proposed a fault diagnosis strategy for cascaded H-bridge multilevel inverter system based on the principle component analysis and the multiclass relevance vector machine. Kamel *et al.* [21] presented fault diagnosis and online monitoring schemes of OC faults for grid-connected single-phase inverters using the adaptive neuro-fuzzy inference system algorithm.

Bayesian network (BN) is a probabilistic graphical model that represents a set of random variables and their conditional dependencies via a directed acyclic graph and is considered to be one of the most useful models in the field of probabilistic knowledge representation and reasoning since it was introduced by Pearl in early 1980s [22]. Recently, BNs are increasingly used in the field of fault detection and diagnosis because it can solve the uncertainty problem. Cho *et al.* [23] proposed a fault detection and isolation method of induction motors using recurrent neural networks and dynamic BNs. The neural network was used to train the data from the system under normal operating conditions and known faulty conditions, and the BNs were employed to produce random residuals. Morgan *et al.* [24] proposed a methodology for detecting and diagnosing faults found in heavy-duty diesel engines based upon spectrometric analysis of lubrication samples, and the output from the classification algorithm was diagnosed using a BN with a specifically designed structure. Cai *et al.* [25] proposed a multisource information fusion-based fault diagnosis methodology for ground-source heat pump system by using the BN method. Namaki-Shoushtari *et al.* [26] proposed a BN-based data-driven method with consideration of process knowledge and training data for control performance diagnosis when some of the abnormality data are sparse or not available in the historical database.

Uncertainty is a big problem in fault diagnosis of inverters. The uncertainty can be caused by various reasons, such as bias and noise of sensors [27]. Motivated by solving these uncertainty problems, a BN-based data-driven fault diagnosis methodology in three-phase inverters for the PMSM drive systems is proposed. Two output line-to-line voltages for different fault modes are measured, the signal features are extracted using the fast Fourier transform (FFT), the dimensions of samples are reduced using the principal component analysis (PCA), and the faults are detected and diagnosed using BNs.

The rest of this chapter is organized as follows. Section 3.2 describes a three-phase voltage source inverter and analyzes the fault modes. Sections 3.3 proposes a BN-based fault diagnosis methodology. Section 3.4 develops a fault diagnosis system and validates it using simulated and experimental data. Sections 3.5 summarizes the chapter.

3.2. System Description and Fault Analysis

Figure 3.2 illustrates a typical three-phase voltage source inverter for the PMSM drive system. It is composed by the parallel connection of three inverter legs. Each leg consists of two power switches T_i (IGBTs or MOSFETs) with corresponding antiparallel connected diodes D_i. These diodes are used to provide a negative current path through the switches. The power switches are controlled by corresponding gate signals g_i. When the gate signal is equal to 1, the switch conducts, and when the gate signal is equal to 0, the switch does not conduct. It is noted that the gate signals in one leg must work in a complementary way with a dead time to prevent a simultaneous conducting of power switches and SC of DC bus voltage. The inverter switching pattern is determined by the modulation strategies, such as PWM, SPWM and space vector pulse width modulation (SVPWM).

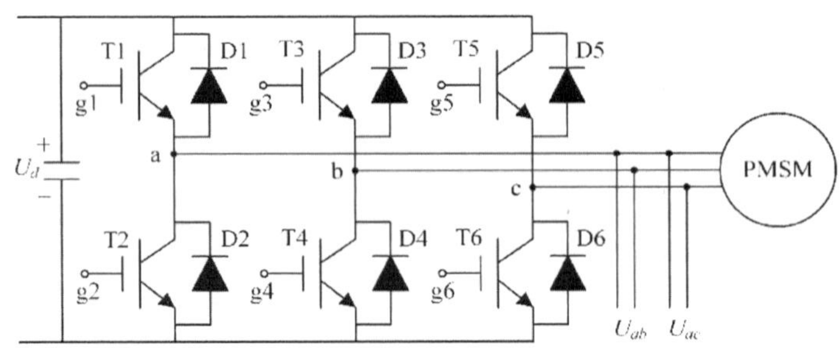

Fig. 3.2. Topology of a typical three-phase inverters.

In the current work, SPWM is used to control the gate signals. SPWM is generated by comparing a voltage reference signal (with amplitude A_r and frequency f_r) and a triangular carrier (with amplitude A_c and frequency f_c). A modulation index m is defined as the ratio of A_r and A_c [12]. It determined the inverter output voltage. According to the different combinations of switching states of six switches, the inverter generates three different line-to-line voltage levels, i.e., $+U_d$, 0, and $-U_d$.

As mentioned above, SC faults can be detected by the standard protection system, whereas OC faults do not trigger any protection systems of inverters. Voltage and current measurement is mandatory for the feedback control and condition monitoring of subsea electric actuators; therefore, it might be practical to use a voltage-based method for fault diagnosis. Voltage-based fault diagnosis has some inherent advantages, such as higher immunity to false alarms, loads, and noise. This current work focuses on the OC faults and uses line-to-line voltage signals for diagnostic purposes.

Line-to-line voltage of inverters is determined by the gate signals g_i and phase current i_j ($j = a, b, c$) [6, 8, 14, 15, 28]. The phase current is conducted by both power switches and antiparallel connected diodes. Taking line-to-line voltage U_{ab} as an example, for a normal operation state, when the gate signals $g_1 = 1$ and $g_3 = 1$, the output currents $i_a > 0$ and $i_b > 0$, the power switches T_1 and T_3 conduct, and the output line-to-line voltage $U_{ab} = 0$. If OC fault occurs in switch T_1, when the gate signals $g_1 = 1$ and $g_3 = 1$, the output currents $i_a > 0$ and $i_b > 0$, the switch T_3 and diode D_2 conduct, and the output line-to-line voltage $U_{ab} = -U_d$.

With different combinations of open-circuit power switches, there are many fault modes. When all the six switches are healthy, the inverter is in normal state, which is defined as a special fault mode. When a single OC fault for one switch occurs, there are six faults modes. When OC faults for two switches occur simultaneously, there are 15 fault modes. Similarly, when OC faults for three switches occur simultaneously, there are 20 fault modes. However, it is rare that three or more switches fail simultaneously. The current work focuses

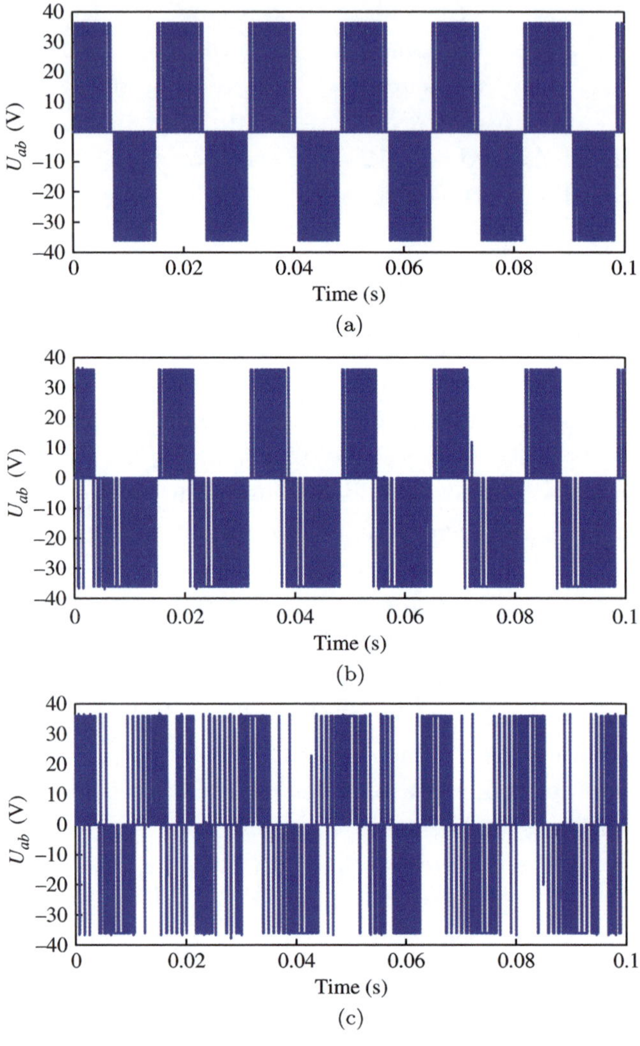

Fig. 3.3. Simulated line-to-line voltage U_{ab}: (a) normal state; (b) OC fault occurred in T1; and (c) OC fault occurred in T1 and T2.

on single OC and double OC faults. Therefore, there are 22 fault modes altogether under both healthy and faulty operating conditions. The simulated and experimental line-to-line voltages under different fault modes are provided in Figs. 3.3 and 3.4, respectively. From these figures, the experimental line-to-line voltages U_{ab} are noisy compared

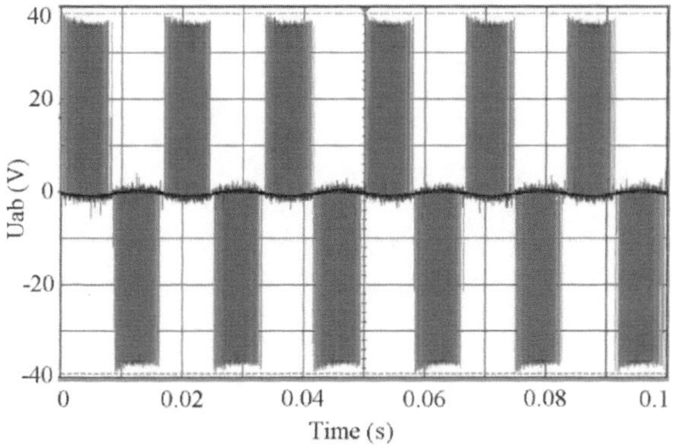

(a)

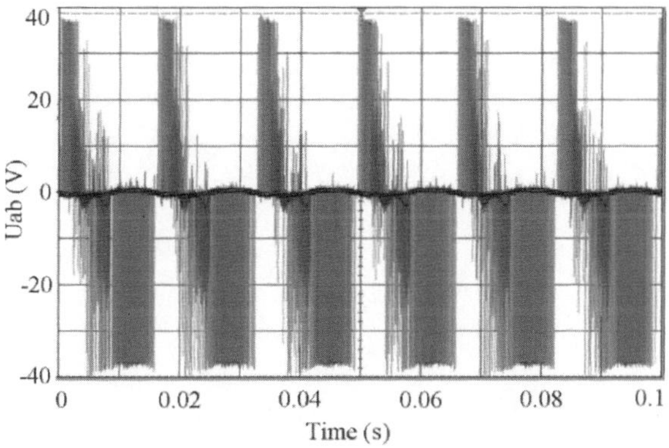

(b)

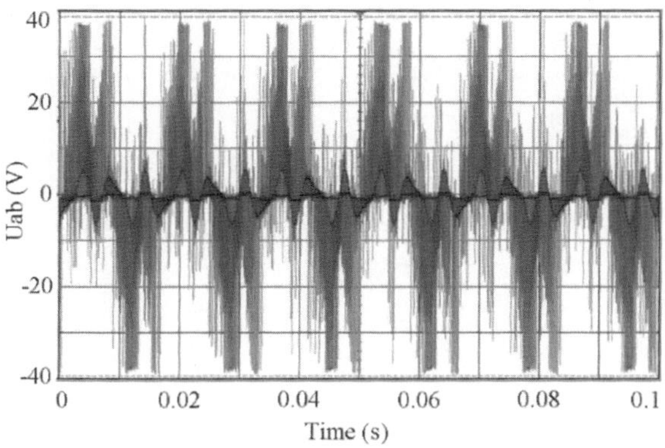

(c)

Fig. 3.4. Experimental line-to-line voltage U_{ab}: (a) normal state; (b) OC fault occurred in T1; and (c) OC fault occurred in T1 and T2.

to the simulated ones, which are caused by disturbance, sensor noise or sensor bias.

3.3. Fault Diagnosis Methodology

3.3.1. *Proposed fault diagnosis methodology*

The flowchart of the proposed fault diagnosis methodology for three-phase inverters is illustrated in Fig. 3.5. Sampled line-to-line voltages from experiments and simulations are used to train and test the developed fault diagnosis system. It is difficult to obtain a large volume of experimental voltages under both healthy and faulty operating conditions. Fortunately, simulated voltages can be easily obtained using reliable simulation tools, such as Matlab/Simulink/Simpower. The signal features of these voltages are extracted using FFT, and the dimensions of samples are reduced using PCA. A large proportion of these data are used for model

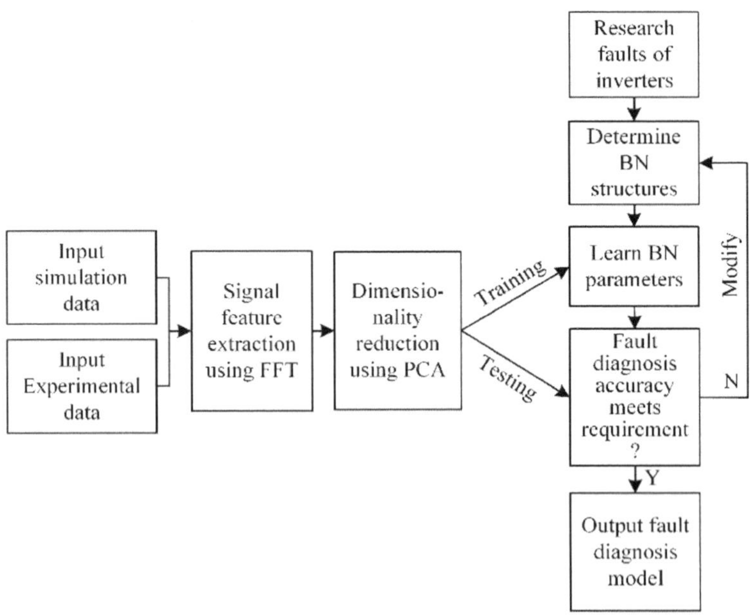

Fig. 3.5. Flowchart of the proposed fault diagnosis methodology.

training, and a small proportion of them are used for model validation. BNs are used to diagnose the faults of inverters. BNs consist of structure and parameter models. The BN structure is determined based on the relationship between faults and fault symptoms. The BN parameter including prior probabilities of root nodes and conditional probabilities of leaf nodes is learned using the simulated and experimental data. The BN-based fault diagnosis model is validated and finally outputted when fault diagnosis accuracy meets requirement.

3.3.2. *Signal feature extraction using FFT*

As shown in Figs. 3.3 and 3.4, the simulated and experimental line-to-line voltages under different fault modes are difficult to distinguish by a computation unit as a human does. Therefore, a signal feature extraction technique is required. Many techniques can be used for signal transformation, such as FFT, wavelet transform, and Hilbert transform. In the current work, FFT is adopted since it has a good identity feature to classify normal and abnormal features.

Assume that the sampled line-to-line voltage $U(n)$ is extended N-periodic, that is, $U(n) = U(n+N)$ for all n. A discrete Fourier transform (DFT) is obtained by decomposing the sequence of voltage values into components of different frequencies. An FFT computes the DFT and produces exactly the same result as that by evaluating the DFT definition directly. The most important difference is that an FFT is much faster than DFT. DFT is defined by the formula [29]

$$F(k) = \sum_{n=0}^{N-1} U(n) e^{-2\pi i k n/N}, \tag{3.1}$$

where $F(k)$ is the frequency-domain output, and k is a positive integer in the range from 0 to $N-1$.

It can also be expressed as follows:

$$F(k) = \sum_{n=0}^{N-1} U(n) \left[\cos\left(2k\pi\frac{n}{N}\right) - i\sin\left(2k\pi\frac{n}{N}\right)\right]. \tag{3.2}$$

The first sample $F(0)$ of the transformed series is the DC component, more commonly known as the average of the sampled line-to-line voltage. The second sample $F(1)$ is the fundamental component, and other samples $F(k)$ are the harmonic components.

It is noted that when OC fault occurs in switches T1 or T2, the magnitude spectrum of line-to-line voltage U_{ab} after FFT is the same. The two faults cannot be distinguished using the magnitude spectrum. Fortunately, the phase spectrum of U_{ab} after FFT is different. In particular, the phase of DC component has a difference of π. Therefore, the phase information of DC component is added into the magnitude spectrum [20]. Hence, the OC faults of switches T1 or T2 can be distinguished. As shown in Fig. 3.6, the DC component involves the phase information; therefore, the magnitude is negative when OC fault occurs in T1. Similarly, the OC failure of switches T3 and T4, and T5 and T6 also leads to the same magnitude spectra of line-to-line voltage U_{ab} and U_{ac}. Since the

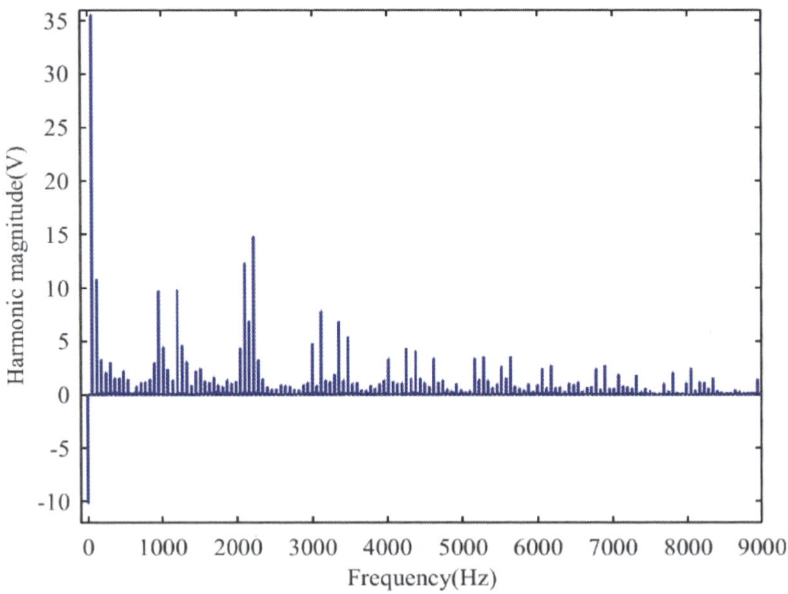

Fig. 3.6. Harmonic magnitude of simulated line-to-line voltage U_{ab} with OC fault occurred in T1.

phase information of DC component is added into the magnitude spectra, the OC faults of T3 and T4, and T5 or T6 can be distinguished.

3.3.3. *Dimensionality reduction using PCA*

PCA is a technique that reduces the dimensionality of a data set consisting of a large number of interrelated variables, while retaining as much as possible of the variation present in the data set. This is achieved by transforming a set of p correlated variables to a new set of uncorrelated variables, the principal components (PCs), which are ordered so that the first few retain most of the variations present in all of the original variables [30]. The transformation is, in fact, an orthogonal rotation in p-space. The fundamental PCA method used in the fault diagnosis of three-phase inverters is illustrated briefly.

Consider an $n \times p$ data matrix, X, where n is the size of samples, representing a different repetition of an experiment, and p is the size of variables, that is, the dimensionality of original data. The transformation is defined by a $p \times p$ load matrix, P, which maps X to an $n \times p$ PC score matrix, T, given by

$$T = X \cdot P. \tag{3.3}$$

The load matrix, P, also known as PC coefficients, is the eigenvector of covariance matrix R for data matrix X:

$$R = E\{[X - E(X)][X - E(X)]^T\}. \tag{3.4}$$

Each column of the load matrix contains coefficients for one PC. The columns are in order of decreasing component variance. Because the first few components account for most of the variation in the original data, a new $n \times k$ score matrix, T', can be expressed by

$$T' = X \cdot P', \tag{3.5}$$

where P' is the first k columns of P, that is a $p \times k$ load matrix. Therefore, PCA reduces the dimensionality of original data greatly, from p to k.

The criterion for choosing k is to select a cumulative percentage of total variation which one desires that the selected PCs contribute, say 90% or 95%. The required number of PCs is then the smallest value of k for which this chosen percentage is exceeded. The cumulative percentage of total variation is defined by [20]

$$t(k) = \frac{\sum_{i=1}^{k} \lambda_i}{\sum_{i=1}^{p} \lambda_i} \times 100\%, \tag{3.6}$$

where λ is the eigenvalue of covariance matrix R for data matrix X.

For the faults of three-phase inverters, line-to-line voltage U_{ab} can indicate the states of switches T1, T2, T3, and T4, but cannot indicate the states of switches T5 and T6. Another line-to-line voltage U_{ac} should be used together to distinguish the faults of all power switches. Taking the FFT data of U_{ab} in Fig. 3.6 as an example, there are 750 harmonic magnitudes within the frequency 9000 Hz, that is, the harmonic order is 750. For distinguishing all faults, the size of variables should be 1500, including the integrated FFT data of U_{ab} and U_{ac}. There are 22 fault modes including normal state (TY1), single OC fault (i.e., from TY2 to TY7), and double OC faults (i.e., from TY8 to TY22). Hence, the variable sequence is a 22 × 1500 data matrix. The dimension of variable is too high to construct the structure of BNs. For example, by transforming the data matrix using PCA with 10 PCs, the dimensionality of original data is reduced to 22 × 10, as shown in Fig. 3.7. The cumulative percentage of total variation is 96.27%.

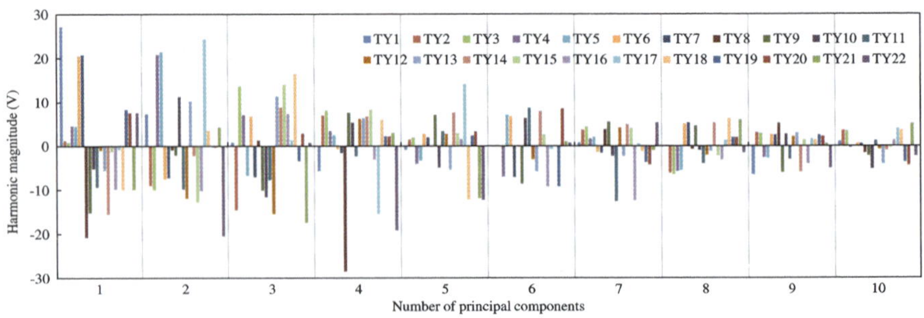

Fig. 3.7. Low-dimensional set of harmonic magnitude for the 22 fault modes.

3.3.4. *Fault diagnosis using BNs*

(1) *Structure of BNs*: The structure of BNs is developed by experts manually or learned using normal and fault data. The first method is adopted in the current work because there are not so many power switches in three-phase inverters. The proposed BNs consist of two layers, i.e., fault layer and fault symptom layer, corresponding to two types of nodes, i.e., fault nodes and fault symptom nodes. The fault layer consists of six fault nodes, representing six power switches from T1 to T6. Each node has two states, i.e., normal (NM) and OC, indicating normal state and open-circuit fault of the corresponding switch given observed evidence, respectively. The fault symptom layer consists of n fault symptom nodes representing n harmonic magnitudes after PCA. The states of each node are the intervals of harmonic magnitude, including positive and negative intervals. The causal relationship between fault nodes and fault symptom nodes is denoted qualitatively using directed arcs. Figure 3.8 gives a typical two-layers BNs for fault diagnosis. The fault symptom layer consists of 10 harmonic magnitudes after PCA. Each node has 30 states, indicating intervals of harmonic magnitude, such as 0–2, 2–4 and so on. The intervals could be different for different inverters, even for different PCs of an inverter system. For the

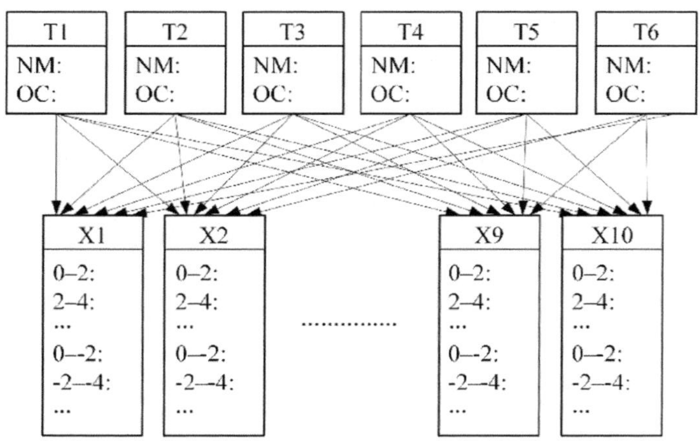

Fig. 3.8. Two-layers BNs for fault diagnosis.

sake of simplicity, the fixed interval of 2 is adopted in the current work.

(2) *Parameter of BNs*: After the network structure is determined, the prior probabilities of root nodes and conditional probabilities of leaf nodes are to be specified. They can be obtained by the experience and judgment of experts or learned using normal and fault data. For the faults of power switches, parameter learning of BNs is adopted with maximum likelihood estimation (MLE) method [31].

Assume that $M = (S, \theta)$ is the proposed BN with known structure S and unknown parameters θ, D is a complete data set of cases, and each case $d \in D$ is complete case, and the probability $P(d \mid M)$ is called the likelihood of M given d. Assume that the cases in D are independent, then the likelihood of M given D is

$$L(M \mid D) = \prod_{d \in D} P(d \mid M). \tag{3.7}$$

It is more convenient to work with the logarithm of the likelihood function, called the log-likelihood, defined as

$$LL(M \mid D) = \prod_{d \in D} \log_2 P(d \mid M). \tag{3.8}$$

The goal of parameter learning is to estimate each conditional probability density function corresponding to every node that maximizes the likelihood function. The likelihood function is greater than or equal to 0, while the log-likelihood function is less than or equal to 0. However, maximizing the likelihood function is equivalent to maximizing the log-likelihood function. Thus, the optimal BN model parameters are estimated from the sequence of training data by maximizing the following log-likelihood-based objective function:

$$\hat{\theta} = \arg\max_{\theta} L(M_\theta \mid D) = \arg\max_{\theta} LL(M_\theta \mid D). \tag{3.9}$$

The PMSM driven by the SPWM inverter is controlled by frequency f_r and modulation index m; therefore, the 22 fault modes and corresponding harmonic magnitudes after PCA with different frequency and modulation index are used for parameter learning. For example, the training data with frequency from 30 to 80 Hz with the interval of 2 Hz and modulation index from 0.6 to 0.9 with the interval of 0.001 could be obtained from simulation and experiment. For the sake of simplicity, a large proportion of training data are generated from simulation, and only a small proportion of them are obtained from the experiments.

It is noted that the proportion of different fault nodes affects the prior probabilities of root nodes. However, the prior probabilities themselves cannot affect the diagnostic performance for the same components. Therefore, a reasonable proportion of normal state, single OC fault, and double OC faults, such as 100:10:1, can be set for training data, indicating that normal state is much more effective than OC faults, and single OC fault is more effective than double OC faults.

(3) *Fault Diagnosis*: After the BNs are constructed and the network model parameters are estimated from the training data, it can be used for fault diagnosis by probabilistic inference with new test data. In the current work, Pearl's belief propagation algorithm is used for inference [32]. Belief propagation is a message-passing algorithm, originally developed for exact inference in polytree networks.

For a variable X with parents U and children Y, the message from node U to child X is denoted by $\pi_X(U)$ and called the causal support from U to X. The message from node Y to parent X is denoted by $\lambda_Y\ (X)$ and called the diagnostic support from Y to X. For some evidence E, the joint marginal for the family of variable X with parents U_i and children Y_i is given by

$$P(XU, E) = \lambda_E(X)\Theta_{X\,|\,U} \prod_i \pi_X(U_i) \prod_j \lambda_{Y_j}(X). \tag{3.10}$$

Here, λ_e (X) is an evidence indicator, where $\lambda_e(x) = 1$ if x is consistent with evidence e and zero otherwise. The causal and diagnostic messages can be defined as

$$\lambda_X(U_i) = \sum_{XU\backslash\{U_i\}} \lambda_E(X)\Theta_{X\,|\,U} \prod_{k\neq i} \pi_X(U_k) \prod_j \lambda_{Y_j}(X), \tag{3.11}$$

$$\pi_{Y_j}(X) = \sum_U \lambda_E(X)\Theta_{X\,|\,U} \prod_i \pi_X(U_i) \prod_{k\neq j} \lambda_{Y_k}(X). \tag{3.12}$$

A node can send a message to a neighbor only after it has received messages from all other neighbors. When a node has a single neighbor, it can immediately send a message to that neighbor. This includes a leaf node X with a single parent U, for which

$$\lambda_X(U) = \sum_X \lambda_E(X)\Theta_{X\,|\,U}. \tag{3.13}$$

It also includes a root node X with a single child Y, for which

$$\pi_Y(X) = \lambda_E(X)\Theta_X. \tag{3.14}$$

During fault diagnosis, new data after FFT and PCA are inputted into fault symptom nodes of BNs, and posterior probabilities of fault nodes are updated using the belief propagation algorithm above. It is noted that the new data should map to adjacent state space by multiplying by load matrix with adjacent frequency and modulation index. Generally, the larger the posterior probability of fault, the higher the possibility of occurrence of the corresponding fault. The fault diagnosis methodology can give a probability of a fault, but not draw a definite diagnostic result. According to engineering experience, two judgment rules are defined to determine the fault diagnosis results as follows.

Rule 1: A single OC fault of power switch with largest posterior probability will be reported if the largest posterior probability is greater than 0.7, or it is 0.5 greater than the second largest one.

Rule 2: A double OC fault of power switch with largest and second largest posterior probabilities will be reported if the two posterior probabilities are both greater than 0.7, or they are both 0.5 greater than the third largest one.

3.4. Developments and Validations

3.4.1. *Simulation and experimental setup*

To evaluate the performance of the BN-based data-driven fault diagnosis method, the simulation and experiment of a PMSM drive system with a three-phase voltage source inverter are conducted. The FFT-based signal feature extraction, PCA-based dimensionality reduction, and BN-based fault diagnosis are all performed using Matlab software, and the functions *tic* and *toc* are used to calculate the time. A DELL Precision T5600 workstation is used for calculation because the calculation time for the same program code is a little different and around a fixed average value, the average values for them are obtained as 0.015, 0.028, and 0.172, respectively. It is noted that these values do not vary greatly when the operating speed varies. It is good enough for fault diagnosis of three-phase inverters for the PMSM drive systems used for a subsea electric actuator. The model of the three-phase voltage source inverter with six power switches is designed using Matlab SimPowerSystems, as shown in Fig. 3.9. The OC faults of switches are simulated by disabling the corresponding gate signal. The line-to-line voltages U_{ab} and U_{ac} for the 22 fault modes are measured and used for training BNs when the frequency varies from 30 to 80 Hz with the interval of 2 Hz, and the modulation index varies from 0.6 to 0.9 with the interval of 0.001. The switching frequency of SPWM is 1 kHz, and the sampling frequencies for experiment and simulation are both 20 kHz, which can reflect the actual waveform well.

The experimental setup of the PMSM drive system is given in Fig. 3.10. A three-phase voltage source inverter with six power switches MOSFET IRF1010 is built. The DRV8301 is a gate driver integrated circuit for three-phase motor drive applications and used to drive the six MOSFETs. Digital signal processor

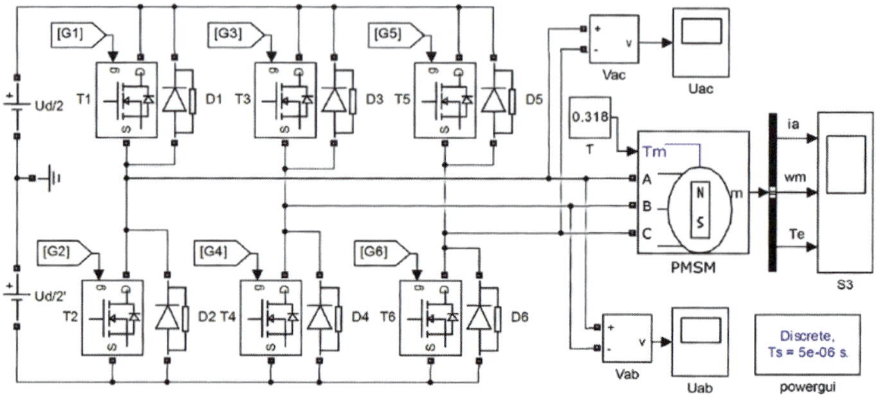

Fig. 3.9. Simulink model of the PMSM drive system.

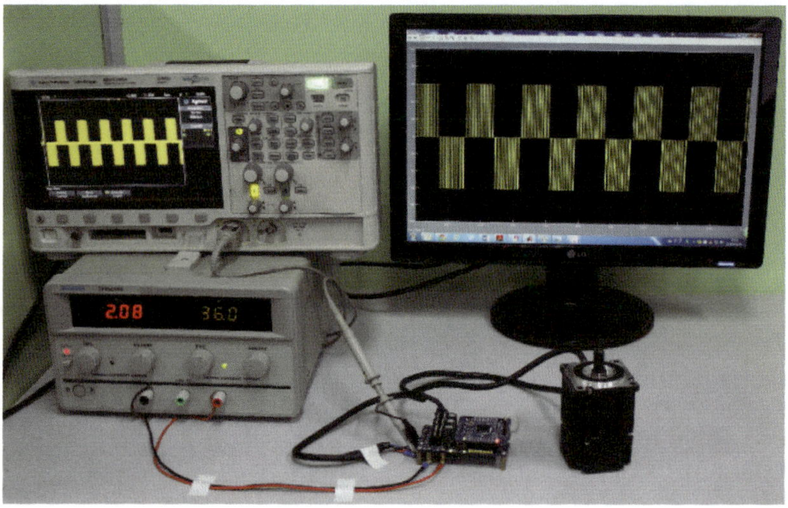

Fig. 3.10. Experimental setup of the PMSM drive system.

TMS320F28035 is used to generate the SPWM signals. The PMSM used is ASMA01L03BAK. Two voltage sensors are installed in the DRV8301 drive card to measure the line-to-line voltages U_{ab} and U_{ac}. The OC faults are inserted by sticking the gate signal of the faulty power switches to the low state. It is difficult to obtain a large volume of experimental voltages under both healthy and faulty

Table 3.1. Main parameters of the PMSM drive system.

Quantity	Value
DC voltage	$U_{dc} = 36$ V
Rated power of PMSM	0.1 kW
Rated torque of PMSM	0.318 Nm
Rated current of PMSM	4 A
Rated speed of PMSM	3000 rpm
Maximum speed of PMSM	4000 rpm
Torque constant of PMSM	0.0866 Nm/Arms
Rotor inertia of PMSM	0.1034 kgm^2 $\times 10^{-4}$
Stator phase resistance of PMSM	0.76 ohms
Armature inductance of PMSM	1.82 mH
Number of pole pairs of PMSM	8
Resistance of MOSFETs	0.001 ohms
Resistance of diodes	0.01 ohms

operating conditions. Therefore, only when the frequency is 60 Hz and the modulation index is 0.8, the line-to-line voltages U_{ab} and U_{ac} for the 22 fault modes are measured and used for training BNs. In actual fault diagnosis, the measured voltages are input to the software system of fault diagnosis unit, including FFT-based signal feature extraction unit, PCA-based dimensionality reduction unit, and BN-based fault diagnosis unit. They are all located in the computer.

The parameters of the PMSM and the inverter used in the simulation and experiment are given in Table 3.1. It is noted that the frequency of the rated speed for the given PMSM is 400 Hz, however, for the purposes of subsea electric actuator control as well as fault diagnosis, the frequency for training BNs varies from 30 to 80 Hz.

3.4.2. *Results*

The validation of the fault diagnosis method is conducted using the test data from simulation and experiment. For all the 22 fault modes, the simulation test data are obtained when the frequency is 60 Hz and the modulation index varies from 0.6005 to 0.9005 with the interval

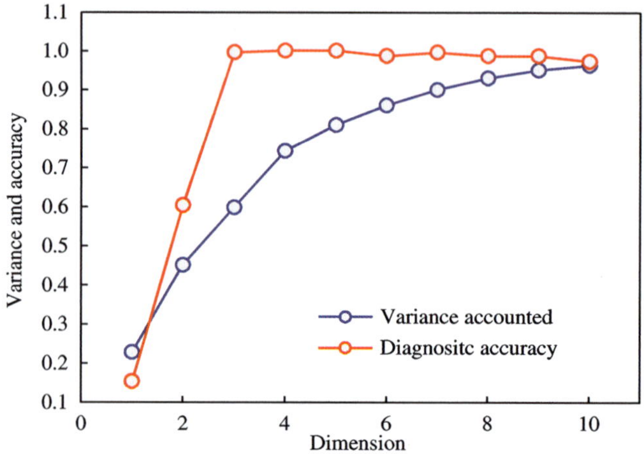

Fig. 3.11. Variance accounted for by first n components and diagnostic accuracy for different PCA dimensions.

of 0.01. Obviously, the test data are totally different from the training data. The cumulative percentage of total variation accounted for by first n components and diagnostic accuracy for different PCA dimensions are shown in Fig. 3.11. It can be seen that with the increase in PCA dimension, the total variation increases continuously from 22.86% to 96.27%. However, the diagnostic accuracy increases almost linearly when the dimension varies from 1 to 3 and then reaches a stable level of almost 100%. This means three PCs are enough for the purpose of fault diagnosis. It is noted that when the PCA dimensions are 3 and 4, the diagnostic accuracies are 99.55% and 100%, respectively; hence, the PCA dimensions of 4 and more are adopted to develop the fault diagnosis program and investigate the effects of sensor noise and bias on the diagnostic accuracy.

A faulty sensor may cause diagnostic performance degradation. Sensor noise and bias are the main types of faults. The white Gaussian noise from 0 to 20 dB and the bias from −1.6 to 1.6 V are added to the measured line-to-line voltage. The effects of sensor noise and bias on the diagnostic accuracy with different PCA dimensions are plotted in Fig. 3.12. As shown in Fig. 3.12(a), with the

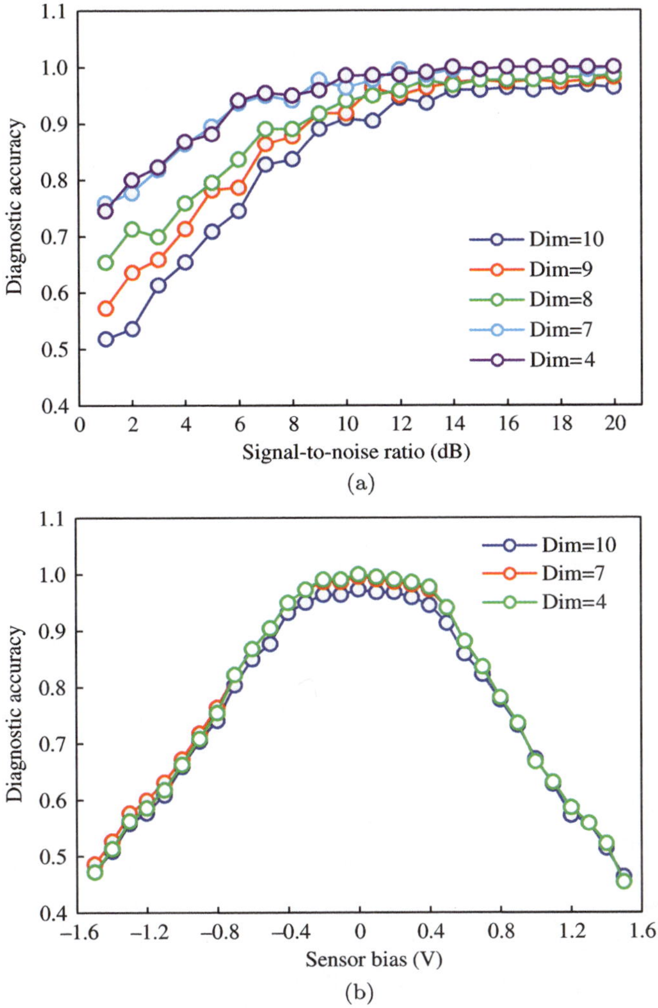

Fig. 3.12. Effects of sensor noise and bias on the diagnostic accuracy for different PCA dimensions: (a) noise and (b) bias.

decrease in noise, the diagnostic accuracies for all PCA dimensions increase. The diagnostic accuracies vary greatly from 1 to 12 dB and reach stable levels when the noise is smaller than 12 dB. In addition, the diagnostic accuracies are almost the same when the PCA dimensions vary from 4 to 7, but they decrease continuously when the dimensions vary from 7 to 10. As shown in Fig. 3.12(b), the relationship between diagnostic accuracy and sensor bias is parabolic type of curve. When the PCA dimension increases from 4 to 10, the diagnostic accuracies decrease slightly. Therefore, the PCA dimension of 4 is the best choice for the OC fault diagnosis of inverters. The results show that the PCA can extract the principle components which represent main fault principle features even when the fault signals are noisy. It can also be seen that the BN-based fault diagnosis methodology has a strong toleration for sensor noise and bias.

In [20], white Gaussian noise with signal-to-noise ratio of 10 dB was added to the simulation samples of the cascaded H-bridge multilevel inverter system fault diagnosis using a PCA and multiclass relevance vector machine approach. The average diagnosis accuracies for testing samples 5 groups, 10 groups, and 20 groups are 97.30%, 98.11%, and 97.97%, respectively. As shown in Fig. 3.12(a), when the signal-to-noise ratio is 10 dB, the diagnosis accuracy is 98.48%, which is better than the previous one.

For all the 22 fault modes, the experimental test data are obtained when the frequencies are 55, 65, and 75 Hz and the modulation indexes are 0.65, 0.75, and 0.85. Therefore, there are nine combinations with different frequencies and modulation indexes, and totally 198 fault modes (this takes lots of time). These data are completely different from the experimental training data. The fault diagnosis is conducted when the PCA dimension is 4 without extra artificial sensor noise and bias. The results show that 196 faults are diagnosed correctly and two faults with double OC are diagnosed incorrectly. Therefore, the experimental diagnostic accuracy for single OC fault is 100%, and the total accuracy is 98.99%.

Taking OC fault of switch T1 as an example, we perform the fault diagnosis using the BN-based data-driven method. The experimental voltage and current waveforms with both healthy and faulty conditions for long-term and short-term waveforms are given in Fig. 3.13. It can be seen that the positive half

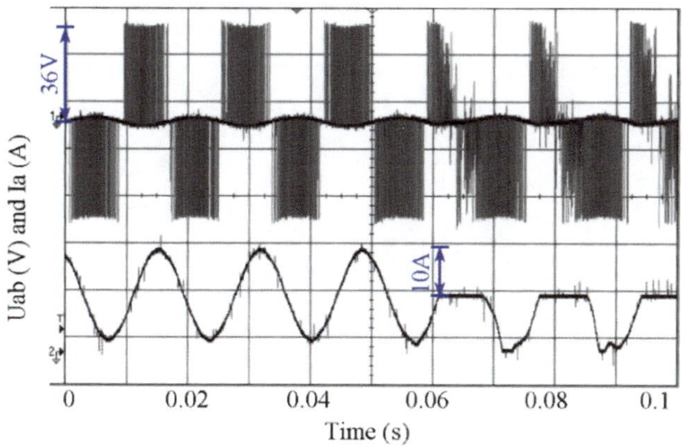

(a)

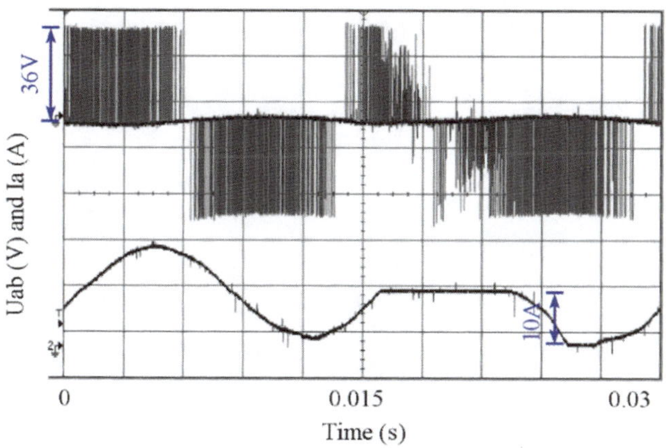

(b)

Fig. 3.13. Experimental line-to-line voltage U_{ab} and phase current I_a with OC fault occurring in T_1. (a) Long-term waveforms and (b) short-term waveforms.

cycle phase current disappear for the OC fault occurring in switch T1.

The FFT-based signal feature extraction results of the line-to-line voltage U_{ab} are shown in oscilloscope directly for better comparison as given in Fig. 3.14. The new data are mapped to adjacent state space by multiplying by load matrix with adjacent frequency and modulation index. When the PCA dimensions is 4, the

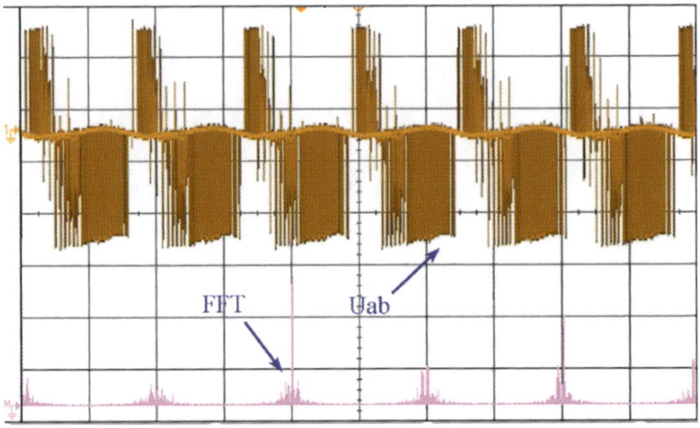

Fig. 3.14. FFT-based signal feature extraction results for the experimental voltage U_{ab} with OC fault occurring in T1.

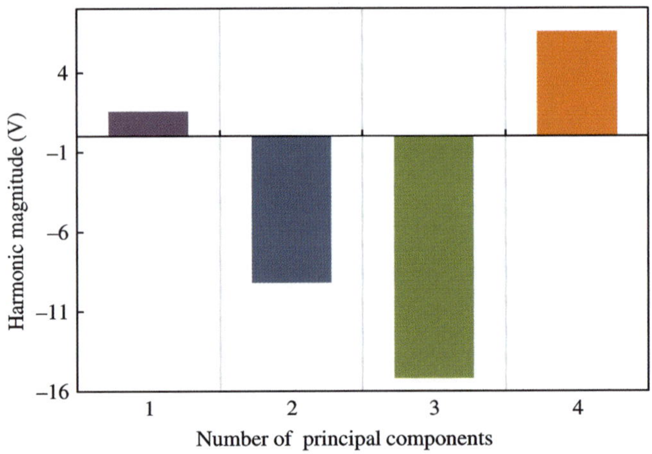

Fig. 3.15. PCA-based dimensionality reduction result for the experimental voltage U_{ab} with OC fault occurring in T1.

dimensionality reduction result is shown in Fig. 3.15. After inputting these data input BN models, as shown in Fig. 3.16, we can see that the posterior probability of fault node T1 is 100% and posterior probabilities of other fault nodes are 0, indicating that OC fault occurs in switch T1.

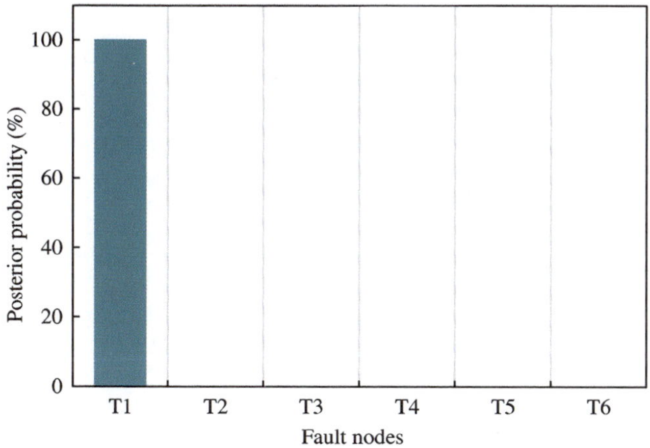

Fig. 3.16. Posterior probabilities of the fault nodes for the experimental OC fault occurring in T1.

3.5. Conclusion

This chapter proposes a BN-based data-driven fault diagnosis methodology in three-phase inverters for PMSM drive systems. Two output line-to-line voltages for different fault modes are measured, the signal features are extracted using FFT, the dimensions of samples are reduced using PCA, and the faults are detected and diagnosed using BNs. The proposed fault diagnosis methodology has been validated in both simulation and experiment, which show that the methodology performs quite well. The results also indicate that the diagnostic accuracy increases almost linearly when the dimension varies from 1 to 3 and then reaches a stable level. Therefore, the PCA dimensions of 4 are adopted to develop the fault diagnosis program. The effects of sensor noise and bias on the diagnostic accuracy with different PCA dimensions are also researched. The results show that the PCA dimension of 4 is the best choice for the OC fault diagnosis of inverters. It can also be seen that the BN-based data-driven fault diagnosis methodology has a better toleration for sensor noise and bias.

The proposed fault diagnosis methodology is also suitable for various power switch-based electric devices, such as multilevel inverter and converter. Future works can be directed at current signal-based fault diagnosis methodology, and the toleration of loads can be investigated using the BN method.

References

[1] D. Abicht, J. V. D. Akker, "The 2nd generation DC all-electronic subsea production control system," In: *Proceedings of Offshore Technology Conference*, Houston, TX, 2011.

[2] Y. T. Song, B. S. Wang, "Survey on reliability of power electronic systems," *IEEE Transactions on Power Electronics*, vol. 28, no. 1, pp. 591–604, 2013.

[3] Z. Gao, C. Cecati, S.X. Ding, "A survey of fault diagnosis and fault-tolerant techniques — Part I: Fault diagnosis with model-based and signal-based approaches," *IEEE Transactions on Industrial Electronics*, vol. 62, no. 6, pp. 3757–3767, 2015.

[4] Z. Gao, C. Cecati, S. X. Ding, "A survey of fault diagnosis and fault-tolerant techniques — Part II: Fault diagnosis with knowledge-based and hybrid/active approaches," *IEEE Transactions on Industrial Electronics*, vol. 62, no. 6, pp. 3768–3774, 2015.

[5] B. G. Park, K. J. Lee, R. Y. Kim, T. S. Kim, J. S. Ryu, D. S. Hyun, "Simple fault diagnosis based on operating characteristic of brushless direct-current motor drives," *IEEE Transactions on Industrial Electronics*, vol. 58, no. 5, pp. 1586–1593, 2011.

[6] Q. T. An, L. Z. Sun, K. Zhao, L. Sun, "Switching function model-based fast-diagnostic method of open-switch faults in inverters without sensors," *IEEE Transactions on Power Electronics*, vol. 26, no. 1, pp. 119–126, 2011.

[7] Q. T. An, L. Sun, L. Z. Sun, "Current residual vector-based open-switch fault diagnosis of inverters in PMSM drive systems," *IEEE Transactions on Power Electronics*, vol. 30, no. 5, pp. 2814–2827, 2015.

[8] S. M. Jung, J. S. Park, H. W. Kim, K. Y. Cho, M. J. Youn, "An MRAS-based diagnosis of open-circuit fault in PWM voltage-source inverters for PM synchronous motor drive systems," *IEEE Transactions on Power Electronics*, vol. 28, no. 5, pp. 2514–2526, 2013.

[9] J. O. Estima, A. J. M. Cardoso, "A new algorithm for real-time multiple open-circuit fault diagnosis in voltage-fed PWM motor drives by the reference current errors," *IEEE Transactions on Industrial Electronics*, vol. 60, no. 8, pp. 3496–3505, 2013.

[10] N. M. A. Freire, J. O. Estima, A. J. M. Cardoso, "Open-circuit fault diagnosis in PMSG drives for wind turbine applications," *IEEE Transactions on Industrial Electronics*, vol. 60, no. 9, pp. 3957–3967, 2013.

[11] N. M. A. Freire, J. O. Estima, A. J. M. Cardoso, "A voltage-based approach without extra hardware for open-circuit fault diagnosis in closed-loop PWM

AC regenerative drives," *IEEE Transactions on Industrial Electronics*, vol. 61, no. 9, pp. 4960–4970, 2014.

[12] M. Alavi, D. W. Wang, M. Luo, "Short-circuit fault diagnosis for three-phase inverters based on voltage-space patterns," *IEEE Transactions on Industrial Electronics*, vol. 61, no. 10, pp. 5558–5569, 2014.

[13] J. H. Zhang, J. Zhao, D. H. Zhou, C. G. Huang, "High-performance fault diagnosis in PWM voltage-source inverters for vector-controlled induction motor drives," *IEEE Transactions on Power Electronics*, vol. 29, no. 11, pp. 6087–6099, 2014.

[14] F. Wu, J. Zhao, "A real-time multiple open-circuit fault diagnosis method in voltage-source-inverter fed vector controlled drives," *IEEE Transactions on Power Electronics*, DOI: 10.1109/TPEL.2015.2422131.

[15] M. Trabelsi, M. Boussak, M.Gossa, "PWM-switching pattern-based diagnosis scheme for single and multiple open-switch damages in VSI-fed induction motor drives," *ISA Transactions*, vol. 51, no. 2, pp. 333–344, 2012.

[16] S. Khomfoi, L. M. Tolbert, "Fault diagnostic system for a multilevel inverter using a neural network," *IEEE Transactions on Power Electronics*, vol. 22, no. 3, pp. 1062–1069, 2007.

[17] S. Khomfoi, L. M. Tolbert, "Fault diagnosis and reconfiguration for multilevel inverter drive using AI-based techniques," *IEEE Transactions on Power Electronics*, vol. 54, no. 6, pp. 2954–2968, 2007.

[18] F. Zidani, D. Diallo, M. E. H. Benbouzid, N. S. Rachid, "A fuzzy-based approach for the diagnosis of fault modes in a voltage-fed PWM inverter induction motor drive," *IEEE Transactions on Industrial Electronics*, vol. 55, no. 2, pp. 586–593, 2008.

[19] M. A. Masrur, Z. Chen, Y. Murphey, "Intelligent diagnosis of open and short circuit faults in electric drive inverters for real-time applications," *IET Power Electronics*, vol. 3, no. 2, pp. 279–291, 2010.

[20] T. Z. Wang, H. Xu, J. G. Han, E. Elbouchikhi, M. E. H. Benbouzid, "Cascaded H-bridge multilevel inverter system fault diagnosis using a PCA and multiclass relevance vector machine approach," *IEEE Transactions on Power Electronics*, vol. 30, no. 12, pp. 7006–7018, 2015.

[21] T. Kamel, Y. Biletskiy, L. C. Chang, "Fault diagnosis and on-line monitoring for grid-connected single-phase inverters," *Electric Power Systems Research*, vol. 126, pp. 68–77, 2015.

[22] J. Pearl, "Bayesian networks: A model of self-activated memory for evidential reasoning," In: *Proceedings of the 7th Conference of the Cognitive Science Society*, University of California, Irvine, CA, 1985.

[23] H. C. Cho, J. Knowles, M. S. Fadali, K. S. Lee,"Fault detection and isolation of induction motors using recurrent neural networks and dynamic Bayesian modeling," *IEEE Transactions on Control Systems Technology*, vol. 18, no. 2, pp. 430–437, 2010.

[24] I. Morgan, H. H. Liu, B. Tormos, A. Sala, "Detection and diagnosis of incipient faults in heavy-duty diesel engines," *IEEE Transactions on Industrial Electronics*, vol. 57, no. 10, pp. 3522–3532, 2010.

[25] B. Cai, Y. Liu, Q. Fan, Y. Zhang, Z Liu, S. Yu, R. Ji, "Multi-source information fusion based fault diagnosis of ground-source heat pump using Bayesian network," *Applied Energy*, vol. 114, pp. 1–9, 2014.

[26] N. S. Omid, B. Huang, "Bayesian control loop diagnosis by combining historical data and process knowledge of fault signatures," *IEEE Transactions on Industrial Electronics*, vol. 62, no. 6, pp. 3696–3704, 2015.

[27] N. Mehranbod, M. Soroush, C. Panjapornpon, "A method of sensor fault detection and identification," *Journal of Process Control*, vol. 15, no. 3, pp. 321–339, 2005.

[28] U. M. Choi, F. Blaabjerg, K. B. Lee, "Study and handling methods of power IGBT module failures in power electronic converter systems," *IEEE Transactions on Power Electronics*, vol. 30, no. 5, pp. 2517–2533, 2015.

[29] J. G. Nash, "Computationally efficient systolic architecture for computing the discrete Fourier transform," *IEEE Transactions on Signal Processing*, vol. 53, no. 12, pp. 4640–4651, 2005.

[30] I. T. Jolliffe, *Principal Component Analysis*, 2nd edn., Springer, 2002.

[31] F. V. Jensen, T. D. Nielsen, *Bayesian Networks and Decision Graphs*, 2nd edn., Springer, 2007.

[32] A. Darwiche, *Modeling and Reasoning with Bayesian Networks*, Cambridge University Press, 2009.

[33] A. K. Aadland, K. Petersen, "Subsea all electric," In: *Proceedings of Offshore Technology Conference*, Houston, TX, 2010.

Chapter 4

A Real-Time Fault Diagnosis Methodology of Complex Systems Using Object-Oriented Bayesian Networks

Bayesian network (BN) is a commonly used tool in probabilistic reasoning of uncertainty in industrial processes, but it requires modeling of large and complex systems in situations, such as fault diagnosis and reliability evaluation. Motivated by the reduction of overall complexities of BNs for fault diagnosis, and the reporting of faults that immediately occur, a real-time fault diagnosis methodology of complex systems with repetitive structures is proposed using object-oriented Bayesian networks (OOBNs). The modeling methodology consists of two main phases: an offline OOBN construction phase and an online fault diagnosis phase. In the offline phase, sensor historical data and expert knowledge are collected and processed to determine the faults and symptoms, and OOBN-based fault diagnosis models are developed subsequently. In the online phase, operator experience and sensor real-time data are placed in the OOBNs to perform the fault diagnosis. According to engineering experience, the judgment rules are defined to obtain the fault diagnosis results.

4.1. Introduction

As modern industrial systems become increasingly complex and large, their reliability and availability suffer fault increases; thus, fault diagnosis has become more important than it was previously [1, 2]. The function of a fault diagnosis system is to rapidly detect and determine the root causes of faults based on various models and algorithms of useful information, such as hardware sensor data and operator experience information. Therefore, a real-time fault diagnosis system is useful in helping a technician to find the fault once it occurs and eliminate the failure immediately.

Over the last several years, some real-time fault diagnosis methods have been developed for various systems and structures. McKenzie *et al.* [3] proposed integrated robust models and device-centered models for real-time fault diagnosis of complex systems and demonstrated the diagnosis technique using power systems. Chen *et al.* [4] proposed an online fault diagnosis approach to estimate the fault section and identify the fault types by using hybrid cause-and-effect networks and fuzzy rule-based approaches. Wong *et al.* [5] proposed an application of extreme learning machine to build a real-time fault diagnostic system, in which data pre-processing techniques were integrated. Zhou *et al.* [6] developed a tree-based, real-time fault diagnosis system for a complex process system. Uncertainty problems were also considered in the proposed methods. Bonvini *et al.* [7] proposed a robust and computationally efficient algorithm for both whole-building and component-level energy fault detection and diagnosis. The algorithm was able to reliably estimate multiple and simultaneous fault conditions, as well as uncertainty. Qian *et al.* [8] developed an expert system that assists engineers and operators in monitoring and diagnosing exceptional cases in the refinery process. Biswas *et al.* [33] developed a real-time, data-driven diagnosis algorithm that runs on a substation and continuously monitors the health condition of a circuit breaker trip coil assembly. Chen *et al.* [34] proposed a fully distributed general anomaly detection scheme that uses graph theory and exploits spatiotemporal correlations of physical processes to conduct real-time anomaly detection for general large-scale networked industrial sensing systems.

Uncertainty is a major problem in the fault diagnosis of complex systems. It can be caused by various reasons, such as limitations of design sensors and observers, inherent uncertainty of observed complex systems, unpredictable factors during observation and diagnosis, and incompleteness and inaccuracy of methods and models. In addition, complex and uncertain relationships exist between faults and fault symptoms because a fault may cause multiple symptoms, and multiple faults may cause a single symptom. Bayesian network (BN) has been considered as one of the most useful models in the area of probabilistic knowledge representation and inference since it was introduced by Pearl in the early 1980s [9, 30]. BNs can be used to solve the uncertainty problems of fault diagnosis.

The use of BNs in the field of fault detection and diagnosis has recently expanded. Lo *et al.* [12] proposed a BN-based fault diagnosis method with a structure that is constructed by mapping the corresponding bond graph models. Tobon-Mejia *et al.* [13] proposed a wear diagnostic and prognostic approach for computer numerical control machine tools by dynamic BNs. Mandal [14] developed new machine learning ensembles for quality spine diagnosis, in which BNs optimized by Tabu search algorithm are used as a classifier. Zhang *et al.* [15] proposed a dynamic fault diagnosis approach of a large and complex system using dynamic uncertain causality graph. Cai *et al.* [16] proposed a BN-based fault diagnosis framework for a ground-source heat pump system by fusing various sensor data and observed information. Liu *et al.* [17] proposed a concept of a node set to improve the efficiency of the exact inference in diagnostic BNs. Sahin *et al.* [35] developed a fault diagnosis system for airplane engines by using BNs and the distributed particle swarm optimization method. Cai *et al.* [32] proposed a novel real-time reliability evaluation approach by combining a BN-based root cause diagnosis phase and a DBN-based reliability evaluation phase. Najafi *et al.* [36] proposed a BN-based diagnostic approach for an air-handling unit. The observed symptoms and various abnormal conditions were matched by using the proposed approach.

During the past two decades, however, only a few studies have reported on the real-time fault diagnosis application of BNs. In addition, some industrial systems are overly complex, and they usually include copies of similar or even almost-identical subsystems. A subsea blowout preventer system, for example, consists of four or six subsea ram preventers and two annular preventers [38], and a subsea production system consists of tens of subsea Christmas trees [29]. To model these complex systems, object-oriented concepts can be adopted. Object-oriented Bayesian network (OOBN) is a suitable tool for building large, complex, hierarchical BNs [10]. Modularity is a significant characteristic of OOBNs, which can greatly simplify program modifications by breaking down a complex system into several smaller subsystems. Moreover, the inference process is particularly efficient because the standard nodes inside an instance are conditionally independent of the rest of the OOBNs.

During the past two decades, OOBNs have been used to describe complex domains in terms of interrelated objects. Weidl *et al.* [18] proposed an OOBN-based process diagnosis method to monitor abnormal conditions, analyze root causes, and make decisions. Weber *et al.* [19] defined dynamic OOBNs and proposed a dynamic OOBN-based reliability modeling and assessment method for a manufacturing process. Jensen *et al.* [20] proposed an OOBN-based evaluation of risk indices method for leg disorders caused by infection, physical, and inherited factors. Kasper *et al.* [21] proposed an OOBN-based method to recognize driving maneuvers. The uncertainties caused by models and measurements are handled using BNs. Khakzad *et al.* [22] proposed an OOBN-based quantitative risk analysis method for offshore drilling operations. The individual BNs were initially developed for accident scenarios and an OOBN was formed by connecting these individual networks. Abramovici *et al.* [23] proposed an OOBN-based method for decision support to improve mass-produced standard products. Huang *et al.* [24] adopted OOBNs to prevent the BN-based fault diagnosis model from becoming overly large and to unburden the calculations.

The application of OOBNs to real-time fault diagnosis methodology is still in the early stage of development, and many critical issues have yet to be investigated, such as constructions of OOBNs. This chapter presents a new approach incorporating OOBN in the study of real-time fault diagnosis methodology of complex systems with repetitive structures. A two-phase modeling methodology, including an offline OOBN construction phase and an online fault diagnosis phase, is developed in the present study. The rest of the chapter is structured as follows. Section 4.2 develops a real-time fault diagnosis methodology, Sec. 4.3 analyzes a case of subsea production system to illustrate the application of the proposed method, and Sec. 4.4 summarizes the work chapter.

4.2. A Proposed Modeling Methodology

4.2.1. *Overview of OOBNs*

A BN consists of qualitative and quantitative parts. The qualitative part is a directed acyclic graph, in which the nodes represent the system variables, while the arcs symbolize the dependencies or the cause-and-effect relationships among the variables. The quantitative part is the conditional probabilistic table (CPT), which represents the relationship between each child node and its parents. In BNs, the leaf nodes have only parent nodes but no child nodes, and root nodes have child nodes but no parent nodes. A complex BN model usually contains some nearly similar BN fragments. A generic BN fragment can be constructed by instantiating these same BN structures for several required times. According to the object-oriented programming paradigm, the generic BN fragment is named class, and each BN fragment produced by instantiating this class is named object [11]. Figure 4.1 is a simple OOBN, which includes instance and usual nodes. The instance node, such as Inst. 1, represents an instance of another BN, notated by a rectangle with rounded edges. They are connected to another instance and usual nodes via input and output nodes. An input node is a circle (or an ellipse) with a dashed line border, such as X. An output node is a circle (or an ellipse) with a bold line border, such as C and D. In OOBNs, an object can be considered as a function that provides a certain input and a probability distribution over a set of variables. Based on standard programming terminology, the input attributes in the class description can be regarded as formal parameters of the corresponding function, while the actual parameters passed to an object are identified as the parent of the input attributes in the surrounding model. For example, X can be considered as a formal parameter, while X1 is the actual parameter passed to the left-most object in Fig. 4.1.

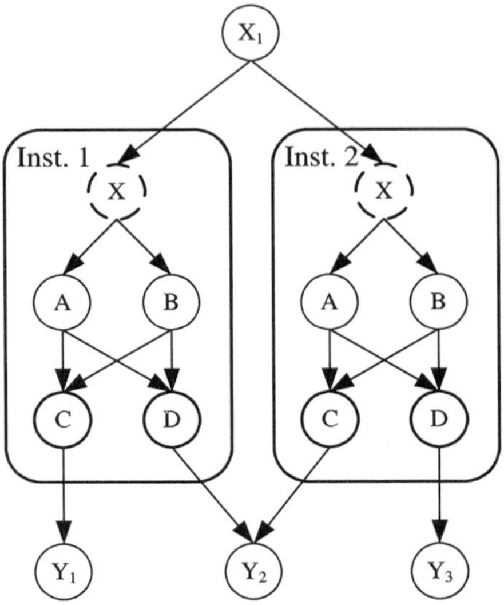

Fig. 4.1. A simple OOBN.

4.2.2. *Modeling methodology*

Generally, some complex industrial systems include repetitive structures, which could be copies of similar or even almost-identical subsystems. For these systems, a generic modeling methodology of real-time fault diagnosis is developed using OOBNs, as shown in Fig. 4.2. The modeling methodology consists of two main phases: an offline OOBN construction phase and an online fault diagnosis phase.

In the offline phase, sensor historical data and expert knowledge are collected and processed to obtain the faults and fault symptoms, which are necessary for constructing the BNs, including the structure model and parameter model. The experts would initially provide the cause-and-effect relationships among the variables. The structure of OOBNs can be constructed using these relationships based on the proposed methodology in Sec. 4.2.3. The experts would then provide the probabilistic relationships between the nodes, and the parameter of OOBNs can be specified based on the proposed methodology in Sec. 4.2.4.

In the online phase, operator experience and sensor real-time data, which act as additional information and fault symptoms, are

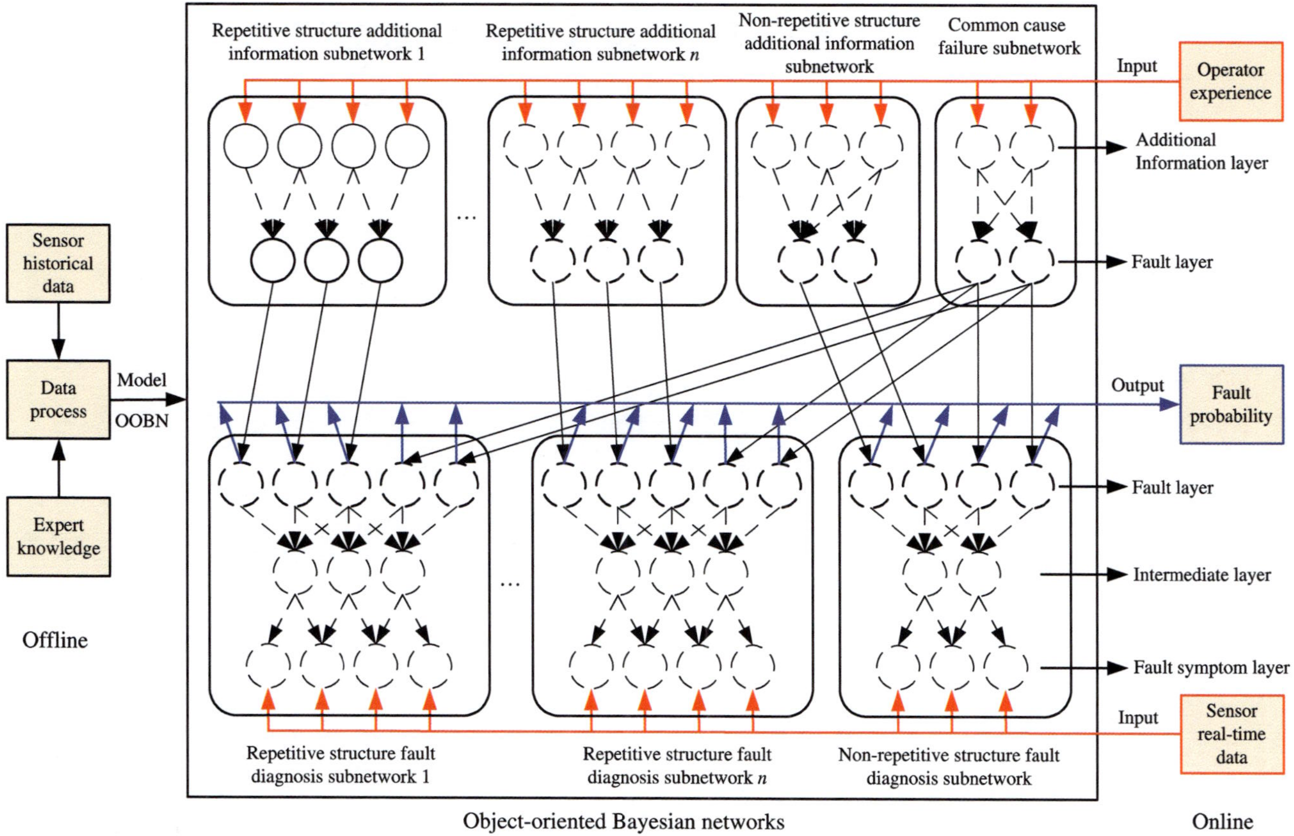

Fig. 4.2. OOBN-based fault diagnosis modeling methodology.

respectively encoded into the OOBNs to perform the fault diagnosis. The quantitative analysis of a BN proceeds along two directions, i.e., forward (or predictive) analysis and backward (or diagnostic) analysis. In the forward analysis, the probabilities of occurrence of any nodes are calculated based on the prior probabilities of root nodes and conditional probabilities of each node. In the backward analysis, the posterior probability of any given set of variables is calculated with some observation (the evidence), which is represented as the instantiation of some of the variables to one of their admissible values. The operator experience, such as repair frequency and weather situation, are considered to be additional information, which can provide more reasonable probabilities of the faults. It is performed by using the forward analysis of BNs. The real-time data detected by temperature and pressure sensors are used as evidence and inputted in fault symptom nodes. Based on these inputs, the posterior probabilities of fault nodes are updated. It is performed through the backward analysis of BNs.

The structure model, parameter model, model validation, and fault diagnosis and verification are important aspects of the proposed methodology. A detailed modeling flowchart is provided in Fig. 4.3, and the details are introduced as follows.

4.2.3. *Structure of OOBNs*

The proposed OOBNs consist of three main parts: repetitive structure subnetwork, non-repetitive structure subnetwork, and common cause failure subnetwork. The repetitive structure subnetwork includes repetitive structure additional information subnetwork and repetitive structure fault diagnosis subnetwork. The repetitive structure subnetwork can be further expanded to more subnetworks, such as repetitive structure additional information subnetwork n and repetitive structure fault diagnosis subnetwork n, which represent the almost-identical subsystems. Similarly, the non-repetitive structure subnetwork includes non-repetitive structure, additional information subnetwork, and non-repetitive structure fault diagnosis subnetwork. Common cause failure subnetwork is an independent subnetwork,

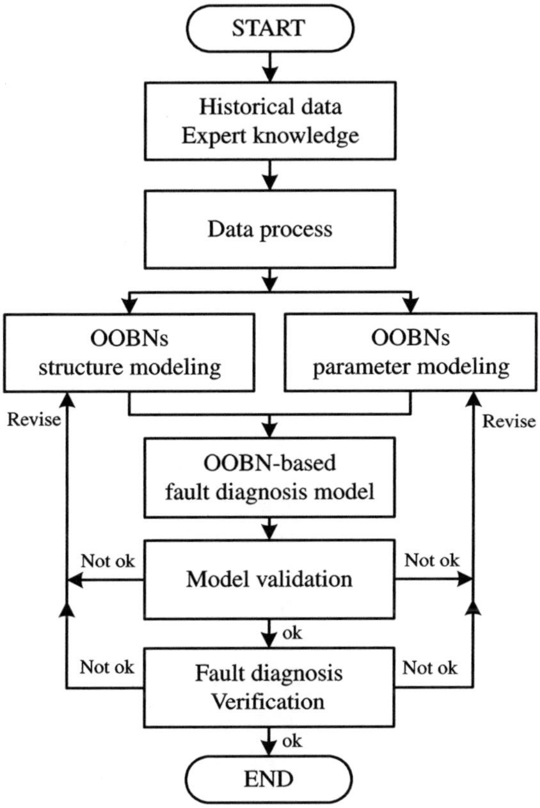

Fig. 4.3. Modeling flowchart of OOBN-based fault diagnosis methodology.

representing the common root faults that cause more than one subsystem failure.

Additional information subnetwork and common cause failure subnetwork have similar network structures, which include two layers, namely, additional information layer and fault layer. The additional information is obtained from the system operator in real time, which can update reasonable probabilities of faults by using the forward analysis of BNs. The fault layer is composed of the possible faults that denote the failure of components in one (non-common cause failure) or more (common cause failure) subsystems. For additional information subnetwork and common cause failure subnetwork, all of the nodes in fault layers are defined as output nodes of OOBNs.

Fault diagnosis subnetworks for both repetitive and non-repetitive structures include three layers, namely, fault layer, intermediate layer, and fault symptom layer. The fault layer is composed of the possible faults that denote the failure of components as described in the preceding paragraph, whereas the intermediate layer is composed of fault category or group, such as hydraulic system, mechanical system, electrical system, hardware issues, or software issues. Each child node in BNs has a maximum of four parent nodes, added by intermediate layers. Because the CPTs of a child node relate to all the states of its parent nodes, the number of parameters required in the CPTs exponentially grows with the number of its parents. A BN with n parent nodes and m states for each parent node requires mn parameters to be able to determine the CPT. The proposed intermediate layer tactically prevents highly complex CPTs, thereby facilitating the generation of diagnosis models. The fault symptoms, which are obtained from the system sensors in real time, represent the evidence of OOBNs and are used to update the posterior probability of faults. For the fault diagnosis subnetwork, all the nodes in the fault layers are defined as input nodes of OOBNs.

4.2.4. *Parameter of OOBNs*

In the OOBNs, the prior probabilities of root nodes in additional information layers and the conditional probabilities of leaf nodes in the fault, intermediate, and fault symptom layers have to be specified. The primary and prior probabilities of additional information nodes are obtained from the expert knowledge in the offline phase, and then they can be updated by the operator based on the real-life situations in the online phase.

The conditional probabilities of fault nodes, namely, the intermediate nodes and fault symptom nodes, are obtained from expert knowledge and historical sensor data in the offline phase. In the CPTs, the exponential growth of the number of parameters is one of the major concerns [31]. The most practical method is to apply noisy-OR and noisy-MAX models to determine the CPTs.

If the nodes have two states, for example, high and low, which would indicate high temperature and low temperature, respectively, these states would be obtained from the temperature sensor. The CPTs can be simplified using noisy-OR. For a noisy-OR model, different causes X_1, $X_2, \ldots, X_n$ are assumed to lead to an effect Y, and all of the variables are assumed to be Boolean. A noisy-OR determines a CPT using n parameters, $q_1, q_2, \ldots, q_n$, one for each parent, where q_i is the probability that Y is false, given that X_i is true and all of the other parents are false [25]. The full CPT can be generated using

$$P(Y = 1|X_1, X_2, \ldots, X_n) = 1 - \prod_{i=1}^{n} q_i. \tag{4.1}$$

If the nodes have more than two states, for example, high, normal, and low, which indicate high pressure, normal pressure, and low pressure, respectively, these can be obtained from the pressure sensor, in which the CPTs can be simplified using noisy-MAX but not noisy-OR. The noisy-MAX is a generalization of the noisy-OR to non-Boolean domains [25]. Suppose, for example, that there are n causes $X_1, X_2, \ldots, X_n$ of Y; by using noisy-MAX, the full CPT can be generated using

$$P(Y \leq y \,|\, X) = \prod_{\substack{i=1 \\ x_i \neq 0}}^{n} \sum_{y'=0}^{y} q_{i,y'}^{x_i} \tag{4.2}$$

$$P(Y = y \,|\, X) = \begin{cases} P(Y \leq 0 \,|\, X) & \text{if } y = 0, \\ P(Y \leq y \,|\, X) - P(Y \leq y-1 \,|\, X) & \text{if } y > 0, \end{cases} \tag{4.3}$$

where X represents parent node configurations of Y; $X = x_1, \ldots, x_n$, and $P(Y = 0 \,|\, X_1 = 0, \ldots, X_n = 0) = 1$.

4.2.5. *Model validation*

Model validation is a significant aspect of the fault diagnosis methodology because it provides reasonable confidence to diagnostic results. It is the best way to develop a real-time fault diagnosis system, monitor its states for a long period of time, and compare the diagnostic results and practical new faults for validation of the proposed methodology. Before the development of a real system, other methods can be used for validation. A three-axiom-based sensitivity analysis is performed for validation of each individual OOBN subnetwork, which includes additional information subnetwork and common cause failure subnetwork. The three axioms proposed by Jones *et al.* [26] must be satisfied to demonstrate that the proposed methodology is reasonable and correct.

Notably, a conflict may occur when two or more pieces of evidences are input into the OOBNs by sensors or even by an operator. The conflict analysis is necessary to detect possible conflicts among pieces of evidences or between them and the OOBN model, i.e., conflict analysis can be used for conflict detection as well as model validation. A conflict measure is used to demonstrate a possible conflict when the product of the probabilities of the individual pieces of evidence is larger than the joint probability of the evidence. Given a set of evidence, $x = \{x_1, \ldots, x_n\}$, under the assumption that pieces of evidence are positively correlated so that $P(x) > \prod_{i=1}^{n} P(x_i)$, the general conflict measure for x can be expressed as [27]

$$\text{conflict}(x) = \text{conflict}(\{x_1, \ldots, x_n\}) = \log \frac{\prod_{i=1}^{n} P(x_i)}{p(x)}. \tag{4.4}$$

This finding shows that a positive value of the conflict measure, conflict(x), demonstrates a possible conflict of evidence or a wrong model. Thus, this evidence-driven conflict analysis can be performed for validation of the complete OOBN-based fault diagnosis models.

4.2.6. *Fault diagnosis and verification*

For a specified system with a certain situation, the operator inputs known experience information, such as repair frequency and weather situation as mentioned, to the additional information layers of additional information subnetwork and common cause failure subnetwork. The real-time data detected by sensors located in the specified system are inputted to the fault symptom nodes of the fault diagnosis subnetwork. The posterior probabilities of fault nodes are updated by the proposed OOBNs. Generally, the larger the difference between prior and posterior probability of the fault, the higher the possibility of the corresponding fault occurrence. The fault diagnosis methodology can provide a probability of a fault but cannot draw a definite diagnostic result. According to engineering experience, two judgment rules are defined to determine the fault diagnosis results as follows:

Rule 1: A fault is reported if the difference between posterior probability and prior probability for a fault node is greater than or equal to 60%.

Rule 2: A warning is reported if the difference between posterior probability and prior probability for a fault node is greater than or equal to 30% and less than 60%.

Similar judgment rules were also developed in BN-based fault detection and diagnosis of variable air volume terminals [39]. According to the rules, the fault diagnosis system reminds the operator to handle the possible faults.

When the additional information, faults and fault symptoms can be obtained from practical engineering or historical data, the fault probabilities can be updated and compared with known information, to verify the proposed methodology. The verification procedure is the same as the fault diagnosis procedure that is described in this chapter.

4.3. Case Study

4.3.1. *Description of subsea production system*

Accidents in the offshore oil and gas industry usually have devastating consequences [40, 41]. Thus, utilizing a fault diagnosis system for oil equipment is necessary to warn operators about a possible failure [42–44]. The fault diagnosis system of a typical oil system, such as a subsea production system in Liuhua 4-1 oil field, is developed to demonstrate the application of the fault diagnosis methodology.

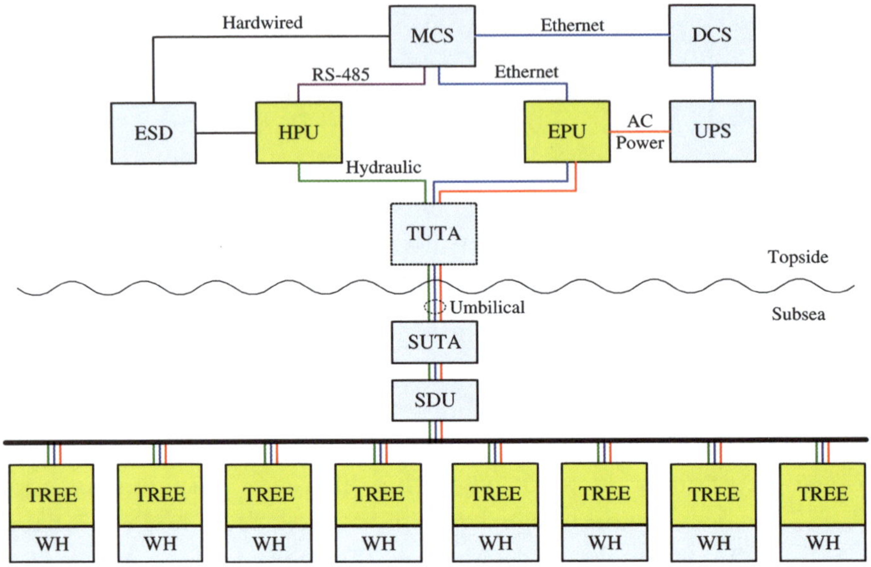

Fig. 4.4. Architecture of the simplified subsea production system of Liuhua 4-1 oil field.

The Liuhua 4-1 oil field is located in the South China Sea, with water depth of approximately 300 m, which is tied back to the nearby Liuhua 11-1 oil field [28]. The development of the Liuhua 4-1 oil field consists of a subsea production manifold with eight subsea wells and eight subsea Christmas trees, which are controlled by a subsea production control system, as shown in Fig. 4.4. The control system is based on the proven standard FMC KS200e electro-hydraulic multiplexed system [28, 29].

The topside components including main control station (MCS), hydraulic power unit (HPU), electrical power unit (EPU), emergency shutdown system (ESD), distributed control system (DCS), uninterrupted power system (UPS), and topside umbilical termination unit (TUTA) are located at the floating production system of the Liuhua 11-1 oil field. These components provide the control signal, hydraulic power, and chemical injections to the subsea production system of the oil field through a 14-km control umbilical system, assigned by a subsea umbilical termination assembly

(SUTA) and a subsea distribution unit (SDU) to each subsea control module (SCM) located at each subsea Christmas tree and wellhead (WH) [28, 29].

The MCS provides an interface between the operators and the subsea equipment through complex control networks. It can be used to monitor and control the subsea wells and manifold, and conduct the data acquisition. The EPU provides conditioned electrical power from the UPS to the topside equipment and subsea equipment. The HPU provides clean hydraulic fluid to the subsea valves located in the subsea Christmas tree through the umbilical system. The TUTA provides the interface between the topside control equipment and the main umbilical system [29]. The SUTA and SDU are the subsea components, and can be standalone or combined on a foundation structure. They receive hydraulic and electrical services from the topside via the production control umbilical system, and they distribute these services to the various subsea production equipment. The subsea Christmas tree is one of the most significant equipments in the subsea production system [45]. This device consists of a stack of subsea valves installed on a subsea wellhead. These subsea trees provide controllable interfaces between the subsea wells and various production facilities.

4.3.2. *Fault diagnosis modeling*

Constructing the OOBN-based fault diagnosis models for the entire subsea production system shown in Fig. 4.4 is challenging because numerous components are involved, and each component has numerous faults and fault symptoms. The purpose of the case is to demonstrate the fault diagnosis methodology; therefore, constructing an OOBN model is sufficient; this model has some repetitive and non-repetitive components with some frequent faults, including eight subsea Christmas trees, one HPU and one EPU, which are marked in yellow in Fig. 4.4. A large proportion of the data obtained from actual expert knowledge and sensor historical data is used for modeling, and another small portion is used for fault diagnosis and verification, as shown in the following section.

For the three equipment, namely, subsea Christmas trees, HPU and EPU, seven class models are developed using Hugin 8.1 software [37], as shown in Fig. 4.5. The figure also includes the subsea Christmas tree additional information (TREE_AI) subnetwork, subsea Christmas tree fault diagnosis (TREE_FD) subnetwork, HPU

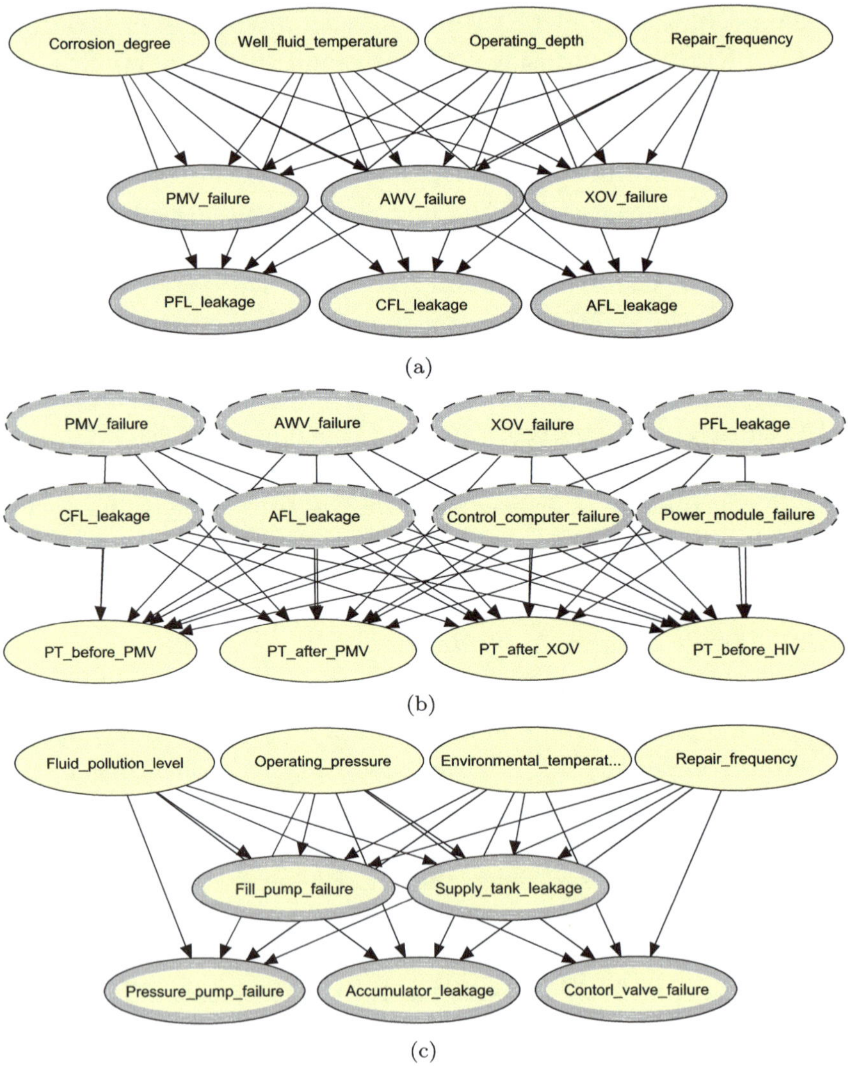

Fig. 4.5. Seven class models including (a) TREE_AI, (b) TREE_FD, (c) HPU_AI, (d) HPU_FD, (e) EPU_AI, (f) EPU_FD, and (g) CCF.

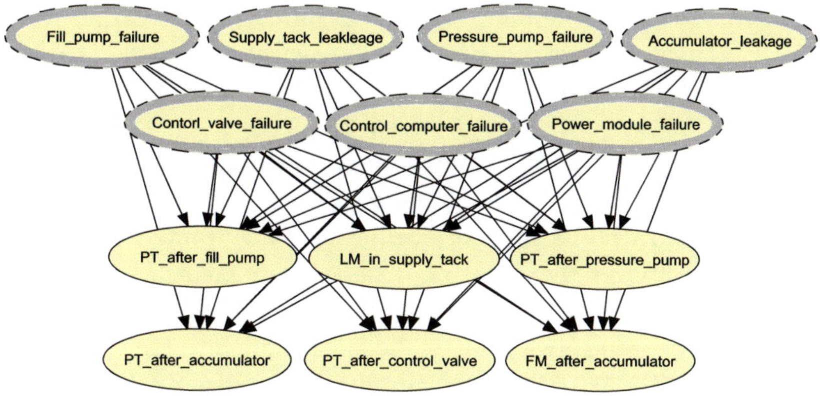

(d)

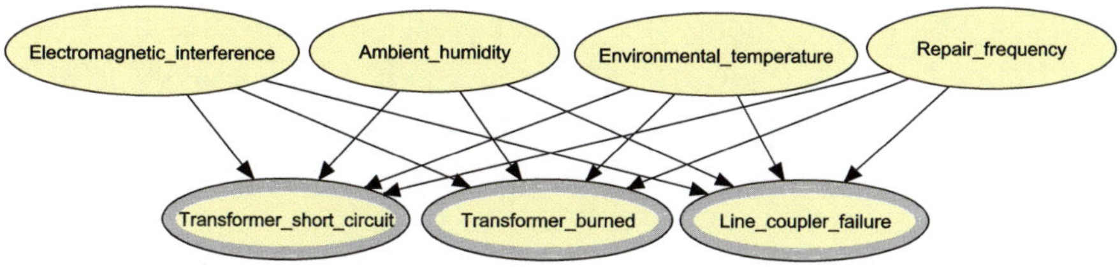

(e)

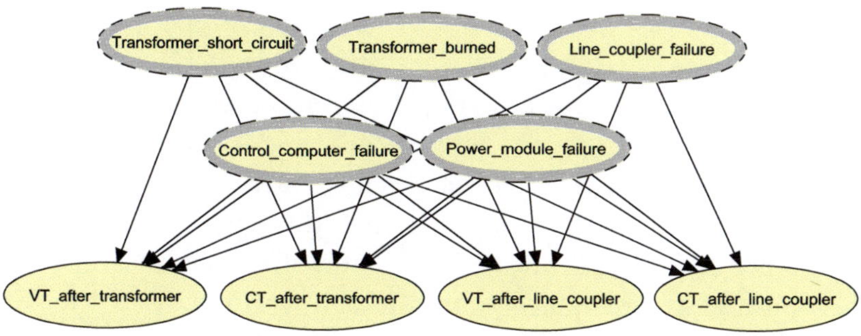

(f)

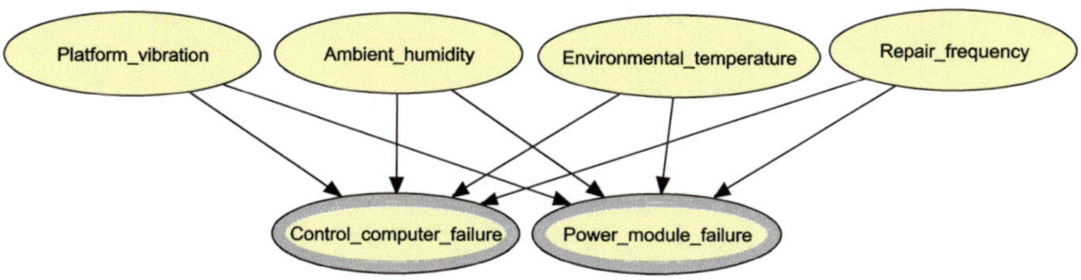

(g)

Fig. 4.5. (*Continued*)

additional information (HPU_AI) subnetwork, HPU fault diagnosis (HPU_FD) subnetwork, EPU additional information (EPU_AI) subnetwork, EPU fault diagnosis (EPU_FD) subnetwork, and common cause failure (CCF) subnetwork.

In the additional information subnetwork, taking TREE_AI shown in Fig. 4.5(a) as an example, we note that four parent nodes are in additional information layer, which includes "corrosion degree", "well fluid temperature", "operating depth", and "repair frequency". Each parent node has two states, which are high and low. Six child nodes are in the failure layer, including "production master valve (PMV) failure", "annulus wing valve (AWV) failure", "crossover valve (XOV) failure", "production flow loop (PFL) leakage", "crossover flow loop (CFL) leakage", and "annulus flow loop (AFL) leakage". Each child node has two states, namely, present and absent. The parent and child nodes are connected via arcs, denoting the cause-and-effect relationship of nodes in the additional information layer and fault layer.

In the fault diagnosis subnetwork, taking TREE_FD shown in Fig. 4.5(b) as an example, we can observe that two layers are constructed because few faults and fault symptoms exist, making the intermediate layer unnecessary. In the fault layer, two other parent nodes exist, namely, "control computer failure" and "power module failure", except the six failure nodes mentioned in the TREE_AI subnetwork. Also, four child nodes are in the fault symptom layer, including "pressure transducer (PT) before PMV", "PT after PMV", "PT after XOV", and "PT before HIV". Each node has three states, namely, normal, high, and low. Also, the parent and child nodes are connected via arcs, denoting the cause-and-effect relationship between the nodes in the fault layer and fault symptom layer.

The relationship between the parent and child nodes for the TREE_AI subnetwork is statistically calculated, as shown in Table 4.1. The CPTs of additional information nodes and fault nodes can be calculated by using the noisy-OR model, which is shown in Eq. (4.1). One modification is that the failure probabilities of the fault nodes are set to 1 but not 0 when all of the states

Table 4.1. Relationship between parent and child nodes for TREE_AI subnetwork.

		Parent node (High)			
Child node	State	Corrosion degree	Well fluid temperature	Operating depth	Repair frequency
PMV failure	Present	0.012	0.015	0.014	0.013
	Absent	0.988	0.985	0.986	0.987
AWV failure	Present	0.012	0.015	0.014	0.013
	Absent	0.988	0.985	0.986	0.987
XOV failure	Present	0.012	0.015	0.014	0.013
	Absent	0.988	0.985	0.986	0.987
PFL leakage	Present	0.014	0.012	0.013	0.013
	Absent	0.986	0.988	0.987	0.987
CFL leakage	Present	0.014	0.012	0.013	0.013
	Absent	0.986	0.988	0.987	0.987
AFL leakage	Present	0.014	0.012	0.013	0.013
	Absent	0.986	0.988	0.987	0.987

for additional information nodes are low. Similarly, the relationship between the parent and child nodes for the TREE_FD subnetwork is also statistically calculated, as shown in Table 4.2. The relationship is found to be uncertain. For instance, when "PMV failure" occurs, the high, normal, and low states of the child node "PT before PMV" are 5%, 10%, and 85%, respectively, but not 0%, 0%, 100%. This uncertainty can be caused by various reasons, such as sensor accuracy and measure uncertainty, and the limitations of design of sensors and observers, as mentioned. In addition, when "AFL leakage" occurs, the high, normal, and low states of the child node "PT before PMV" are 4%, 12%, and 84%, respectively, but not 0%, 0%, 100%. We can observe that the two kinds of faults cause the same symptom, indicating that complex and uncertain relationships exist between faults and fault symptoms. A similar situation was also found in the fault diagnosis systems of the ground-source heat pump [16] and variable air volume terminals [39] when various sensors were used. The CPTs of the fault nodes and fault symptoms can be calculated by using the noisy-MAX model shown in Eqs. (4.2) and (4.3).

Table 4.2. Relationship between parent and child nodes for TREE_FD subnetwork.

		Parent node (Present)							
Child node	State	PMV failure	AWV failure	XOV failure	PFL leakage	CFL leakage	AFL leakage	Control computer failure	Power module failure
PT before PMV	High	0.05	0.02	0.05	0.02	0.02	0.04	0.01	0.03
	Normal	0.10	0.92	0.91	0.08	0.88	0.12	0.03	0.06
	Low	0.85	0.06	0.04	0.90	0.10	0.84	0.96	0.91
PT after PMV	High	0.86	0.07	0.01	0.04	0.10	0.12	0.01	0.05
	Normal	0.10	0.87	0.05	0.12	0.79	0.82	0.89	0.90
	Low	0.04	0.06	0.94	0.84	0.11	0.06	0.10	0.05
PT after XOV	High	0.10	0.03	0.94	0.03	0.05	0.05	0.06	0.91
	Normal	0.89	0.06	0.04	0.95	0.13	0.84	0.12	0.05
	Low	0.01	0.91	0.02	0.02	0.82	0.11	0.82	0.04
PT before HIV	High	0.10	0.09	0.05	0.07	0.10	0.02	0.02	0.92
	Normal	0.85	0.82	0.86	0.02	0.88	0.89	0.87	0.05
	Low	0.05	0.09	0.09	0.91	0.02	0.09	0.11	0.03

The entire OOBNs for the fault diagnosis of the subsea production system are developed based on the constructed seven class models, as shown in Fig. 4.6.

For the repetitive structure, the objects TREE_AI_n and TREE_FD_n ($n = 1, \ldots, 8$) are produced by instantiating the classes TREE_AI and TREE_FD eight times, denoting the additional information and fault diagnosis models of the eight subsea Christmas trees. For the non-repetitive structure, the objects HPU_AI, HPU_FD, EPU_AI, EPU_FD and CCF are produced by instantiating the corresponding classes only once. Then, the output nodes in the objects of additional information subnetwork are connected with the input nodes in the objects of the fault diagnosis subnetwork via arcs, finally constructing the entire OOBN. As shown in Fig. 4.6(a), only the input and output nodes are presented for each instance node, making the BNs less complex and more tractable. In addition, collapsing the instance nodes can make the network less cluttered, as shown in Fig. 4.6(b).

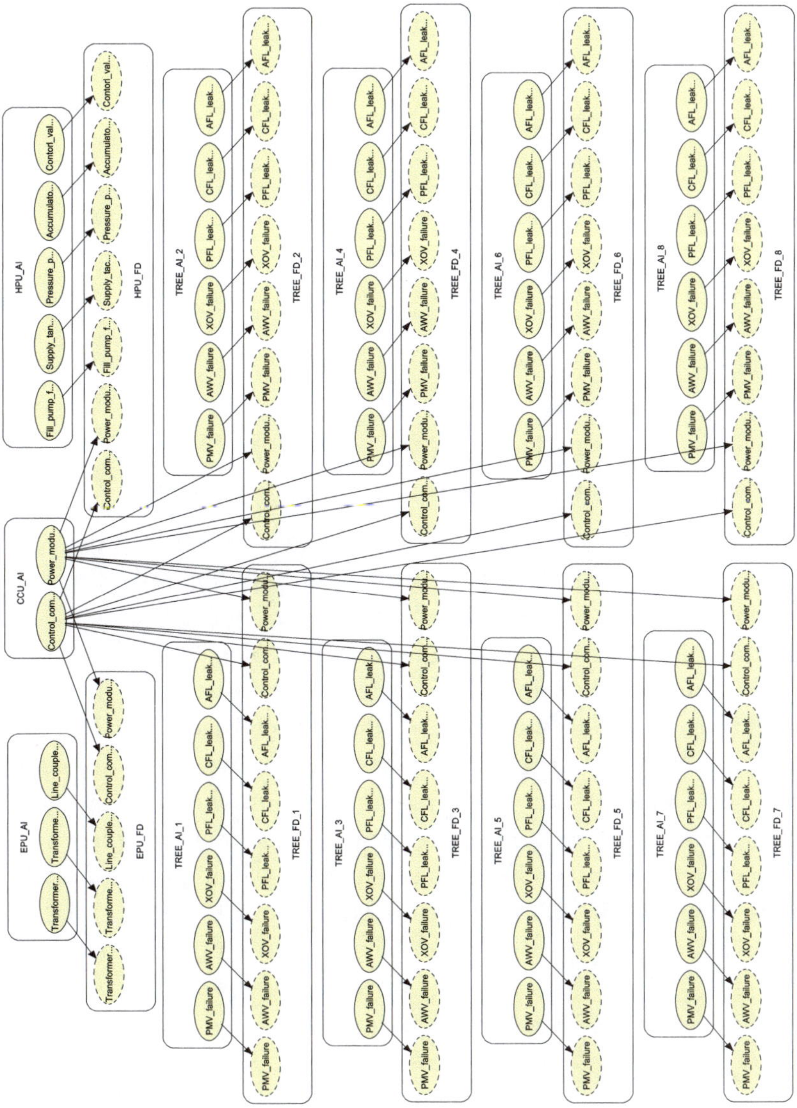

(a)

Fig. 4.6. Entire OOBN for fault diagnosis of subsea production system (a) non-collapsed form and (b) collapsed form.

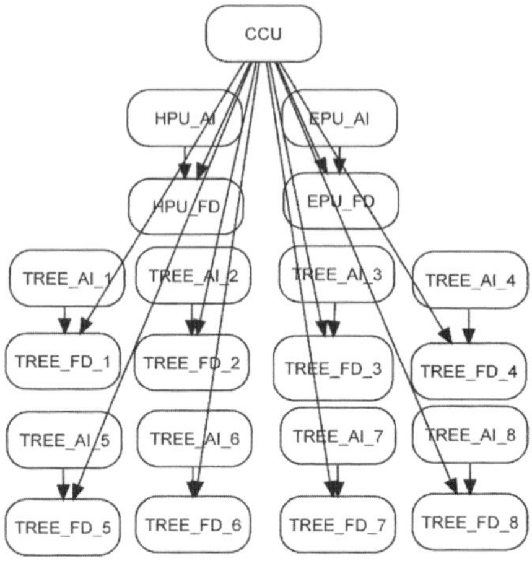

(b)

Fig. 4.6. (*Continued*)

4.3.3. *Results and discussion*

A sensitivity analysis was conducted to validate the additional information subnetworks and the common cause failure subnetwork. The models should satisfy at least three axioms [26]. Taking CCF shown in Fig. 4.5(g) as an example, when the "platform vibration" was set to 100% high from 40%, the probabilities of "control computer failure" and "power module failure" increased to 3.32% and 3.16% from 2.51% and 2.47%, respectively. The change and "ambient humidity" that were set to 100% high caused further increases of 4.08% and 3.86% in the failure probabilities. Similarly, when the "environmental temperature" was also set to 100% high, the failure probabilities were 4.95% and 4.85%. Lastly, when all of the parent nodes were set to 100% high, the failure probabilities increased to 5.87% and 5.77%. The exercise of increasing each influencing node satisfied these axioms stated in [26], thereby validating the CCF subnetwork models. Using the same method, we also validated the TREE_AI, HPU_AI, and EPU_AI subnetworks, and the results showed that these models were correct.

Except for the larger proportion of the data used in modeling, a small proportion of the data obtained from actual expert knowledge

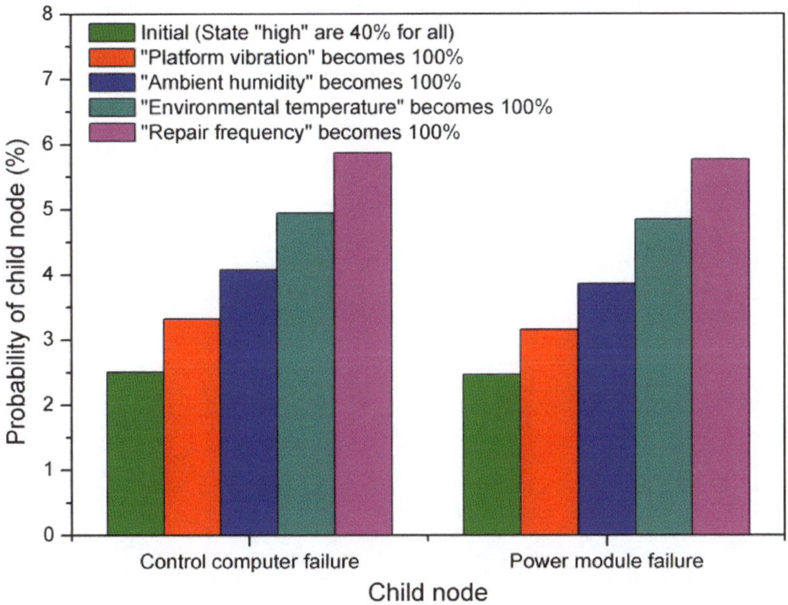

Fig. 4.7. Sensitivity analysis of CCF subnetwork.

and sensor historical data was used for conflict analysis, fault diagnosis, and verification in four cases. For these cases, the same operator experience (i.e., expert knowledge in this work) is inputted to the additional information layers of additional information subnetwork and common cause failure subnetwork. These include "well fluid temperature", "environmental temperature", "ambient humidity", and "environmental temperature", "platform vibration", which are high for the Christmas tree, HPU, EPU, and CCF, respectively, whereas all the others are low. The sensor real-time data (i.e., sensor historical data in this study) listed in the second column of Table 4.3 were inputted to the fault symptom layers of the fault diagnosis subnetwork as evidence. The inference algorithm adopted in this case study is from the Junction Tree algorithm.

In Case No. 1, three pressure transducers in subsea Christmas tree No.1 detected abnormal data, which included "PT before PMV" as low, "PT after PMV" as low, and "PT before HIV" as low. At the same time, two pressure transducers in subsea Christmas tree No. 4

Table 4.3. Three fault diagnosis cases.

Case	Evidence	Conflict measure	Fault report	Fault warning	True fault
No. 1	TREE1-PT before PMV(Low) TREE1-PT after PMV(Low) TREE1-PT before HIV(Low) TREE4-PT before PMV(Low) TREE4-PT after PMV(High)	−8.4014	TREE1-PFL leakage (98.81% − 1.2% = 97.61%) TREE4-PMV failure (90.32% − 1.5% = 88.82%)	None	TREE1-PFL leakage TREE4-PMV failure
No. 2	HPU-PT after fill pump(Low) HPU-LM in supply tack(Low) HPU-PT after accumulator(High) HPU-PT after control valve(Low) HPU-FM after accumulator(Low)	−10.3059	HPU-Control valve failure (99.93% − 1.4% = 98.53%) HPU-Fill pump failure (73.47% − 1.2% = 72.27%)	None	HPU-Control valve failure HPU-Fill pump failure
No. 3	TREE8-PT before PMV(Low) TREE8-PT after XOV(Low)	−1.9307	EPU-Line coupler failure (99.62% − 2.29% = 97.33%)	TREE8-AFL leakage (40.23% − 1.2% = 39.03%)	TREE8-AFL leakage TREE8-AWV failure
	EPU-VT after line coupler(Low) EPU-CT after line coupler(Low)			TREE8-AWV failure (33.58% − 1.5% = 32.08%)	EPU-Line coupler failure
No. 4	TREE8-PT before PMV(Low) TREE8-PT after XOV(Low)	3.5009	EPU-Line coupler failure (98.69% − 2.29% = 96.40%)	TREE8-AFL leakage (40.23% − 1.2% = 39.03%)	TREE8-AFL leakage TREE8-AWV failure
	EPU-VT after line coupler(Low) EPU-CT after line coupler(Normal)			TREE8-AWV failure (33.58% − 1.5% = 32.08%)	EPU-Line coupler failure

also detected abnormal data, which include “PT before PMV” as low and “PT after PMV” as high. The conflict measure is calculated based on Eq. (4.4) using the Hugin software. The negative value of −8.4014 indicates no conflicting evidence and that the developed fault diagnosis models were correct. The posterior probabilities of faults were also updated along with the evidence. The differences in posterior probability and prior probability for the “PFL leakage” and “PMV failure” were 97.61% and 88.82%, respectively, thus, the two faults should be reported in accordance with *Rule* 1. Compared to the true faults in Table 4.3, the fault diagnosis results were correct, indicating that the proposed methodology worked as designed. Particularly, the difference between the posterior probability and the prior probability was large for a single fault in one structure. Therefore, the single fault in one structure was considered to be simple to diagnose.

In Case No. 2, the abnormal data are detected in one single structure, that is, HPU. “PT after fill pump” is low, “LM in supply tack” is low, “PT after accumulator” is high, “PT after control valve” is low, and “FM after accumulator” is also low. The conflict measure value of −10.3059 also indicates the free conflict and correctness of the models. In addition, two faults in a single structure are diagnosed because the differences of posterior probability and prior probability for the “control valve failure” and “fill pump failure” are both larger than 60%. Compared to the true faults in Table 4.3, the fault diagnosis results were also correct.

In Case No. 3, two pressure transducers in subsea Christmas tree No. 8 detected abnormal data, which included low “PT before PMV” and low “PT after XOV”. The voltage and current transducers in the EPU also detected abnormal data, which included low “VT after line coupler” and low “CT after line coupler”. The conflict measure value of −1.9307 indicated no conflicting evidence. According to the difference of the posterior probability and the prior probability of 97.33%, diagnosing the fault “line coupler failure” in the EPU could be performed easily. For the subsea Christmas tree No. 8, however, the differences in the posterior probability and prior probability for “AFL leakage” and “AMV failure” were 39.03% and 32.08%, respectively; thus, a warning should be reported in accordance with *Rule* 2. Compared to the true faults in Table 4.3, the fault warnings were actual faults, indicating that the results were correct. We can infer from the case that multiple-simultaneous faults in a single structure might not be simple to diagnose, and the operator is required to handle the possible faults further.

Case No. 4 with the fault symptom "CT after line coupler" set to normal is an artificial modification of Case No. 3. Although the fault diagnosis models provide the correct result, the positive conflict measure of 3.5009 indicated the existence of conflict in the evidence gathered. Thus, the operator should check the hardware sensors to eliminate failure. The modified case indirectly showed that the proposed methodology was correct.

4.4. Conclusion

The present study proposed a real-time fault diagnosis methodology of complex systems with repetitive structures. The uncertainty problem of fault diagnosis was solved using BNs, and the complicated models were reduced using OOBNs. Once faults occur, the proposed OOBN-based fault diagnosis system can report the faults and warnings. For a specified system with a certain situation, the operator can input some known experience information to the additional information layers of the additional information and common cause failure subnetworks. At the same time, the sensors located in the specified system input real-time data to the fault symptom layers of the fault diagnosis subnetwork. Then, the OOBNs update the posterior probabilities of faults to perform fault diagnosis.

The application of the proposed methodology is demonstrated by using a subsea production system, including eight subsea Christmas trees, one hydraulic power unit, and one electrical power unit. The OOBN-based fault diagnosis models were then developed. The three-axiom-based sensitivity analysis was conducted to validate the proposed additional information subnetwork and common cause failure models, and the evidence-driven conflict analysis was performed to verify the complete fault diagnosis methodology. Four fault diagnosis cases were researched, and the fault report and fault warning were consistent with the true faults, indicating that the proposed methodology was correct in diagnosing both the single fault and multiple-simultaneous faults.

References

[1] X. Dai, Z. Gao, "From model, signal to knowledge: A data-driven perspective of fault detection and diagnosis," *IEEE Transactions on Industrial Informatics*, vol. 9, no. 4, pp. 2226–2238, 2013.

[2] L. M. Bartlett, E. E. Hurdle, E. M. Kelly, "Integrated system fault diagnostics utilising digraph and fault tree-based approaches," *Reliability Engineering & System Safety*, vol. 94, no. 6, pp. 1107–1115, 2009.

[3] F. D. McKenzie, A. J. Gonzalez, R. Morris, "An integrated model-based approach for real-time on-line diagnosis of complex systems," *Engineering Applications of Artificial Intelligence*, vol. 11, no. 2, pp. 279–291, 1998.

[4] W. H. Chen, C. W. Liu, M. S. Tsai, "On-line fault diagnosis of distribution substations using hybrid cause-effect network and fuzzy rule-based method," *IEEE Transactions on Power Delivery*, vol. 15, no. 2, pp. 710–717, 2000.

[5] P. K. Wong, Z. Yang, C. M. Vong, J. Zhong, "Real-time fault diagnosis for gas turbine generator systems using extreme learning machine," *Neurocomputing*, vol. 128, pp. 249–257, 2014.

[6] Z. Zhou, M. Zhuang, X. Lu, L. Hu, G. Xia, "Design of a real-time fault diagnosis expert system for the EAST cryoplant," *Fusion Engineering and Design*, vol. 87, no. 12, pp. 2002–2006, 2012.

[7] M. Bonvini, M. D. Sohn, J. Granderson, M. Wetter, R. G. Jungst, M. A. Piette, "Robust on-line fault detection diagnosis for HVAC components based on nonlinear state estimation techniques," *Applied Energy*, vol. 124, pp. 156–166, 2014.

[8] Y. Qian, L. Xu, X. Li, L. Lin, A. Kraslawski, "LUBRES: An expert system development and implementation for real-time fault diagnosis of a lubricating oil refining process," *Expert Systems with Applications*, vol. 35, no. 3, pp. 1252–1266, 2008.

[9] J. Pearl, "Bayesian networks: A model of self-activated memory for evidential reasoning," In: *Proceedings of the 7th Conference of the Cognitive Science Society*, University of California, Irvine, CA, 1985.

[10] D. Koller, A. Pfeffer, "Object-oriented Bayesian network," In: *Proceedings of the Thirteenth Annual Conference on Uncertainty in Artificial Intelligence*, Providence, RI, 1997.

[11] F. V. Jesen, T. D. Nielsen, *Bayesian Networks and Decision Graphs*, 2nd edn., Springer, 2007.

[12] C. H. Lo, Y. K. Wong, A. B. Rad, "Bond graph based Bayesian network for fault diagnosis," *Applied Soft Computing*, vol. 11, no. 1, pp. 1208–1212, 2011.

[13] D. A. Tobon-Mejia, K. Medjaher, N. Zerhouni, "CNC machine tool's wear diagnostic and prognostic by using dynamic Bayesian networks," *Mechanical Systems and Signal Processing*, vol. 28, pp. 167–182, 2012.

[14] I. Mandal, "Developing new machine learning ensembles for quality spine diagnosis," *Knowledge-Based Systems*, vol. 73, pp. 298–310, 2015.

[15] Q. Zhang, S. Geng, "Dynamic uncertain causality graph applied to dynamic fault diagnoses of large and complex systems," *IEEE Transactions on Reliability*, vol. 64, pp. 910–927, 2015.

[16] B. Cai, Y. Liu, Q. Fan, Y. Zhang, Z. Liu, S. Yu, R. Ji, "Multi-source information fusion based fault diagnosis of ground-source heat pump using Bayesian network," *Applied Energy*, vol. 114, pp. 1–9, 2014.

[17] D. Liu, Y. Huang, Q. Yu, J. Chen, H. Jia, "A search problem in complex diagnostic Bayesian networks," *Knowledge-Based Systems*, vol. 30, pp. 95–103, 2012.

[18] G. Weidl, A. L. Madsen, S. Israelson, "Applications of object-oriented Bayesian networks for condition monitoring, root cause analysis and decision support on operation of complex continuous processes," *Computers & Chemical Engineering*, vol. 29, no. 9, pp. 1996–2009, 2005.

[19] P. Weber, L. Jouffe, "Complex system reliability modelling with Dynamic Object Oriented Bayesian Networks (DOOBN)," *Reliability Engineering & System Safety*, vol. 91, no. 2, pp. 149–162, 2006.

[20] T. B. Jensen, A. R. Kristensen, N. Toft, N. P. Baadsgaard, S. Østergaard, H. Houe, "An object-oriented Bayesian network modeling the causes of leg disorders in finisher herds," *Preventive Veterinary Medicine*, vol. 89, pp. 237–248, 2009.

[21] D. Kasper, G. Weidl, T. Dang, G. Breuel, A. Tamke, A. Wedel, W. Rosenstiel, "Object-oriented Bayesian networks for detection of lane change maneuvers," *IEEE Intelligent Vehicles Symposium Proceedings*, vol. 4, no. 1, pp. 673–678, 2012.

[22] N. Khakzad, F. Khan, P. Amyotte, "Quantitative risk analysis of offshore drilling operations: A Bayesian approach," *Safety Science*, vol. 57, pp. 108–117, 2013.

[23] M. Abramovici, A. Lindner, "Knowledge-based decision support for the improvement of standard products," *CIRP Annals Manufacturing Technology*, vol. 62, no. 1, pp. 159–162, 2013.

[24] N. Huang, R. McMurran, G. Dhadyalla, R. P. Jones, "Probability based vehicle fault diagnosis, Bayesian network method," *Journal of Intelligent Manufacturing*, vol. 19, no. 3, pp. 301–311, 2008.

[25] W. Li, P. Poupart, P. V. Beek, "Exploiting structure in weighted model counting approaches to probabilistic inference," *Journal of Artificial Intelligence Research*, vol. 40, pp. 729–765, 2011.

[26] B. Jones, I. Jenkinson, Z. Yang, J. Wang, "The use of Bayesian network modelling for maintenance planning in a manufacturing industry," *Reliability Engineering & System Safety*, vol. 95, no. 3, pp. 277–367, 2010.

[27] U. Kjærulff, A. L. Madsen, *Bayesian Networks and Influence Diagrams*, 2nd edn., Springer, 2013.
[28] Z. Yue, Y. Liu, J. Mao, J.Yang, "Field development overview of Liuhua 11-1 and Liuhua 4-1 oil field," In: *The Offshore Technology Conference*, Houston, TX, 2013.
[29] Y. Bai, Q. Bai, *Subsea Engineering Handbook*, Elsevier, 2012.
[30] A. Bobbio, L. Portinale, M. Minichino, E. Ciancamerla, "Improving the analysis of dependable systems by mapping fault trees into Bayesian networks," *Reliability Engineering & System Safety*, vol. 71, no. 3, pp. 249–260, 2001.
[31] Y. F. Wang, M. Xie, K. S. Chin, X. J. Fu, "Accident analysis model based on Bayesian network and evidential reasoning approach," *Journal of Loss Prevention in the Process Industries*, vol. 26, pp. 10–21, 2013.
[32] B. Cai, Y. Liu, Y. Ma, Z. Liu, Y. Zhou, J. Sun, "Real-time reliability evaluation methodology based on dynamic Bayesian networks: A case study of a subsea pipe ram BOP system," *ISA Transactions*, vol. 58, pp. 595–604, 2015.
[33] S. S. Biswas, A. K. Srivastava, D. Whitehead, "A real-time data driven algorithm for health diagnosis and prognosis of a circuit breaker trip assembly," *IEEE Transactions on Industrial Electronics*, vol. 62, no. 6, pp. 3822–3831, 2015.
[34] P. Chen, S. Yang, J. A. McCann, "Distributed real-time anomaly detection in networked industrial sensing systems," *IEEE Transactions on Industrial Electronics*, vol. 62, no. 6, pp. 3832–3842, 2015.
[35] F. Sahin, M. C. Yavuz, Z. Arnavut, O. Uluyol, "Fault diagnosis for airplane engines using Bayesian networks and distributed particle swarm optimization," *Parallel Computing*, vol. 33, no. 2, pp. 124–143, 2007.
[36] M. Najafi, D. M. Auslander, P. L. Bartlett, P. Haves, M. D. Sohn, "Application of machine learning in the fault diagnostics of air handling units," *Applied Energy*, vol. 96, pp. 347–358, 2012.
[37] Hugin Expert A/S, Hugin version 8.1, Available at http://www.hugin.com. Accessed Feb. 2016.
[38] B. Cai, Y. Liu, Z. Liu, X. Tian, Y. Zhang, J. Liu, "Performance evaluation of subsea blowout preventer systems with common-cause failures," *Journal of Petroleum Science and Engineering*, vol. 90–91, pp. 18–25, 2012.
[39] F. Xiao, Y. Zhao, J. Wen, S. Wang. "Bayesian network based FDD strategy for variable air volume terminal," *Automation in Construction*, vol. 41, pp. 106–118, 2014.
[40] BOEMRE, Report regarding the causes of the April 20, 2010, The Bureau of Ocean Energy Management Regulation and Enforcement, Sep. 14, 2011.
[41] W. F. Harlow, B. C. Brantley, R. M. Harlow, "BP initial image repair strategies after the Deepwater Horizon spill," *Public Relations Review*, vol. 37, pp. 80–83, 2011.
[42] E. Altamiranda, E. Colina, "Intelligent supervision and integrated fault detection and diagnosis for subsea control systems," In: *OCEANS 2007 — Europe*, pp. 1–6, IEEE, 2007.

[43] L. P. Chze, "Importance of condition and performance monitoring to maximize subsea production system availability," In: *SPE/IATMI Asia Pacific Oil & Gas Conference and Exhibition*, Society of Petroleum Engineers, 2015.

[44] J. D. Friedemann, A. Varma, P. Bonissone, N. Iyer, "Subsea condition monitoring: A path to increased availability and increased recovery," In: *Intelligent Energy Conference and Exhibition*, Society of Petroleum Engineers, 2008.

[45] Y. Kang, M. Duan, B. Chen, "Optimization design of subsea Christmas tree pipeline," *International Journal of Energy Engineering*, vol. 1, pp. 12–18, 2011.

Chapter 5

A Dynamic Bayesian Network-Based Fault Diagnosis Methodology Considering Transient and Intermittent Faults

Transient and intermittent faults of complex electronic systems are difficult to diagnose. As the performance of electronic products degrades over time, the results of fault diagnosis could be different at different times given the identical fault symptoms. A dynamic Bayesian network (DBN)-based fault diagnosis methodology in the presence of transient and intermittent faults for the electronic systems is proposed. DBNs are used to model the dynamic degradation process of electronic products, and Markov chains are used to model the transition relationships of four states, i.e., no fault, transient fault, intermittent fault, and permanent fault. Our fault diagnosis methodology can identify the faulty components and distinguish the fault types. Four fault diagnosis cases of Genius modular redundancy control system are investigated to demonstrate the application of this methodology.

5.1. Introduction

Modern industrial systems, such as subsea blowout preventer (BOP) and subsea oil production systems, are mainly controlled by various electronic systems, such as programmable logic controllers, programmable automation controllers, and industrial computers. Faults in these electronic systems degrade the performance of industrial systems greatly; therefore, fault diagnosis plays an important role in detecting and isolating faults [1–4].

Permanent faults (PFs), together with two types of non-permanent faults, i.e., transient faults (TFs) and intermittent faults (IFs), cause most failures of various electronic products and systems

[5, 6]. PFs are malfunctions that always occur no matter when the system is examined. On the contrary, TFs and IFs are sporadic faults that are not easily repeatable because of their complicated behavioral patterns. They do not manifest themselves all the time and disappear in an unpredictable manner. They are difficult to detect because they are not easy to verify, replicate, or localize to a specific failure location, mode, or mechanism. The costs caused by TF and IFs could be much higher than that by PFs because the maintenance team has to run many tests to detect these faults [7, 8]. A common example of TF/IF occurs when a computer "hangs up". Clearly, a "failure" has occurred. However, if the computer is rebooted, it often works again.

Several methods have been developed to detect and diagnose TF and IF of various systems. Singh *et al.* [7] proposed an off-board, data-driven method for analyzing the fleet-wide operating parameter identifier data and investigating the IFs of electronic control units.

Zhou *et al.* [9] proposed an innovative ontology-based fault propagation analysis method to analyze the TF propagation effects in networked control systems. Contant *et al.* [10] proposed a modeling methodology of IFs in the context of discrete event system models by defining new notions of diagnosability associated with faults and reset events. Abreu and van Gemund [11] proposed a fault diagnosis framework to diagnose multiple IFs by using a maximum likelihood estimation method. Mahapatro and Khilar [12] formulated the IF detection and diagnosis of node failure in wireless sensor networks as an optimization problem and proposed a two-lbests-based multiobjective particle swarm optimization algorithm for solving the multiobjective problem. Monekosso and Remagnino [13] described a data-driven method to detect and mask PFs and TFs by using the principal component analysis and canonical correlation analysis techniques. Deng *et al.* [14] proposed a discrete event systems' method to distinguish PFs and IFs. Liu *et al.* [15] proposed an experimental method for detecting IFs in a field programmable gate array of ball grid array packages by injecting current into the input/output port and detecting the voltage signal. Yu and Wang [16] proposed a model-based fault diagnosis and prognosis method for a vehicle steering system and constructed a fault discriminator to distinguish the multiple faults of different types, including abrupt faults, incipient faults, and IFs. Cui *et al.* [17] proposed a novel method for transient and IF detection by using the Hilbert transform-based instantaneous power direction detection with intermittent spike identification. Wang and Liang [18] presented a method for the identification of multiple TFs based on the adaptive spectral kurtosis analysis of the vibration signal from a single sensor.

The Bayesian network (BN) is considered to be one of the most useful models in the field of probabilistic knowledge representation and reasoning since it was introduced in early 1980s [19]. Recently, BNs are increasingly used in the field of fault detection and diagnosis. Liu *et al.* [20] proposed an adaptive sensor allocation strategy for process monitoring and diagnosis using BNs. Bennacer *et al.* [21] proposed a hybrid method that combines BNs and case-based reasoning for overcoming the usual limits of fault diagnosis techniques and reducing human intervention. Zhu *et al.* [22] proposed three element-oriented fault diagnosis models for transmission lines, busbars, and transformers of the transmission power system using BNs. Khanafer *et al.* [23] proposed automated diagnosis in troubleshooting for universal mobile telecommunication system networks by using BNs. Barua and Khorasani [24, 25] proposed a systematic, transparent, and hierarchical fault diagnosis framework for a satellite formation flight and investigated the verification and validation of the proposed methodology. Ricks and Mengshoel [26] proposed a comprehensive fault diagnosis approach for dynamic and hybrid domains with uncertainty using BNs and arithmetic circuits and validated them using electrical power system data. Liu *et al.* [27] proposed a fault diagnosis framework for a solar-assisted heat pump system with incomplete data and expert knowledge by using BNs. Xu *et al.* [28] proposed a novel integrated system health management-oriented intelligent diagnostics methodology based on data mining, which uses a robust diagnostic BN to identify faults with uncertainty in a dynamic environment.

DBNs are a long-established extension to ordinary BNs and allow the explicit modeling of changes over time. It is a powerful tool that allows modeling sequential data, and becomes popular thanks to its modeling, graphical representation, and inference capabilities [29]. The technique has an advantage of representing uncertain knowledge in dynamic systems and the inference can be done fast. DBNs are used in the field of reliability evaluation, risk analysis, as well as fault detection and diagnosis. Cai *et al.* [30] proposed a DBN-based real-time reliability evaluation methodology, and a subsea pipe ram BOP system was used to illustrate the methodology. They also proposed a multiphase DBNs methodology for the determination of safety integrity levels [31]. Codetta-Raiteri and Portinale [32] described the modeling features and inference capabilities of DBNs in designing and implementing an innovative method to fault detection, identification, and recovery for an autonomous spacecraft. Tobon-Mejia *et al.* [29] performed the CNC machine tool's wear diagnostic and prognostic by using DBNs. A hidden Markov model is the simplest DBN, and was also used for fault diagnosis recently [33, 34].

As the performance of electronic products degrades over time, the results of fault diagnosis could be different at different times given the identical fault symptoms. Earlier chapters did not propose a fault diagnosis method for solving this problem. DBNs are used to model the dynamic degradation process of electronic systems for fault diagnosis. The main contribution of this study is to represent a DBN-based fault diagnosis methodology, which can identify the faulty component and distinguish the fault types, including TF, IF, and PF.

The chapter is structured as follows. Section 5.2 describes the different types of faults for electronic systems. Section 5.3 proposes the DBN-based fault diagnosis methodology in the presence of TF and IF. Section 5.4 investigates four cases of Genius modular redundancy (GMR) control systems to demonstrate the application of the proposed methodology. Section 5.5 summarizes the chapter.

5.2. Faults Description

Generally, three types of faults exist in complex electronic systems, i.e., PF, TF, and IF, as shown in Fig. 5.1. A PF is a persistent failure,

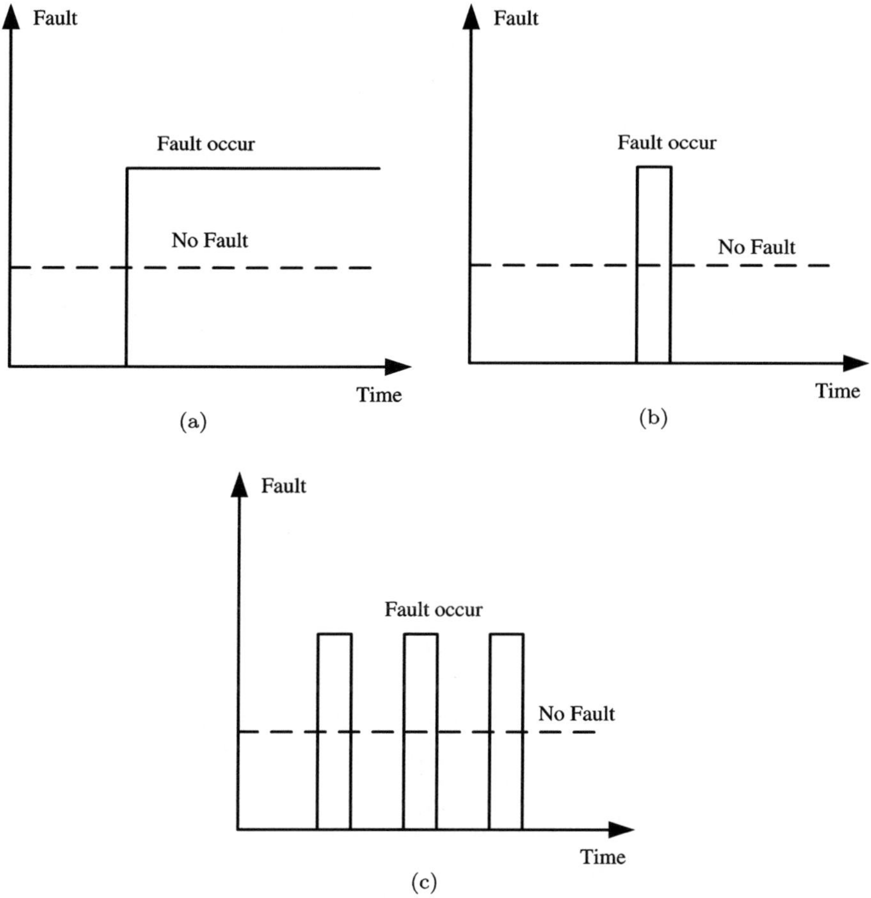

Fig. 5.1. Three types of faults: (a) PF; (b) TF; and (c) IF.

and it continues to exist until the faulty component is repaired or replaced. A TF is a temporary, one-time failure that causes a change of system and may not cause a permanent damage. TFs tend to be completely random in nature. An IF is a repeatable failure that appears and disappears with the changes in operation conditions. It is cumulative and might be a prelude to a PF. TF and IF can recover their abilities to perform their required functions without being subjected to any external corrective action [11, 12, 15, 16, 35]. TFs do not cause a permanent damage of systems, and we can ignore them; however, IFs cause repeatable failure of systems, and we must replace or repair the faulty components with IFs. It is necessary to differentiate these two types of faults.

Permanent faults of electronic systems are usually caused by overvoltage, overcurrent, or overheating, subsequent aging or burning, etc. For example, a capacitor in PLC burns because of overvoltage, which leads to a PF of the PLC. TFs of electronic systems are usually caused by disturbance signals, electromagnetic radiation, overheating, input power variations, etc. For example, an inverter of power systems fails because of input power variation and recovers subsequently when the input power stabilizes. IFs of electronic systems are usually caused by vibration, coefficient of thermal expansion mismatch, stress relaxation, etc. For example, vibrations can cause IFs of connectors and wire bonds. Coefficient of thermal expansion mismatch can cause IFs of wire bonds because of temperature changes [5, 8].

The current study does not aim to diagnose the root causes of TF and IF, but aims to identify the faulty components and distinguish the fault types of electronic systems. BN-based fault diagnosis is a typical data-driven method and can be used to distinguish these faults based on historical statistical data by using various inference algorithms.

5.3. Modeling Methodology

Given identical fault symptoms at different times, the diagnosis results may be different because the performance of electronic products degrades continuously over time. A general DBN-based fault diagnosis methodology is developed in the current study. All the faults and fault symptoms are merged into DBNs to simulate what engineers do in actual fault diagnosis. It aims at identifying the component faults and distinguishing the fault types, including TF, IF, and PF. They are modeled using DBNs and Markov. In particular, the dynamic degradation process is modeled using DBNs, and the relationship of three types of faults are modeled using Markov, and finally DBN models and Markov models are combined together. Because DBNs and Markov models are graphical representations, the theoretical analysis and technical process are described using graphical models. The details of the proposed DBNs are introduced as follows.

5.3.1. *DBNs structure modeling*

The proposed DBN model contains two layers of events, namely, faults and fault symptoms, as shown in Fig. 5.2. The nodes, F_n, at the fault layer represent the fault of each electronic product in an electronic system, such as "power source (PS) fault" and "communication module fault". Each node has four states, i.e., no fault (NF), transient fault (TF), intermittent fault (IF), and permanent fault (PF), indicating the fault types. The nodes, S_n, at the fault symptom layer represent the symptoms reported by engineers or operators, such as "program missing" and "lose all communication". Each node has four states, i.e., zero time, one time, two or more times, and permanent failure. The times indicate the cumulative number of various fault symptoms recorded in a short period of time, such as one week or 10 days. If a fault occurs and never recovers before repair, the node state will be permanent failure.

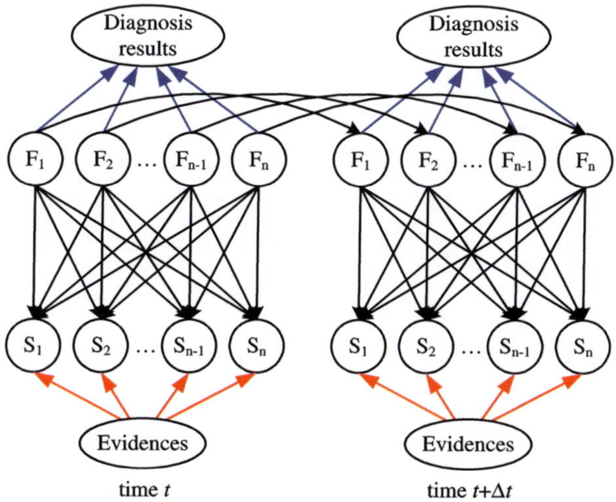

Fig. 5.2. DBN-based fault diagnosis model.

In static BNs, the causal relationships between the fault layer and fault symptom layer are connected by intra-slice arcs at a specific time slice t. In DBNs, the dynamic degradation process is illustrated by using inter-slice arcs, which connects the corresponding fault nodes between adjacent time slices t and $t + \Delta t$. All the information required to predict a state at time $t + \Delta t$ is contained in the description at time t, and no information about earlier times is required, which means that the process has the Markov property.

5.3.2. *DBN parameter modeling*

DBN parameter model is composed of an intra-slice parameter model and an inter-slice parameter model. The intra-slice parameter model is the conditional probability between fault nodes and fault symptom nodes. There are usually two methods for determining the conditional probability. One is probability elicitation by experts and subsequent filling-up of remaining relationships with various algorithms, such as noisy-OR and noisy-Max [36]. However, this method might bring errors for fault diagnosis results because of the independent assumption of noisy models. The other method is parameter learning using full-set fault and symptom data. The disadvantage of the method is that obtaining full-set data is difficult and sometimes impractical.

The inter-slice parameter model is the probability of fault nodes between time slices t and $t + \Delta t$. Since the degradation process has the Markov property, the probability is determined by using the Markov state transition relationship based on the assumption that all transition rates including failure rate and repair rate are constant for electronic products [37, 38]. The state transition diagram in terms of the four-state Markov model is provided in Fig. 5.3. A component is in its good function state initially, namely, NF state. As time goes by, it goes to the degraded states, namely, TF state, IF state, and PF state, upon degradation process. According to the model, the NF state makes a transition to TF state, IF state, and PF state with failure rates λ_1, λ_2, and λ_3, respectively. A TF of the component can develop into an IF and a PF, or recover to NF. Therefore, the TF state makes a transition to IF state, PF state, and NF state with

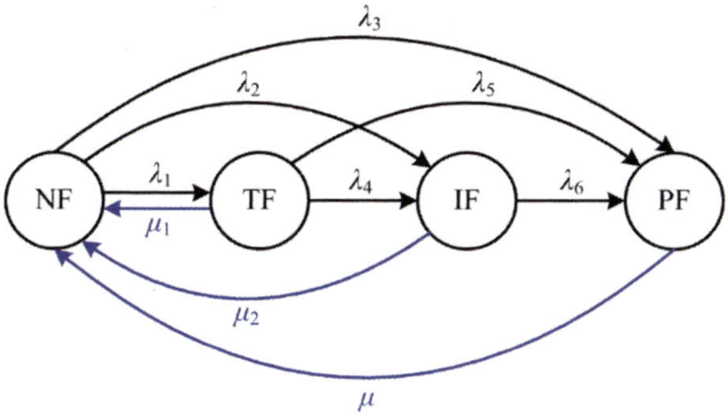

Fig. 5.3. State transition diagram with TF and IFs.

failure rates λ_4, λ_5 and repair rate μ_1, respectively. Similarly, an IF of the component can develop into a PF, or recover to NF. Therefore, the IF state makes a transition to the PF state and the NF state with failure rate λ_6 and repair rate μ_2, respectively. When a PF occurs, the component should be repaired. Therefore, the PF state makes a transition to the NF state with a repair rate μ.

Given the current time t and the next time $t+\Delta t$, the transition relationships between consecutive nodes in the presence of TF and IFs are provided in Table 5.1 [39]. For example, TF state can make a transition to IF state with the failure rate λ_4, and then the transition probabilistic relationships between consecutive nodes from TF state to IF state can be obtained using $\mathrm{e}^{-(\lambda_4+\lambda_5+\mu_1)\Delta t}$.

It is difficult to collect all the transition rates for all the states of each component; therefore, based on practical engineering experiences, some assumptive equations are developed for simplicity using the basic parameter, i.e., failure rate λ of a component. The failure rates λ_1, λ_2, λ_3, λ_4, λ_5, and the repair rates μ_1, μ_2 are calculated by using the equations as follows:

$$\lambda_1 = x \cdot \lambda, \tag{5.1}$$

$$\lambda_2 = y \cdot \lambda, \tag{5.2}$$

$$\lambda_3 = z \cdot \lambda, \tag{5.3}$$

$$\lambda_4 = a \cdot \lambda_1, \tag{5.4}$$

$$\lambda_5 = b \cdot \lambda_1, \tag{5.5}$$

Table 5.1. Transition relationships between consecutive nodes.

	$t + \Delta t$			
t	NF	TF	IF	PF
NF	$e^{-(\lambda_1+\lambda_2+\lambda_3)\Delta t}$	$\frac{[1-e^{-(\lambda_1+\lambda_2+\lambda_3)\Delta t}]\lambda_1}{\lambda_1+\lambda_2+\lambda_3}$	$\frac{[1-e^{-(\lambda_1+\lambda_2+\lambda_3)\Delta t}]\lambda_2}{\lambda_1+\lambda_2+\lambda_3}$	$\frac{[1-e^{-(\lambda_1+\lambda_2+\lambda_3)\Delta t}]\lambda_3}{\lambda_1+\lambda_2+\lambda_3}$
TF	$\frac{[1-e^{-(\lambda_4+\lambda_5+\mu_1)\Delta t}]\mu_1}{\lambda_4+\lambda_5+\mu_1}$	$e^{-(\lambda_4+\lambda_5+\mu_1)\Delta t}$	$\frac{[1-e^{-(\lambda_4+\lambda_5+\mu_1)\Delta t}]\lambda_4}{\lambda_4+\lambda_5+\mu_1}$	$\frac{[1-e^{-(\lambda_4+\lambda_5+\mu_1)\Delta t}]\lambda_5}{\lambda_4+\lambda_5+\mu_1}$
IF	$\frac{[1-e^{-(\mu_2+\lambda_6)\Delta t}]\mu_2}{\mu_2+\lambda_6}$	0	$e^{-(\mu_2+\lambda_6)\Delta t}$	$\frac{[1-e^{-(\mu_2+\lambda_6)\Delta t}]\lambda_6}{\mu_2+\lambda_6}$
PF	$1 - e^{-\mu\Delta t}$	0	0	$e^{-\mu\Delta t}$

$$\mu_1 = c \cdot \lambda_1, \tag{5.6}$$

$$\lambda_6 = m \cdot (\lambda_2 + \lambda_4), \tag{5.7}$$

$$\mu_2 = n \cdot (\lambda_2 + \lambda_4), \tag{5.8}$$

where λ is the total failure rate, and x, y, z, a, b, c, m, and n are transition coefficients. The sums of x, y and z, a, b and c, and m and n are all less than or equal to 1. The total failure rate λ and repair rate μ can be collected from the reference reviews and expert judgments, and the transition coefficient can be determined by multiexperts.

5.3.3. *Fault diagnosis*

At different times, the cumulative numbers of various fault symptoms in a short period of time, such as one week or ten days, are recorded. They are input to the fault symptom layer at the specified time as evidences. Fault diagnosis is conducted through the backward analysis of DBNs, which involves the computation of the posterior probability of any given set of variables from given observations (evidence) represented as the instantiation of some variables to one of their admissible values. Generally, the larger the posterior probability of fault, the higher the possibility of the corresponding faults. According to engineering experience, two judgment rules are defined to determine the fault diagnosis results as follows.

Rule 1: A fault will be reported if the difference of posterior probability and prior probability for a fault node is greater than or equal to 60%.

Rule 2: A warning will be reported if the difference of posterior probability and prior probability for a fault node is or greater than or equal to 30% and less than or equal to 60%.

According to the rules, the fault diagnosis system will remind the operators to handle the possible faults.

5.4. Case Study

5.4.1. *Description of GMR control systems*

GMR is a flexible system specifically designed for industrial control applications including applications with safety-related requirements.

It has been applied to produce fire and gas systems that conform to the requirements of IEC 61508, such as the subsea BOP control system [37]. Based on GE Fanuc's globally available hardware, a GMR system consists of a number of modular subsystems, integrated to form a flexible and powerful whole. The number and type of subsystems are completely scaleable to the requirements of the application. GMR can provide variable redundancy from the input devices through one, two, or three PLC central processing unit (CPU) processors to the output devices. This flexibility means less critical inputs and outputs may be configured for simplex or duplex operation while maintaining the triplicated elements for critical control. A typical GMR system consists of input devices gathering data from multiple or single sensors, multiple PLCs running the same application program, and groups of Genius blocks controlling shared output loads. Communications between the blocks and PLCs and among the PLCs themselves are provided by redundant Genius buses [40–44].

Take the GMR control system used in a subsea BOP system as an example. It consists of three series 90–70 PLCs and four sets of Genius buses. The PLC primary racks are connected to Genius distributed input/output (I/O) subsystems located in toolpusher's panel, driller's panel, and hydraulic power unit, via two sets of independent dual Genius buses and 12 Genius bus controllers (GBCs). Similarly, the PLC expansion rack 1 is connected to Genius distributed I/O subsystems, located in the blue and yellow subsea electronic modules, via two sets of independent triple Genius buses and 18 GBCs. The PLC expansion rack 1 is connected to the CPU through an 18-twisted-pair parallel I/O cable with one end connected to the lower connector on the bus transmitter module (BTM) in the PLC primary rack, and the other end connected to the top connector on a bus receiver module (BRM) installed in slot 1 in the PLC expansion rack 1. Each PLC individually monitors and controls the Genius distributed I/O subsystems for triple or dual redundancy of analog input (AI) module and digital output (DO) module.

In the work, the faults and fault symptoms of the GMR control system used in multiple sets of subsea BOP systems are collected for constructing the DBN-based fault diagnosis models. The details are introduced as follows.

5.4.2. *Fault diagnosis modeling*

The DBNs for fault diagnosis of a GMR control system are constructed using Netica software, as shown in Fig. 5.4. Netica is a powerful, easy-to-use, and complete commercial software program for working with BNs and DBNs, and can perform various kinds of inference by using the fastest and most modern algorithms, such as junction trees.

The nodes in fault layer are determined firstly. It includes eight nodes, i.e., "power source (PS) fault", "CPU fault", "Ethernet controller (ETH) fault", "BTM fault", "BRM fault", "GBN fault", "AI fault", and "DO fault". The nodes in the fault symptom layer are determined subsequently. It also includes eight nodes, i.e., "system halted", "program does not work", "blue screen of death", "program missing", "inaccurate temperature test", "lose all communication", "solenoid valve lose control", and "lose subsea information". The fault nodes and fault symptom nodes are connected via intra-slice arcs, indicating the causal relationships.

The DBNs are extended from BNs for every time interval Δt, and Δt could be one hour, one week, and even one month. A great number of time slices correspond to a smaller value of Δt, and, hence, a longer time at which Netica runs. In the current work, the time interval Δt is one month.

The prior probabilities of states NF, TF, IF, and PF for all fault nodes in initial time are set to 97%, 1%, 1%, and 1%, respectively. The conditional probability between fault nodes and fault symptom nodes in static BNs is determined using parameter learning. Parameter learning is the automatic learning of the specific relationships nodes have with their parents using case data, once it has already been determined which nodes are the parents of each node. The full-set fault and symptom case data are determined by three electrical engineers in the subsea BOP manufacturer, Rongsheng Machinery Manufacture Ltd. of Huabei Oilfield, Hebei. They use the GMR system to develop the 3000 m subsea BOP control system, and are

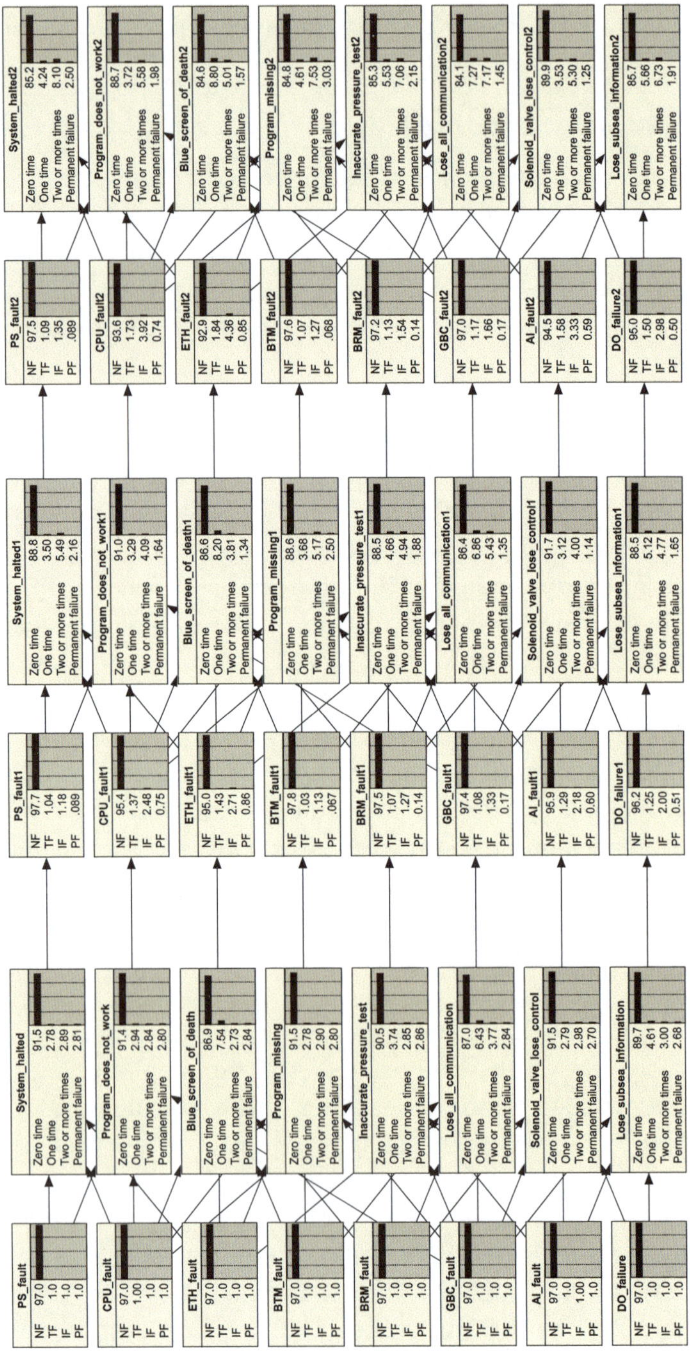

Fig. 5.4. Fault diagnosis models of GMR control systems.

Table 5.2. Failure rates and repair rate of components.

Component	Failure rate (h^{-1})	Repair rate (h^{-1})
PS	0.527×10^{-6}	0.200
CPU	4.514×10^{-6}	0.143
ETH	5.211×10^{-6}	0.167
BTM	0.399×10^{-6}	0.333
BRM	0.805×10^{-6}	0.333
GBC	0.966×10^{-6}	0.250
AI	3.580×10^{-6}	0.042
DO	3.030×10^{-6}	0.042

familiar with the faults and fault symptoms of the GMR control system.

The transition relationships between consecutive fault nodes from time t to time $t + \Delta t$ are determined using equations in Table 5.1 and Eqs. (5.1)–(5.8). The total failure rate and repair rate of each component are provided in Table 5.2. The failure rates are obtained from product manuals of GMR systems [45], and the repair rates are obtained from the three electrical engineers mentioned above. When the transition coefficients x, y, z, a, b, c, m, and n are difficult to determine, we have to appeal to the three electrical engineers for assistance, and the coefficients are set to 10%, 40%, 20%, 10%, 10%, 70%, 20%, and 50%, respectively.

Four cases obtained from practical engineering experience are used for fault diagnosis and modeling validation. In a short period of time, such as one week in the current work, the cumulative number of various fault symptoms are recorded, and node states are provided as evidences. In Case 1, the states of nodes "system halted", "program does not work", and "blue screen of death" are one time, two or more times, and zero times in the first year, respectively. In Case 2, the states of nodes "solenoid valve lose control" and "lose subsea information" are both one time in the fifth year. In Case 3, the states of nodes "program missing" and "lose all communication" are both permanent failure in the first year. In Case 4, the states of nodes "program missing" and "inaccurate pressure test" are both permanent failure in the fifth year.

Although these evidences are found at a specified time, such as first year and fifth year, they are input to the fault symptom layer of DBNs at six different times, i.e., zeroth year, first year, third year, fifth year, eighth year and tenth year for comparison. Fault diagnosis is conducted through backward analysis of DBNs, and the posterior probabilities of fault nodes are calculated by using Netica software.

5.4.3. *Results and discussion*

In Case 1, as shown in Fig. 5.5, the absolute differences of posterior probabilities and prior probabilities of IF for PS and GBC are larger than 30% at certain times, while the absolute differences of TF, IF, and PF for all the other nodes are always smaller than 30%, which indicates that the components PS and GBC may be faulty and the other components are normal, such as CPU, ETH, BTM, BRM, AI, and DO.

For the PS in Fig. 5.5(a), the absolute difference of IF is significantly larger than those of TF and PF, indicating that the possible fault should be an IF. It is noted that with an increase in time, the difference of IF decreases, that is it is larger than 60% in the first 5 years, and smaller than 60% after the eighth year. According to the rules defined in Sec. 5.3.3, an IF of PS will be reported if the fault symptoms are found in the first 5 years, and a warning of IF of PS will be reported if the fault symptoms are found after the eighth year. For the GBC in Fig. 5.5(b), with an increase in time, the difference of IF increases continuously, that is it is smaller than 30% in the first 8 years, and larger than 30% in the tenth year. According to the rules, NF of GBC will be reported if the fault symptoms are found in the first 8 years, and a warning of IF of GBC will be reported if the fault symptoms are found in the tenth year. In addition, the fault diagnosis result that an IF of PS occurs in the first year is consistent with the actual fault, indicating that the proposed methodology is correct.

In Case 2, as shown in Fig. 5.6, the absolute differences of posterior probabilities and prior probabilities of TF for GBC and DO are larger than 30% at certain times, while the absolute differences of

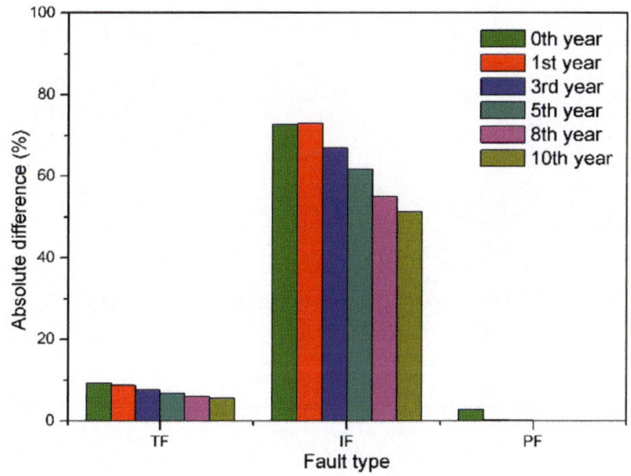

(a)

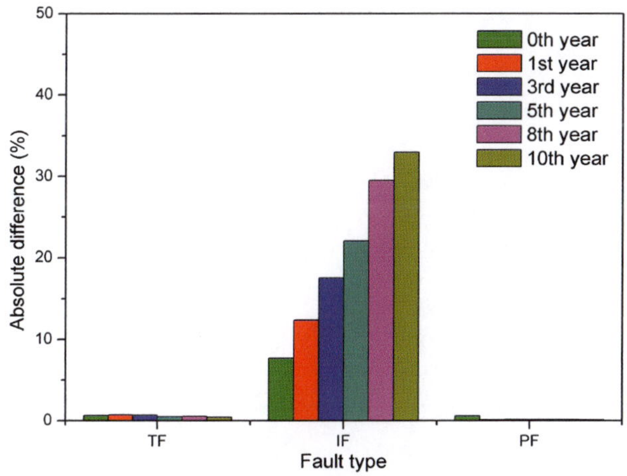

(b)

Fig. 5.5. Fault diagnosis results of Case 1: (a) PS and (b) GBC.

TF, IF, and PF for all the other nodes are always smaller than 30%, which indicates that the components GBC and DO may be faulty and the other components are normal.

For the GBC in Fig. 5.6(a), the absolute difference of TF is significantly larger than those of IF and PF, indicating the possible fault should be a TF. It is noted that with an increase in time, the difference of TF decreases continuously, that is it is larger than 30%

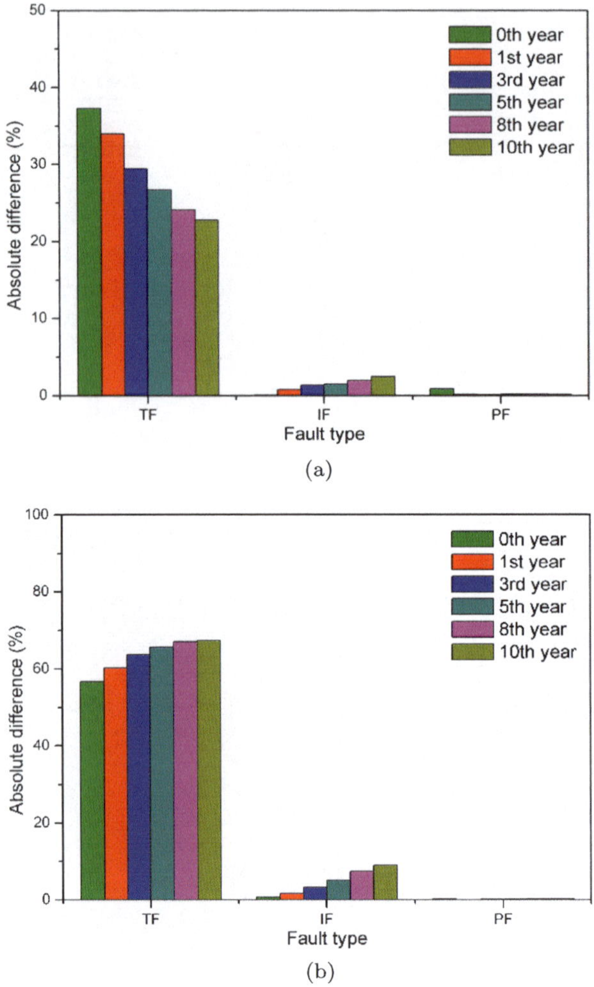

Fig. 5.6. Fault diagnosis results of Case 2: (a) GBC and (b) DO.

in the first year, and smaller than 30% after the third year. According to the rules, a TF of GBC will be reported if the fault symptoms are found in the first year, and NF will be reported if the fault symptoms are found after the third year. For the DO in Fig. 5.5(b), with an increase in time, the difference of IF increases, that is it is smaller than 60% in the initial time, and larger than 60% after the first year. According to the rules, a warning of TF of DO will be reported if the

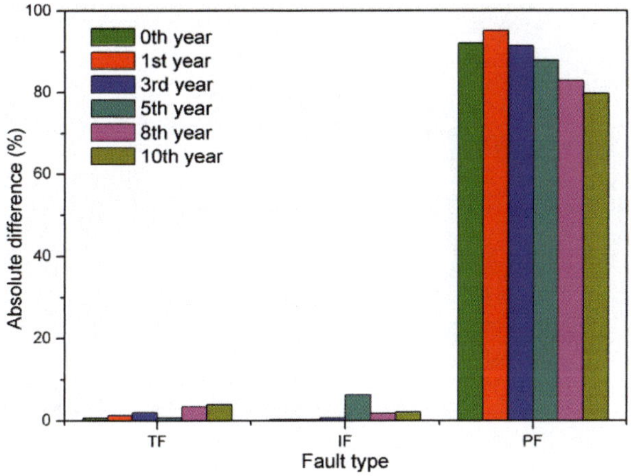

Fig. 5.7. Fault diagnosis results of Case 3 for ETH.

fault symptoms are found in the initial time, and a TF of DO will be reported if the fault symptoms are found after the first year. In addition, the fault diagnosis result that TF of DO occurs in the fifth year is consistent with the actual fault, indicating that the proposed methodology is correct.

In Case 3, as shown in Fig. 5.7, the absolute differences of posterior probabilities and prior probabilities of PF for ETH are significantly larger than those of TF and IF, indicating that the possible fault of ETH should be a PF. The absolute differences of TF, IF, and PF for all the other nodes are very small, indicating that all the other components are normal. As shown in Fig. 5.7, the absolute difference of PF from the initial time to the tenth year are greater than 60%. According to the rules, a PF of ETH will be reported if the fault symptoms are found in the first 10 years. In addition, the fault diagnosis result that PF of ETH occurs in the first year is consistent with the actual fault, indicating that the proposed methodology is correct.

In Case 4, as shown in Fig. 5.8, the absolute differences of posterior probabilities and prior probabilities of PF for CPU and GBC are larger than 30% at certain times, while the absolute differences of TF, IF, and PF for all the other nodes are always

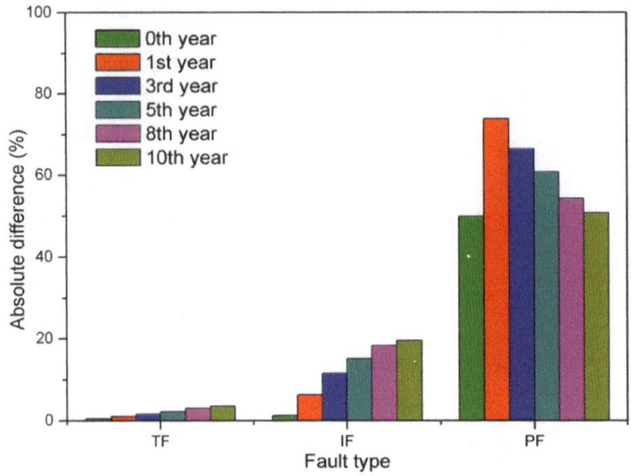

(a)

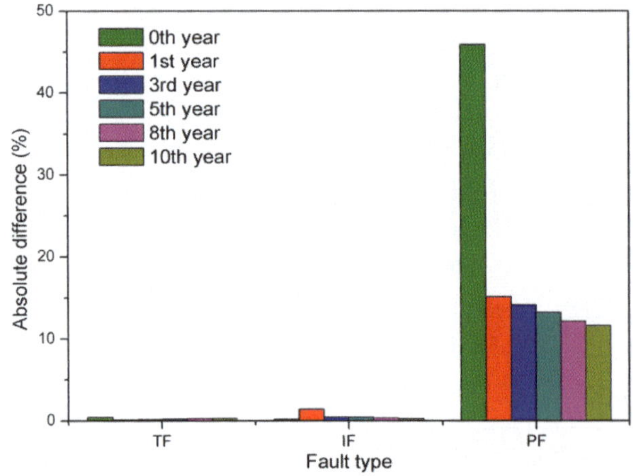

(b)

Fig. 5.8. Fault diagnosis results of Case 4: (a) CPU and (b) GBC.

smaller than 30%, which indicates that the components CPU and GBC may be faulty and the other components are normal.

For the CPU in Fig. 5.8(a), the absolute difference of PF is significantly larger than those of TF and IF, indicating the possible fault should be a PF. It is noted that the value is larger than 60% from the first year to the fifth year, and is between 30% and 60% in the initial time and after the eighth year. According to the rules, a PF of CPU will be reported if the fault symptoms are found from the first year to the fifth year, and a warning of PF of CPU will be reported if the fault symptoms are found at other times. For GBC in Fig. 5.8(b), with an increase in time, the difference of PF decreases, that is it is greater than 30% in the initial time, and is smaller than 30% after the first year. According to the rules, a warning of PF of GBC will be reported if the fault symptoms are found in the initial time, and NF of GBC will be reported if the fault symptoms are found after the first year. It is noted that there is a "sudden change" in the absolute difference of PF from initial time to the first year for both CPU and GBC, and the absolute difference increases for CPU and decreases for GBC. This is because the prior probabilities of states PF for all fault nodes in initial time are set to 1%, but the CPU has higher possibility of failure and the GBC has lower possibility of failure when they work after initial time, making the absolute difference different. When the permanent failures of "program missing" and "inaccurate pressure test" occur after the initial time, the fault diagnosis results show higher fault probability for CPU and lower fault probability for GBC because of their different degradation processes. In addition, the fault diagnosis result that PF of CPU occurs in the first year is consistent with the actual fault, indicating that the proposed methodology is correct.

5.5. Conclusion

This chapter proposes a DBN-based fault diagnosis methodology in the presence of TF and IF for electronic systems. It aims at identifying the faulty components and distinguishing the fault types, including TF, IF, and PF. DBNs are used to model the dynamic degradation process of electronic products, and the Markov process is used to model the transition relationships of four states for each node, i.e., NF, TF, IF, and PF. According to engineering experiences, two judgment rules are defined to determine the fault diagnosis results. Four fault diagnosis cases of the GMR control system are investigated to prove the effectiveness of this methodology. The results show that for given fault symptoms at a specified time, the fault diagnosis results diagnosed by using the proposed DBN models are consistent with the actual faults, indicating that the methodology is correct. Given identical fault symptoms at different times, the diagnosis results may be different because the performance of electronic products degrades continuously over time.

For a specified electronic system, a DBN-based fault diagnosis system could be developed according to the proposed methodology. The engineers can record the cumulative number of various fault symptoms of this electronic system in a short period of time, such as one week or ten days at different times, and then input these data to the fault diagnosis system. The fault diagnosis results can help engineers to identify the component faults and distinguish the fault types more accurately at different times.

References

[1] Q. Liu, S. J. Qin, T. Chai, "Decentralized fault diagnosis of continuous annealing processes based on multilevel PCA," *IEEE Transactions on Automation Science and Engineering*, vol. 10, no. 3, pp. 687–698, 2013.

[2] D. Lefebvre, "Fault diagnosis and prognosis with partially observed Petri nets," *IEEE Transactions on Systems, Man, and Cybernetics: Systems*, vol. 44, no. 10, pp. 1413–1424, 2014.

[3] E. Gascard, Z. Simeu-Abazi, "Modular modeling for the diagnostic of complex discrete-event systems," *IEEE Transactions on Automation Science and Engineering*, vol. 10, no. 4, pp. 1101–1123, 2013.

[4] Y. Ren, A. Wang, H. Wang, "Fault diagnosis and tolerant control for discrete stochastic distribution collaborativecontrol systems," *IEEE Transactions on Systems, Man, and Cybernetics: Systems*, vol. 45, no. 3, pp. 462–471, 2015.

[5] H. Qi, S. Ganesan, M. Pecht, "No-fault-found and intermittent failures in electronic products," *Microelectronics Reliability*, vol. 48, no. 5, pp. 663–674, 2008.

[6] A. Kodali, S. Singh, K. Pattipati, "Dynamic set-covering for real-time multiple fault diagnosis with delayed test outcomes," *IEEE Transactions on Systems, Man, and Cybernetics: Systems*, vol. 43, no. 3, pp. 547–562, 2013.

[7] S. Singh, H. S. Subramania, S. W. Holland, J. T. Davis, "Decision forest for root cause analysis of intermittent faults," *IEEE Transactions on Systems, Man, and Cybernetics, Part C (Applications and Reviews)*, vol. 42, no. 6, pp. 1818–1827, 2012.

[8] R. Bakhshi, S. Kunche, M. Pecht, "Intermittent failures in hardware and software," *Journal of Electronic Packaging, Transactions of the ASME*, vol. 136, no. 1, pp. 011014-1–011014-5, 2014.

[9] C. Zhou, X. Huang, X.Naixue, Y. Qin, S. Huang, "A class of general transient faults propagation analysis for networked control systems," *IEEE Transactions on Systems, Man, and Cybernetics: Systems*, vol. 45, no. 4, pp. 647–661, 2015.

[10] O. Contant, S. Lafortune, D. Teneketzis, "Diagnosis of intermittent faults," *Discrete Event Dynamic Systems: Theory and Applications*, vol. 14, no. 2, pp. 171–202, 2004.

[11] R. Abreu, A. J. C. van Gemund, "Diagnosing multiple intermittent failures using maximum likelihood estimation," *Artificial Intelligence*, vol. 174, no. 18, pp. 1481–1497, 2010.

[12] A. Mahapatro, P. M. Khilar, "Detection and diagnosis of node failure in wireless sensor networks: a multiobjective optimization approach," *Swarm and Evolutionary Computation*, vol. 13, pp. 74–84, 2013.

[13] D. N. Monekosso, P. Remagnino, "Data reconciliation in a smart home sensor network," *Expert Systems with Applications*, vol. 40, no. 8, pp. 3248–3255, 2013.

[14] G. Deng, J. Qiu, G. Liu, K. Lyu, "A discrete event systems approach to discriminating intermittent from permanent faults," *Chinese Journal of Aeronautics*, vol. 27, no. 2, pp. 390–396, 2014.

[15] C. Liu, J. Wang, A. Zhang, H. Ding, "Research on the fault diagnosis technology of intermittent connection failure belonging to FPGA solder-joints in BGA package," *Optik*, vol. 125, no. 2, pp. 737–740, 2014.

[16] M. Yu, D. Wang, "Model-based health monitoring for a vehicle steering system with multiple faults of unknown types," *IEEE Transactions on Industrial Electronics*, vol. 61, no. 7, pp. 3574–3586, 2014.

[17] T. Cui, X. Dong, Z. Bo, A. Juszczyk, "Hilbert-transform-based transient/intermittent earth fault detection in noneffectively grounded distribution systems," *IEEE Transactions on Power Delivery*, vol. 26, no. 1, pp. 143–151, 2011.

[18] Y. Wang, M. Liang, "Identification of multiple transient faults based on the adaptive spectral kurtosis method," *Journal of Sound and Vibration*, vol. 331, no. 2, pp. 470–486, 2012.

[19] J. Pearl, "Bayesian networks, a model of self-activated memory for evidential reasoning," In: *Proceedings of the 7th Conference of the Cognitive Science Society*, University of California, Irvine, CA, 1985.

[20] K. Liu, X. Zhang, J. Shi, "Adaptive sensor allocation strategy for process monitoring and diagnosis in a Bayesian network," *IEEE Transactions on Automation Science and Engineering*, vol. 11, no. 2, pp. 452–462, 2014.

[21] L. Bennacer, Y. Amirat, A. Chibani, A. Mellouk, L. Ciavaglia, "Self-diagnosis technique for virtual private networks combining Bayesian networks and case-based reasoning," *IEEE Transactions on Automation Science and Engineering*, vol. 12, no. 1, pp. 354–366, 2015.

[22] Y. Zhu, L. Huo, J. Lu, "Bayesian networks-based approach for power systems fault diagnosis," *IEEE Transactions on Power Delivery*, vol. 21, no. 2, pp. 634–639, 2006.

[23] R. M. Khanafer, B. Solana, J. Triola, R. Barco, L. Moltsen, Z. Altman, P. Lázaro, "Automated diagnosis for UMTS networks using Bayesian network approach," *IEEE Transactions on Vehicular Technology*, vol. 57, no. 4, pp. 2451–2461, 2008.

[24] A. Barua, K. Khorasani, "Hierarchical fault diagnosis and health monitoring in satellites formation flight," *IEEE Transactions on Systems, Man, and Cybernetics, Part C (Applications and Reviews)*, vol. 41, no. 2, pp. 223–239, 2011.

[25] A. Barua, K. Khorasani, "Verification and validation of hierarchical fault diagnosis in satellites formation flight," *IEEE Transactions on Systems, Man, and Cybernetics, Part C (Applications and Reviews)*, vol. 42, no. 6, pp. 1384–1399, 2012.

[26] B. Ricks, O. J. Mengshoel, "Diagnosis for uncertain, dynamic and hybrid domains using Bayesian networks and arithmetic circuits," *International Journal of Approximate Reasoning*, vol. 55, no. 5, pp. 1207–1234, 2014.

[27] Z. Liu, Y. Liu, D. Zhang, B. Cai, C. Zheng, "Fault diagnosis for a solar assisted heat pump system under incomplete data and expert knowledge," *Energy*, vol. 87, pp. 41–48, 2015.

[28] J. Xu, K. Sun, L. Xu, "Data mining-based intelligent fault diagnostics for integrated system health management to avionics," *Proceedings of the Institution of Mechanical Engineers, Part O: Journal of Risk and Reliability*, vol. 229, pp. 3–15, 2015.

[29] D. A. Tobon-Mejia, K. Medjaher, N. Zerhouni, "CNC machine tool's wear diagnostic and prognostic by using dynamic Bayesian networks," *Mechanical Systems and Signal Processing*, vol. 28, pp. 167–182, 2012.

[30] B. Cai, Y. Liu, Y. Ma, Z. Liu, Y. Zhou, J. Sun, "Real-time reliability evaluation methodology based on dynamic Bayesian networks: A case study of a subsea pipe ram BOP system," *ISA Transactions*, vol. 58, pp. 595–604, 2015.

[31] B. Cai, Y. Liu, Q. Fan, "A multiphase dynamic Bayesian networks methodology for the determination of safety integrity levels," *Reliability Engineering & System Safety*, vol. 150, pp. 105–115, 2016.

[32] D. Codetta-Raiteri, L. Portinale, "Dynamic Bayesian networks for fault detection identification, and recovery in autonomous spacecraft," *IEEE Transactions on Systems, Man, and Cybernetics: Systems*, vol. 45, no. 1, pp. 13–24, 2015.

[33] A. D. Kenyon, V. M. Catterson, S. D. J. McArthur, J. Twiddle, "An agent-based implementation of hidden Markov models for gas turbine condition monitoring," *IEEE Transactions on Systems, Man, and Cybernetics: Systems*, vol. 44, no. 2, pp. 186–195, 2014.

[34] S. Ntalampiras, Y. Soupionis, G. Giannopoulos, "A fault diagnosis system for interdependent critical infrastructures based on HMMs," *Reliability Engineering & System Safety*, vol. 138, pp. 73–81, 2015.

[35] A. Correcher, E. García, F. Morant, E. Quiles, L. Rodríguez, "Intermittent failure dynamics characterization," *IEEE Transactions on Reliability*, vol. 61, no. 3, pp. 649–658, 2012.

[36] A. Zagorecki, M. J. Druzdzel, "Knowledge engineering for Bayesian networks: How common are noisy-MAX distributions in practice?" *IEEE Transactions on Systems, Man, and Cybernetics: Systems*, vol. 43, no. 1, pp. 186–195, 2013.

[37] V. B. Prasad, "Computer networks reliability evaluations and intermittent faults," In: *Proceedings of the 33rd Midwest Symposium on Circuits and Systems*, vol. 1, pp. 327–330, Calgary, 1990.

[38] V. B. Prasad, "Markovian model for the evaluation of reliability of computer networks with intermittent faults," In: *IEEE International Symposium on Circuits and Systems*, vol. 4, pp. 2084–2087, 1991.

[39] T. Kohda, W. Cui, "Risk-based reconfiguration of safety monitoring system using dynamic Bayesian network," *Reliability Engineering & System Safety*, vol. 92, pp. 1716–1723, 2007.

[40] B. Cai, Y. Liu, Z. Liu, F. Wang, X. Tian, Y. Zhang, "Development of an automatic subsea blowout preventer stack control system using PLC based SCADA," *ISA Transactions*, vol. 51, no. 1, pp. 198–207, 2012.

[41] GE Fanuc Automation, "Genius modular redundancy flexible triple modular redundant (TMR) system: User's manual," GFK-0787B, 1995.

[42] GE Fanuc Automation, "Genius modular redundancy: User's manual," GFK-1277E, 2007.

[43] GE Fanuc Automation, "Genius modular redundancy for fire and gas applications," GFK-1649A, 1999.

[44] GE Fanuc Automation, "Genius modular redundancy flexible triple modular redundant (TMR) system: Technical product overview," GFT-177A, 1998.

[45] GE Fanuc Automation, "Genius modular redundancy manual: Triple voting system," GFA-525CN, 2007.

Chapter 6

An Integrated Safety Prognosis Model for Complex System Based on Dynamic Bayesian Network and Ant Colony Algorithm

In a complex industrial system, most of the single faults have multiple propagation paths, so any local slight deviation is able to propagate, spread, accumulate, and increase through system fault causal chains. It will finally result in unplanned outages and even catastrophic accidents, which lead to huge economic losses, environmental contamination, or human injuries. In order to ensure system's intrinsic safety and increase operational performance and reliability in the long term, this study proposes an Integrated Safety Prognosis Model (ISPM) considering the randomness, complexity, and uncertainty of fault propagation.

ISPM is developed based on dynamic Bayesian networks to model the propagation of faults in a complex system, integrating *a priori* knowledge of the interactions and dependencies among subsystems, components, and the environment of the system, as well as the relationships between fault causes and effects. So, the current safety state and potential risk of system can be assessed by locating potential hazard origins and deducing corresponding possible consequences. Furthermore, ISPM is also developed to predict the future degradation trend in terms of future reliability or performance of the system, and provide proper proactive maintenance plans. Ant colony algorithm is introduced in the ISPM by comprehensively considering two factors — probability and severity of faults — to perform the quantitative risk estimation of the underlining system. The feasibility and benefits of ISPM are investigated with a field case study of the gas turbine compressor system. According to the outputs given by the ISPM in the application, proactive maintenance, safety-related actions, and contingency plans are further discussed and then implemented to keep the system in a high reliability and safety level in the long term.

6.1. Introduction

The assessment and prediction of the safety situation of a complex system is a very critical challenge due to technical difficulty as well as dynamic environment and also because it has a significant impact on economy, environment, and society performance. In complex industrial systems, operating, regulating, maintenance activities, and external incidents occur dynamically and multiple entities (e.g., persons, subsystems, components, and environment) in same or different subsystems interact in a complex manner. So, single faults may have multiple propagation paths, which will finally lead to catastrophic accidents.

System safety management nowadays is of vital importance to keep the system safe at an acceptable level, the key issues of which focus on both of how to reduce the probability of fault occurrence and decrease the loss of fault consequence. The implementation of such requirements can be studied in terms of the determination of the fault root causes, possible consequence, estimated risk, and timing of various maintenance activities (i.e., repair or replacement of parts), which are considered in a safety prognosis scheme. Safety prognosis is usually carried out based on the historical and current conditions and determines whether and when the underlying system is in need of maintenance or risk control taking into consideration both fault probability and the level of its severity. With the assistance of prognosis, a pre-warning alarm can be set when the predicted values fall within the warning region (or beyond the safety threshold). This provides adequate time for safety engineers to make a proactive maintenance plan, inspect the hardware of a system, conduct a repair on the defect, and even make a contingency plan before the catastrophic failure occurs.

Nevertheless, safety prognosis has been a difficult task and has attracted much attention of researchers in the field [5, 6, 11, 12, 14, 22]. The approaches to prognosis fall into three main categories: statistical approaches, artificial intelligent approaches, and model-based approaches [11]. Model-based approaches utilize physics specific, explicit mathematical models of the underling entities. Such approaches can be more effective than other model-free approaches if a correct and accurate model is built. However, explicit mathematical modeling may not be feasible for complex systems since changes in structural dynamics and operating conditions can affect the mathematical model, and it would be very difficult or even impossible to build mathematical models for all real-life conditions.

Statistical prognostic approaches require the component failure history data or event data. They involve collecting statistical information from a large number of component samples to indicate the survival duration of a component before a failure occurs, and use these statistical parameters to predict the Remaining Useful Life (RUL) of individual components [23]. A few statistical models in survival analysis, such as Hidden Semi-Markov Model (HSMM) [1], Proportional Hazard Model (PHM) [20], PIM [16, 24], Proportional Covariate Model (PCM) [21] and Extended Hazard Regression Model (EHRM) [15], are useful tools for RUL estimation. Iung [10] and Muller [17, 18] also included maintenance policies in the consideration of the machine prognostic process, in order to provide decision support for maintenance actions.

Artificial Intelligence (AI) approaches utilize large amounts of historical failure data or condition data to build a prognostic model which learns the system behavior, instead of building models based on comprehensive system physics and human expertise. Some of the published researches using these approaches can be found in [3, 7, 9, 14, 25]. In the literature, a neural network-based model for prognosis is regularly used because of its flexibility in generating the appropriate model. Tran *et al.* [23] proposed a multistep ahead prediction methodology for forecasting the machines' operating conditions using regression trees and neuron-fuzzy systems. Heng *et al.* [5] used a feed-forward neural network to estimate the future survival probability of a monitored item, given the corresponding condition monitoring indices.

The above-mentioned methods have advanced the development of safety prognostics. However, several issues need to be further investigated when they are applied to complex industrial systems:

(1) In order to control fault occurrence probability, system reliability or performance prediction is an effective way to track the

degradation of system in the future and make a proper proactive maintenance plan to avoid failure or reduce the fault influence range. However, the existing literature [5, 9, 13] largely focuses on the degradation mechanism and RUL from the component or single equipment deterioration point of view; for example, ball bearing, pump, and steam generator. Very few models have taken into account not only the interdependencies among the components and subsystems, but also the impact of degradations and the influence of exogenous variables on the degradation processes. The main reason for this is that to decrease the model's complexity and avoid combinational explosion, two hypotheses, according to which there is no simultaneous occurrence of failure and the statistical independence between events, are assumed [25]. However, such hypotheses are no longer valid when components have common causes or when components have several failure modes, which make a negative impact on the rationality of the reliability prediction for complex system.

(2) On the other hand, in a complex industrial system, most single faults have multiple propagation paths, and different propagation paths may lead to different consequences, some of which may gradually recover by its self-control system, while others may further cause adjacent components' failure and eventually lead to catastrophic accidents by the fault coupling mechanism. So, in order to reduce the loss of fault consequence, safety assessment and risk evaluation are the main issues in safety prognosis. Traditional risk estimation model [4] partially depended on subjective factors simply based on the product of fault probability and severity, so as to make major revisions according to the field condition when applied to the specific engineering system. Meanwhile, not all of the compatible fault propagation pathways derived from traditional risk analysis would happen in the real world, since a considerable part of them have a very low occurrence probability [27], which usually makes it difficult for safety engineers to find the actual fault root cause and thus miss the best time to repair. So, there is an urgent need to develop an

effective and systematic quantitative risk evaluation mechanism in the safety prognosis framework.

This work presents an approach for addressing the above challenges. Taking into consideration the relation among safety assessment, risk evaluation, and prediction, an Integrated Safety Prognosis Model (ISPM) is proposed based on dynamic Bayesian networks (DBNs) and ant colony algorithm. Utilizing both the knowledge of the system structure as well as flow process and monitoring data, ISPM is developed to analyze the potential hazards of the system (e.g., functional faults, component failures, human mistakes, external destruction), reflecting its hidden degraded states, possible hazard origins (root causes), and consequences with corresponding probabilities. With the knowledge of consequence severity, the risk of hazard and its propagation path are calculated by ant colony algorithm in ISPM, which provides instructions for decision making for safety-related actions and contingent plans.

Furthermore, on the basis of safety assessment and risk evaluation, ISPM aims to foresee how a system (or component) will evolve from its current degraded state until its failure and then until the system's breakdown (performance level), analyzing the impact of degradation on the component itself and on the other entities of the system to predict system failures and its RUL. A field case study for the gas turbine compressor system presents how to apply the ISPM to a real industrial system and to investigate its rationality and validity.

The rest of this chapter is organized as follows. In Section 6.2, the dynamic Bayesian network and its inference scheme used in ISPM are described. In Section 6.3, the integrated safety prognosis model is introduced and then algorithms and workflows are also presented. In Section 6.4, a real case study is organized explicitly to illustrate the overall flow of this research in detail. Section 6.5 presents the application results of safety assessment, risk evaluation, and prediction given by ISPM, and a further discussion of the proactive maintenance plan. Conclusions are givens in Section 6.6.

6.2. Dynamic Bayesian Networks

A Bayesian network (BN) as a probability-based knowledge representation method is appropriate for the modeling of causal processes with uncertainty. A BN is a Directed Acyclic Graph (DAG) whose nodes represent random variables and links define probabilistic dependences between variables. These relationships are quantified by associating a conditional probability table with each node, given any possible configuration of values for its parents.

The static Bayesian network can be extended to a DBN model by introducing relevant temporal dependencies that capture the dynamic behaviors of the domain variables between representations of the static network at different times. Two types of dependencies can be distinguished in a DBN: contemporaneous dependencies and non-contemporaneous dependencies. Contemporaneous dependencies refer to arcs between nodes that represent variables within the same time period. Non-contemporaneous dependencies refer to arcs between nodes that represent variables at different times. Figure 6.1 shows a typical DBN graph for an industrial complex system, where nodes "X" represent the degradation states of different components in each subsystem, and nodes "S" represent corresponding observable internal variables which are usually outputs of sensors in the monitoring system (observed system).

Therefore, considering the architecture in Fig. 6.1, a DBN is a way to extend BN to model probability distributions over semi-infinite collections of random variables $\{X_1, X_2, \ldots\}$. A DBN is defined to be a pair $(B_1, B_{\rightarrow})$, where B_1 is a BN which defines the

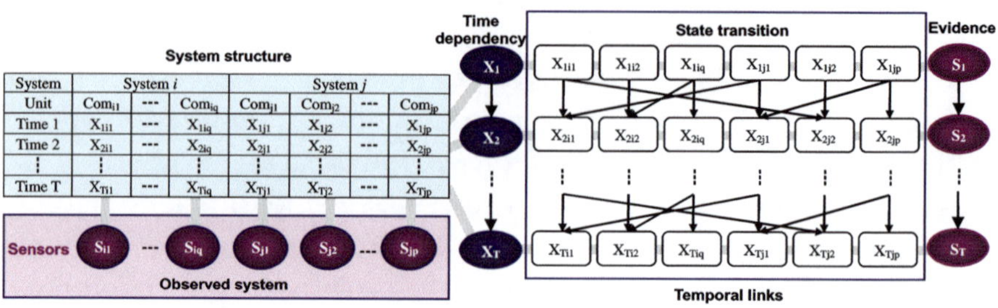

Fig. 6.1. DBN architecture for complex system.

prior $P(X_1)$, and $B_{\rightarrow}$ is a two-slice temporal Bayesian net (2TBN) which defines $P(X_t|X_{t-1})$ by means of a DAG as follows:

$$P(X_t|X_{t-1}) = \prod_{i=1}^{N} P(X_t^i|\text{Pa}(X_t^i)), \tag{6.1}$$

where X_t^i is the ith node at time t and $\text{Pa}(X_t^i)$ are the parents of X_t^i in the graph.

The nodes in the first slice of a 2TBN do not have any parameters associated with them, but each node in the second slice of the 2TBN has an associated conditional probability distribution (CPD) for continuous variables or conditional probability table (CPT) for discrete variables, which defines $P(X_t^i|\text{Pa}(X_t^i))$ for all $t > 1$. The parents of a node $\text{Pa}(X_t^i)$ can either be in the same time slice or in the previous time slice.

The arcs between slices are from up to down (in Fig. 6.1), reflecting the causal flow of time. If there is an arc from X_{t-1}^i to X_t^i, this node is called persistent. The arcs within a slice are arbitrary. Directed arcs within a slice represent "instantaneous" causation. In this chapter, the parameters of the CPTs used by the proposed model are assumed time-invariant; i.e., the model is time-homogeneous.

The semantics of a DBN can be defined by "unrolling" the 2TBN until there are T time-slices. The resulting joint distribution is then given by

$$P(X_{1:T}) = \prod_{t=1}^{T}\prod_{i=1}^{N} P(X_t^i \text{Pa}(X_t^i)). \tag{6.2}$$

Several inference methods for a discrete-state DBN can be used, i.e., forward–backward (FB) algorithm, unrolled junction tree, and the frontier algorithm [18]. In this chapter, the forward–backward method is used for Bayesian inference in the proposed model.

6.3. Proposed Integrated Safety Prognosis Model

To improve the effectiveness and accuracy of safety management, the proposed Integrated Safety Prognosis Model (ISPM) studies how

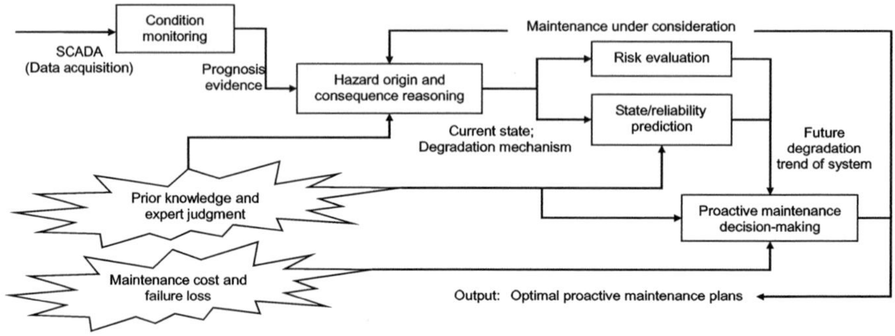

Fig. 6.2. Overall workflow of ISPM.

subsystems, component, humans, and environment interact with each other based on the P&ID system, the knowledge of the system and process function, historical database, and Supervisory Control And Data Acquisition (SCADA) monitoring system. The overall workflow of ISPM is shown in Fig. 6.2, and its application architecture is defined by seven modeling steps of the complex system (Fig. 6.3), that is the Hazard and Operability (HAZOP) model, degradation model, dynamic Bayesian network model (DBN model), monitoring model, assessment model, risk evaluation model, and prediction model. The main corresponding calculation steps in Fig. 6.3 are illustrated in detail in Fig. 6.4.

6.3.1. *HAZOP model development*

The HAZOP model is developed to (1) identify system variables which can represent or indicate system safety condition, flow process, and performance state; (2) determine the state space of each variable with corresponding numerical range; (3) build explicit relationships based on system knowledge of cause and effect as well as system structure and process function (i.e., interdependencies) among above variables with different attributions, for example, exogenous variables, system internal variables, variables of failure modes in terms of hidden-state variables, as well as hazard reason/consequence variables.

For a complex industrial system, the HAZOP study starts with the system hierarchical analysis shown in Fig. 6.5 by examining

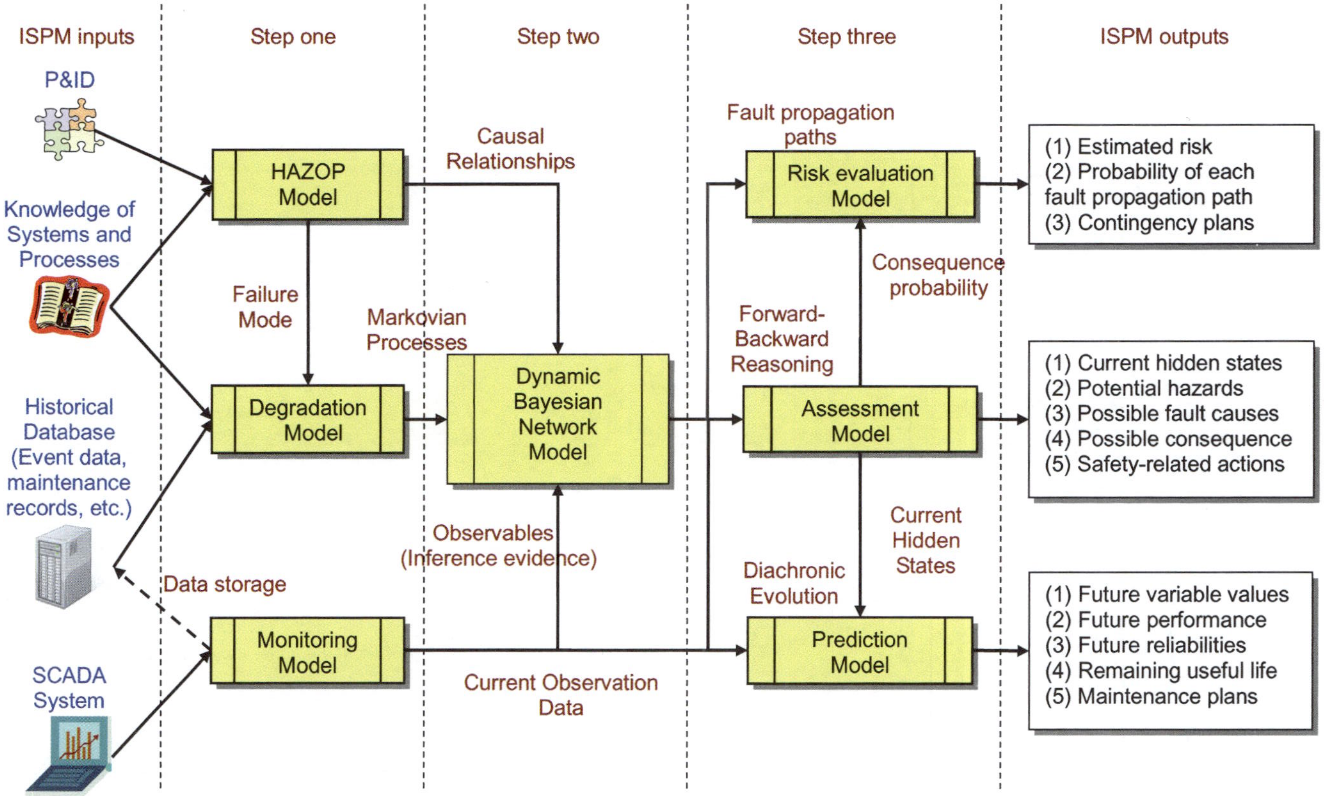

Fig. 6.3. Application architecture of ISPM.

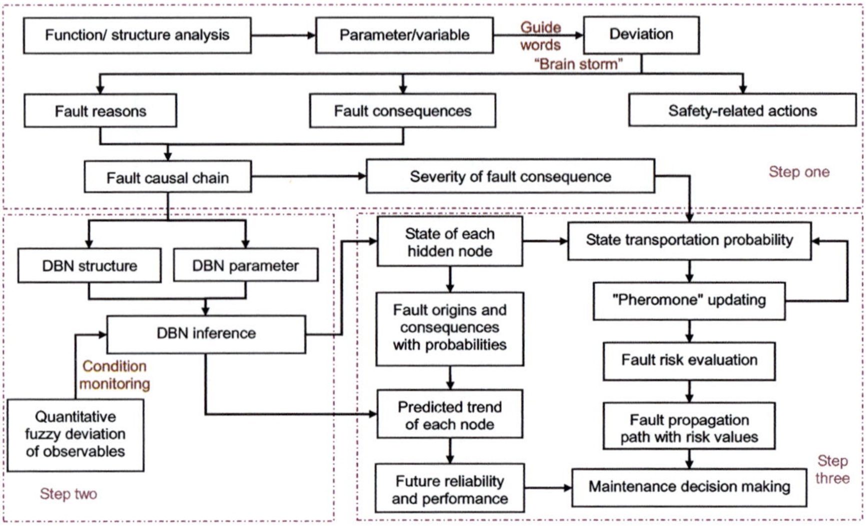

Fig. 6.4. Main calculation steps in ISPM.

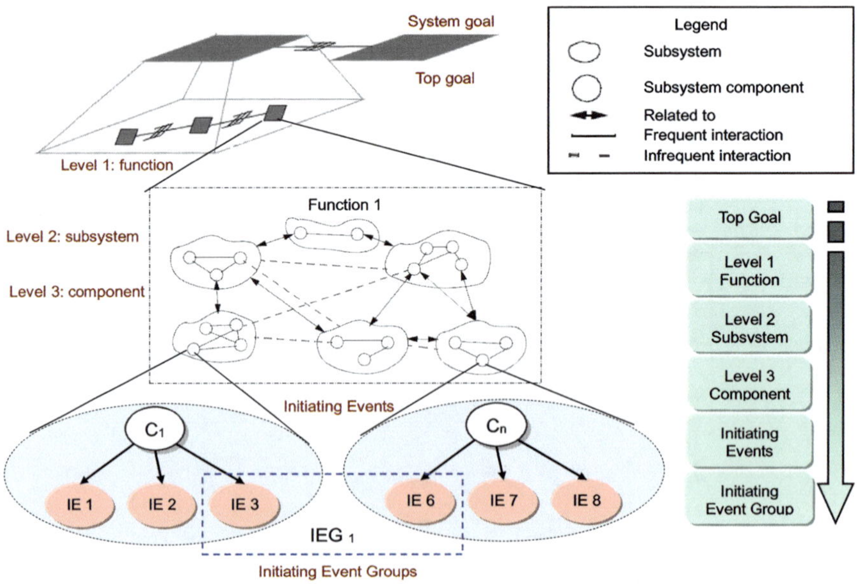

Fig. 6.5. Relationship of function subsystem component and initiating events.

P&ID systematically with knowledge of system/process support and function. According to initiating events (or initiating events group), the system process is divided into sessions called "analysis nodes", and then meaningful deviations in every analysis node are generated by combining process parameters and HAZOP guidewords. All conceivable deviations from design intention in the system are identified, and all the possible abnormal causes and adverse consequences of those deviations are determined, i.e., cause–effect relationships and the fault propagation path. The state space of each variable with different attribution is usually defined based on both historical data and expert judgment when certain data is unavailable. Several terms are considered as follows:

- *Intentions*: They define how the part is expected to function.
- *Deviations*: They refer to departures from the design intention which are discovered by the systematic application of the guide words.
- *Causes*: These include the reasons why deviations might occur.
- *Consequences*: These are the results of the deviations.
- *Guide words*: These refer to simple words which are used to qualify the intention and hence deviations. The list of guide words includes, but is not limited to, NO/NOT, MORE, LESS, AS WELL AS, PART OF, REVERSE.

During the development process of the HAZOP model, the failure modes, internal variables, exogenous factors, the explicit cause–effect relationships are determined. The main outputs of the HAZOP model will be used as the input to the DBN model and degradation model in the next stage.

6.3.2. *Degradation model development*

According to the failure modes with corresponding state spaces obtained by the HAZOP model in the last stage, this stage consists of modeling all physical mechanisms of deterioration, i.e., fouling, wearing, cracking, and corrosion. Some entities may continue to operate in a degraded mode following certain types of faults. The system may continue to perform its function but not at a specified operating level. For example, a lubricating oil pump may experience a problem in some components. If it is important to distinguish the degraded state from that of a complete failure, then Markov or Semi-Markov analysis can be utilized if constant failure rates can be estimated from a historical event database. Simplified models such as Markov chain with finite discrete states can be used to reduce the entire complexity of the system and the calculation of the whole algorithm, on condition that this simplification has little negative impact on the accuracy of practical industrial application. Meanwhile, the transition probability distribution (or transition matrix) of each degradation process is either calculated based on historical inspection and maintenance data (records) or determined by expert knowledge.

By considering a Markov process $\{X(t), t \geq 0\}$ with state space $S = \{0, 1, 2, \ldots, r\}$ and stationary transition probabilities, the transition probabilities of the Markov process $P_{ij}(t) = \Pr(X(t) = j|X(0) = i)$ for all $i, j \in S$ may be arranged as a matrix:

$$P(t) = \begin{pmatrix} P_{00}(t) & P_{01}(t) & \cdots & P_{0r}(t) \\ P_{10}(t) & P_{11}(t) & \cdots & P_{1r}(t) \\ \vdots & \vdots & \ddots & \vdots \\ P_{r0}(t) & P_{r1}(t) & \cdots & P_{rr}(t) \end{pmatrix}. \tag{6.3}$$

In this modeling stage, dependent failures are extremely important and must be given adequate treatment so as to minimize gross overestimation of performance. In general, dependent failures are defined as events in which the probability of each failure is dependent on the occurrence of other failures. If a set of dependent events $\{E_1, E_2, \ldots, E_n\}$ exists, then the probability of each failure in the set depends on the occurrence of other failures in the set.

Let a system be composed of n identical components, each with a constant failure rate λ. Let $\lambda^{(i)}$ denote the failure rate due to independent failures, and let $\lambda^{(c)}$ denote the failure rate due to common cause failures. Assuming independence of the two failure causes, the total failure rate λ of the component can be written as the sum of the two failure rates $\lambda = \lambda^{(i)} + \lambda^{(c)}$, and let $\beta = \frac{\lambda^{(c)}}{\lambda^{(i)}+\lambda^{(c)}}$, then $\lambda^{(c)} = \beta\lambda$, $\lambda^{(i)} = (1-\beta)\lambda$. Therefore, common cause failure (CCF) parametric models can be used to obtain a more accurate assessment of CCF probabilities in systems with a higher level of redundancy. Other methods, e.g., Multiple Greek Letters model, α-factor model, and Binomial failure rate model can also be used in more complicated situations.

The main outputs of the degradation model, including Markov deterioration processes with finite state space and corresponding parameters, will be used in the DBN model.

6.3.3. *DBN model development*

In this stage, a DBN is developed to simultaneously integrate the structure of the system, the variables, and causal mechanisms (or interdependencies) analyzed by the HAZOP model, and the dynamic deterioration process as a result of the degradation model. There are three procedures in this step which are illustrated explicitly in Fig. 6.6.

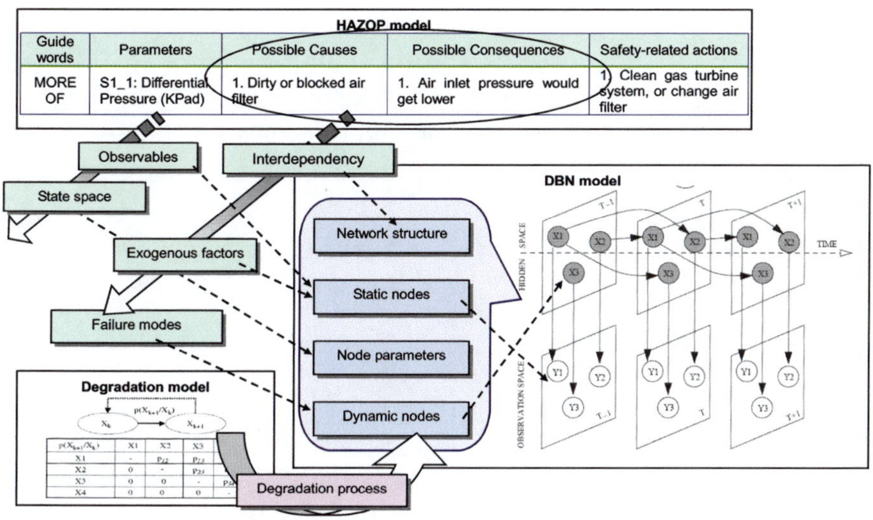

Fig. 6.6. Development of DBN model.

(1) Identification of the DBN variables (nodes). According to the results of the HAZOP model, all the variables in the HAZOP model are associated with nodes in BNs. Meanwhile, each deterioration mechanism analyzed by the degradation model is associated with a hidden-state variable in DBN (dynamic nodes), whereas variables in the HAZOP model which can be observed by a monitoring system are associated with observable variables in DBN (static nodes), and exogenous factors in the HAZOP model are associated with hidden-state variables in DBN (static nodes).
(2) Development of the network structure by creating the directed edges from the node corresponding to fault causes to the node representing its consequence, which is indicative of a conditional dependency between the variables it links.
(3) Determination of DBN parameters, i.e., the definition of all the CPTs. CPTs can be determined by learning the parameters on the database or depending on an expert's judgment.

The DBN model integrating both HAZOP and degradation models describes the evolution of the states of some systems (or subsystems, components) in terms of their joint probability distribution. At each instance in time t, the states of the system depend only on the states at the previous time slice $(t-1)$ and possibly current time instance. Furthermore, the DBN model is a probability distribution function on the sequence of T hidden-state variables $X = \{x_0, \ldots, x_{T-1}\}$ and the sequence of T observables $Y = \{y_0, \ldots, y_{T-1}\}$ that has the following factorization, which satisfies the requirements for DBNs that state x_t depends only on state x_{t-1}:

$$\Pr(X, Y) = \prod_{t=1}^{T-1} \Pr(x_t|x_{t-1}) \cdot \prod_{t=0}^{T-1} \Pr(y_t|x_t) \cdot \Pr(x_0). \tag{6.4}$$

The learning of ISPM in this step needs to define the following DBN parameters using historical data and prior knowledge: the state transition pdf $\Pr(x_{t+1}|x_t)$, the observation pdf $\Pr(y_t|x_t)$, and the initial state distribution $\Pr(x_0)$. The DBN model will be used for safety assessment, risk evaluation, and prediction model in the next step (step three).

6.3.4. *Monitoring model development*

Based on online condition and process monitoring system, the monitoring model is dedicated to providing the current state of the observable variables as evidence defined in the previous DBN model.

According to the condition and process monitoring, each observable variable value is obtained from field sensors based on local Programmable Logic Controller (PLC) or SCADA systems (Fig. 6.7). Then the variable value is mapped into probabilities of each state space defined in the HAZOP model by fuzzy methods, which is explicated in detail in the previous work [27]. Corresponding posterior probabilities are then calculated to supply hard or soft evidences for DBN inference. On the other hand, the real-time data are stored in a safety database consisting of real-time database, fault sample database, normal operating condition database, and unknown or potential fault database [8]. It is developed to update the parameters in degradation and DBN models which help to improve the accuracy of safety assessment, risk evaluation, and prediction.

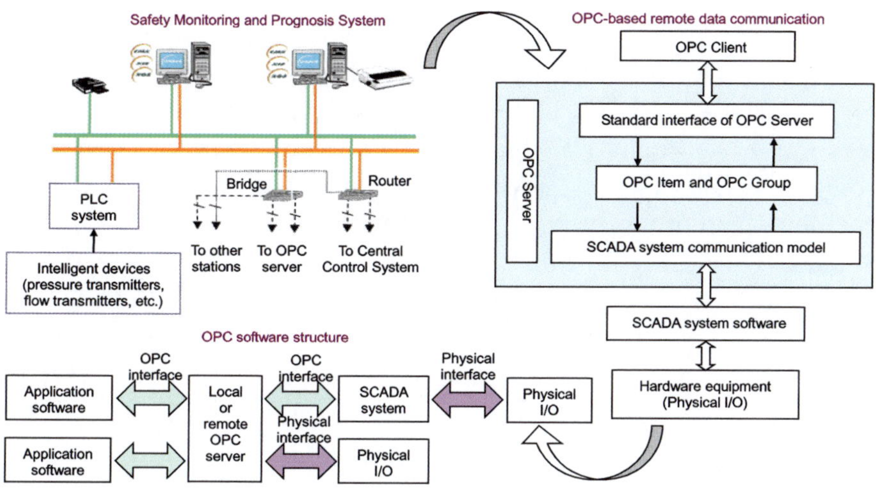

Fig. 6.7. Safety monitoring system with OPC technology based on SCADA system.

6.3.5. *Assessment model development*

In this stage of step three, the basic aim is to determine the hidden states in the DBN model based on current observation data obtained by the monitoring model as inference evidence, and then discover possible hazard reasons, consequences, and decide on proper safety-related actions. In other words, assessment with DBNs consists of fixing the values of the observed variables and computing the posterior probabilities of the unobserved variables.

The assessment model finally gives the probability of each hidden state of the system, i.e., the degradation condition of each component or subsystems, the probability of each potential exogenous hazard reasons, as well as possible consequences. According to the results of the assessment model, a decision can be made to eliminate hazard origins, repair degraded entities, and establish a proper safety control plan.

The problem of inference in assessment model can be posed as the problem of finding $\Pr(X_0^T|Y_0^T)$, where Y_0^T denotes a finite set of T consecutive observations, $Y_0^T = \{y_1, \ldots, y_T\}$, and X_0^T is the set of the corresponding hidden variables, which are depicted as dark rings in Fig. 6.6. While the white ring in Fig. 6.6 indicates that the distribution of x_t is to be estimated based on observations Y_0^T.

In this chapter, FB algorithm is used for inference in the assessment model. The basic idea of the FB algorithm is to recursively compute $\alpha_t(i) = P(X_t = i|y_{1:t})$ in the forwards pass, to recursively compute $\beta_t(i) = P(y_{t+1:T}|X_t = i)$ in the backwards pass, and then to combine them to produce the final estimation $\gamma_t(i) = P(X_t = i|y_{1:T})$:

$$
\begin{aligned}
P(X_t = i|y_{1:T}) &= \frac{1}{P(y_{1:T})} P(y_{t+1:T}|X_t = i, y_{1:t}) P(X_t = i, y_{1:t}) \\
&= \frac{1}{P(y_{1:T})} P(y_{t+1:T}|X_t = i) P(X_t = i|y_{1:t}) \qquad (6.5)
\end{aligned}
$$

or

$$\gamma_t \propto \alpha_t \cdot \beta_t. \qquad (6.6)$$

6.3.6. *Risk evaluation model development*

Automatic optimization mechanism based on ant colony algorithm is introduced for system risk evaluation in ISPM with integration of fault occurrence probability and its severity of each node in the DBN model. According to the state transition probability and the cycle "pheromone" update algorithm in ant colony algorithm, the risk level of each fault propagation path is calculated, which is further used for decision-making of safety pre-warning scheme or contingency plan.

The basic ant colony algorithm used in risk evaluation model here is composed of the transition rule and the pheromone update rule. In m ants' routing process, each ant calculates state transition probability according to the artificial pheromone trail (associated with the fault probability of each node in the DBN model) and the heuristic information (associated with the consequence severity of each node in the DBN model) on the fault propagation path. The probability of the kth ant making the transition from node i to node j is given by

$$p_{ij}^{k}(t) = \frac{[\tau_{ij}(t)]^{\alpha}[\eta_{ij}(t)]^{\beta}}{\sum_{s=1}^{n}[\tau_{is}(t)]^{\alpha}[\eta_{is}(t)]^{\beta}}, \tag{6.7}$$

where $\tau_{ij}(t)$ is the quantity of pheromone laid on the path (i, j) at time t, $\eta_{ij}(t)$ is the heuristic information (also called visibility) and is defined as the quantity $1/d_{ij}$ (d_{ij} is the path length between node i and node j), s is the selected node of the kth ant in the next searching step, α and β control the relative importance of the pheromone trail and visibility, respectively.

An epoch is defined to be every n iterations, when each ant has completed a tour. In order to prevent pheromone intensity being ignored due to too much residual pheromone, after each epoch, the pheromone intensity trails (i.e., system temporary estimated risk value) are updated based on ant-cycle model [2] according to the following equations:

$$\tau_{ij}(t+n) = (1-\rho)\tau_{ij}(t) + \Delta\tau_{ij}(t), \tag{6.8}$$

$$\Delta\tau_{ij}(t) = \sum_{k=1}^{m} \Delta\tau_{ij}^{k}(t), \tag{6.9}$$

$$\Delta\tau_{ij}^{k}(t) = \frac{Q}{L_k}, \tag{6.10}$$

where $\rho \in [0, 1)$ is the evaporation rate and $\Delta\tau_{ij}^{k}(t)$ is the quantity of pheromone laid on the path (i, j) by the kth ant between time t and $(t+n)$; Q is a constant and L_k is the tour length of the kth ant.

In complex industrial system, as mentioned before, most of single faults have multiple propagation paths, so any local slight deviation is able to propagate, spread, accumulate, and increase through fault causal chains. In the ISPM, the risk evaluation model helps to determine which fault propagation path would happen most possibly, and which path would have the largest estimated risk, considering the system's current running situation. So, the safety contingent plan can be made according to the output of the risk evaluation model, which includes but not limited to the organization of fault isolation, emergency rescue technical support, and comprehensive emergency rescue teams.

6.3.7. *Prediction model development*

In the last stage of the ISPM development, the prediction model deals with predicting future observations which indicate system performance as well as future hidden states which indicate component reliability based on the current observation data from the monitoring model, estimated current hidden-states from the assessment model, and also diachronic evolution mechanism from the DBN model. A one-step prediction can be started as the following inference problem [$\Pr(x_{t+1}|Y_0^T)$ or $\Pr(y_{t+1}|Y_0^T)$]:

$$\Pr(x_{t+1}|Y_0^T) = \frac{\sum_{x_t} \Pr(x_{t+1}|x_t)\alpha_t(x_t)}{\sum_{x_t} \alpha_t(x_t)} \tag{6.11}$$

similarly

$$\Pr(y_{t+1}|Y_0^T) = \frac{\sum_{x_{t+1}} \alpha_{t+1}(x_{t+1})}{\sum_{x_t} \alpha_t(x_t)}. \tag{6.12}$$

The result from the prediction model will be used for the decision-making procedure: if the future estimated situation is a state considered “safe” or “successful” to fulfil the goal of the system, no action should be planned. In a contradictory way, if the degradation degree can be the cause of risk or insufficient performance, an action must be started immediately (in safety case) or at a scheduled time (predictive maintenance).

In summary, the results from the assessment, risk evaluation, and prediction models in step three of the proposed ISPM development play a fundamental role for advanced safety management of complex industrial system providing proper and accurate safety pre-warnings as well as effective proactive maintenance and contingent plans. In this way, the proposed ISPM is conducive to both reducing hazard happening probability and the loss of failures or accidents, in other words, improving the system’s intrinsic safety level.

6.4. Application to Gas Turbine Compressor System

The proposed ISPM is applied to a real system to assess the safety state and predict the reliability and performance of a gas turbine compressor system, which is an important facility in long distance gas pipeline systems. The information about the gas turbine compressor system is shown in Table 6.1. Based on the principle of HAZOP model development, the gas turbine compressor system consists of seven main subsystems shown in Fig. 6.8. In this chapter, the former six subsystems are considered and analyzed in detail, while the centrifugal compressor is relatively independent, and can be analyzed alone in the same way.

The results of DBN model integrating the HAZOP model and the degradation model are shown in the appendix, in which Table A.1 records all information of dynamic nodes in the DBN model, and Table A.2 records all information about static nodes. Taking the first subsystem (air system) for example, in the HAZOP model, the main HAZOP study records of the air system are shown in Table 6.2. In the degradation model, the degradation processes are studied, and two degradation modes are determined, which are “air filter degradation

Table 6.1. Information of the system.

Gas turbine		Centrifugal compressor	
Company	SOLAR	Company	MAN TURBO
Type	Titan 130	Type	RVO50/04
NGP	11220 rpm	Max. inlet flow	80000 m^3/h
NPT	8856 rpm	First critical speed	4428 rpm
Power	14000 kW	Max. continuous speed	8856 rpm
Inlet airflow	83000 kg/h	Efficiency	77%–84%
Heat efficiency	33.3%	Inlet pressure	4.5 Mpa
Axial compressor stage	14	Exhaust pressure	6.4 Mpa
Compression ratio	16	Pressure ratio	1.86
Variable inlet Guide vanes	Six stages (adjustable)	Impeller	Four stages
Fuel nozzle	21	Max. temperature	193 (°C)
T5 (°C)	760 (°C)	impeller Diameter	500 mm

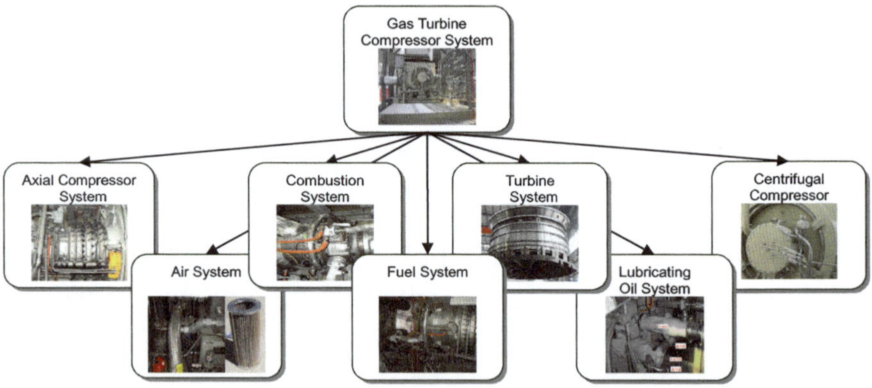

Fig. 6.8. Gas turbine compressor system and main subsystems.

(AFD)" and "ventilation system degradation (VSD)". The transition probabilities of the Markov process (shown in Tables 6.3 and 6.4) of the above two degradation modes are calculated based on both historical data stored in databases and expert judgment according to engineering practices in the field. The DBN model is then developed

Table 6.2. HAZOP study records of air system (subsystem 1).

Guide word	Deviation	Possible cause	Possible consequences	Action required (safety measurements)
MORE OF	S1_1 differential pressure of inlet filter (kPaD)	Dirty or blocked air filter	Air inlet pressure would get lower	Clean gas turbine system, or change air filter
	S1_3 cabinet temperature (°C)	1. Malfunction in the fan on cabinet 2. Dirty air filter 3. Blocked air exhaust path 4. External fire accident	1. It would make the whole facility heat up, and impact on the function of each component 2. Temperature of inlet air and the power consumption would increase 3. Corrosion process of internal metal material would be accelerated 4. It would cause fire	1. Change fan 2. Change inlet air filter 3. Check and clean air exhaust path
	S1_5 inlet differential pressure (kPa)	Air inlet path has worn or been fouled	Output power would decrease	Repair air inlet path

(*Continued*)

Table 6.2. (*Continued*)

Guide word	Deviation	Possible cause	Possible consequences	Action required (safety measurements)
LESS OF	S1_2 cabinet pressure (kPa)	1. Partially opened cabinet doors, causing a mass of air leakage 2. Malfunction in the fan on cabinet 3. Dirty air inlet filter	1. Efficiency of the system would decrease 2. Barotropic state of the cabinet would be destroyed	1. Check and close the cabinet door 2. Change fan 3. Change air filter
	S1_4 T1 temperature (°C)	Environment temperature is on the low side	1. If air contains ice, the components would be destroyed 2. Power loss would increase	Make air inlet temperature higher or append air heater

Table 6.3. D1_1 air filter degradation (AFD).

D1_1	AFD_{k+1} Normal	Fouled	Blocked
AFD_k = Normal	0.9959	0.0031	0.0010
AFD_k = Fouled	0	0.9978	0.0022
AFD_k = Blocked	0	0	1

Table 6.4. D1_2 ventilation system degradation (VSD).

D1_2	VSD_{k+1} Normal	Degraded	Failure
VSD_k = Normal	0.99969	0.00026	0.00005
VSD_k = Degraded	0	0.99986	0.00014
VSD_k = Failure	0	0	1

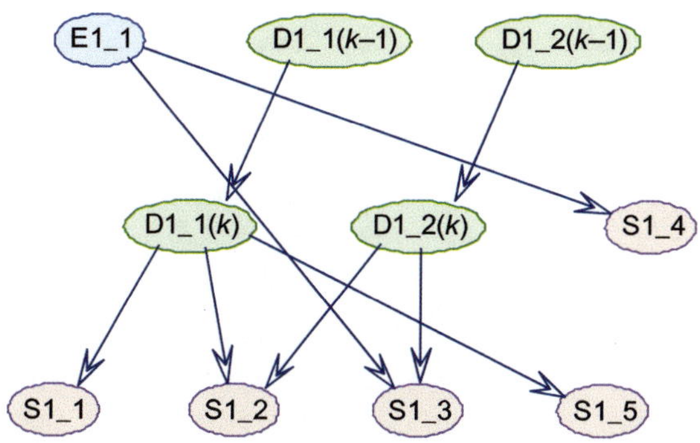

Fig. 6.9. Dynamic network structure of air system.

integrating the HAZOP and degradation models of the air system, and the DBN structure is shown in Fig. 6.9, according to which the CPTs of all the observations are shown in Tables 6.5–6.9 determined by historical databases and maintenance records. The node IDs (e.g., S1_1) in Tables 6.5–6.9 and figures are explained in Tables A.1 and A.2.

The monitoring model (Fig. 6.10) gives updated information about each observable variable as inference evidence, which is used

Table 6.5. CPT of Pr(S1_1 | D1_1).

D1_1	S1_1 = Normal	On the high side	Superhigh
Normal	0.7490	0.1975	0.0535
Fouled	0.2635	0.3979	0.3386
Blocked	0.0664	0.3253	0.6083

Table 6.6. CPT of Pr(S1_2 | D1_1, D1_2).

D1_1	D1_2	S1_2 = Normal	On the low side	Ultra low
Normal	Normal	0.9553	0.0447	0
Fouled	Normal	0.7557	0.2443	0
Blocked	Normal	0.5945	0.3539	0.0516
Normal	Degraded	0.2502	0.6063	0.1435
Fouled	Degraded	0.1527	0.7496	0.0977
Blocked	Degraded	0.1109	0.7453	0.1438
Normal	Failure	0.0563	0.7950	0.1487
Fouled	Failure	0	0.7016	0.2984
Blocked	Failure	0	0.5935	0.4065

Table 6.7. CPT of Pr(S1_3 | E1_1, D1_2).

E1_1	D1_2	S1_3 = Normal	On the high side
Normal	Normal	1	0
On the low side	Normal	1	0
On the high side	Normal	0.9819	0.0181
Normal	Degraded	0.7622	0.2378
On the low side	Degraded	0.8224	0.1776
On the high side	Degraded	0.7468	0.2532
Normal	Failure	0.4013	0.5987
On the low side	Failure	0.4511	0.5489
On the high side	Failure	0.3548	0.6452

Table 6.8. CPT of Pr(S1_4 | E1_1).

E1_1	S1_4 = Normal	On the low side	On the high side
Normal	0.9787	0.0121	0.0092
On the low side	0.1536	0.8464	0
On the high side	0.1461	0	0.8539

Table 6.9. CPT of Pr(S1_5 | D1_1).

D1_1	S1_5 = Normal	On the high side
Normal	0.8469	0.1531
Fouled	0.4464	0.5536
Blocked	0.2507	0.7493

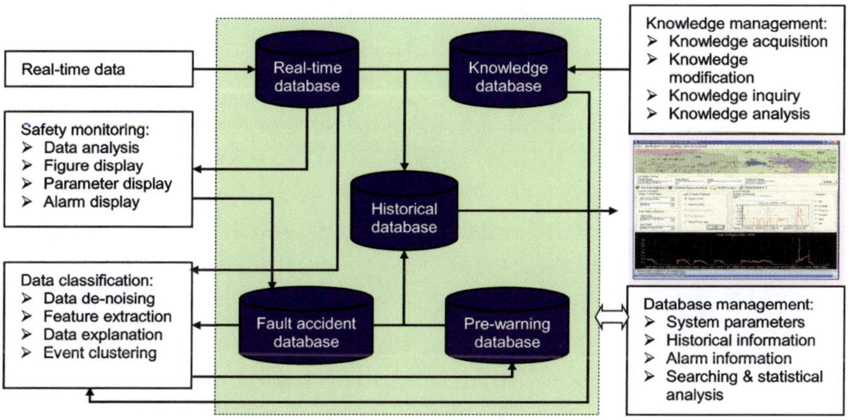

Fig. 6.10. Structure and function of monitoring model in ISPM.

in the assessment model to estimate system's hidden states, possible hazard reasons, and consequences. The risk evaluation model takes advantage of the possible hazard reasons and consequences to deduce possible fault propagation paths with corresponding estimated risk. While the prediction model based on the results of safety assessment outputs the future trend of both observable variables and hidden states, which are used to calculate future system performance, reliability, and RUL. A maintenance plan is then made according to the results of assessment and prediction, which helps the system to keep safe and healthy in the long run.

6.5. Results and Discussion

6.5.1. *The results of safety assessment*

During the online monitoring of the gas turbine compressor system at the beginning of May 2008, the average lubricating oil temperature was 57 (°C) and the lubricating oil pressure was 0.195 MPa, all of which have deviated from the normal value interval. Based on the principle of the assessment model presented in Sec. 6.3.5, the analysis processes are as follows (where the meaning of node ID (e.g., S7_2) is explicated in Tables A.1 and A.2):

(1) The observable node deviations are fuzzily quantified as posterior probability. The normal value range of lubricating oil temperature at machine running state is [35, 55] (°C) (shown in Table A.2), while the measured lubricating oil temperature was 57 (°C), then the posterior probability of observable node S7_2 is calculated as {0, 0.1546, 0.7939, 0.0515} according to the state space: {On the low side; Normal; On the high side; Superhigh}. In the same way, the normal value range of the lubricating oil pressure is [0.210, 0.449] (MPa) (shown in Table A.2), then the posterior probability of the observable node S7_1 is calculated as {0.0593, 0.8536, 0.0871, 0} according to the state space: {Ultra low; On the low side; Normal; On the high side}. The abnormal observations belong to the lubricating oil system, whose dynamic network structure is shown in Fig. 6.11. The probabilities of the normal state of D7_1, D7_2, D7_3, D7_4, and D7_5, which can be considered as reliabilities of such entities, are then estimated as 0.3140, 0.6583, 0.7244, 0.5395, and 0.9016, respectively, by inference of the assessment model. The probabilities of normal state of node D7_1 and node D7_4 have lowest values, which mean

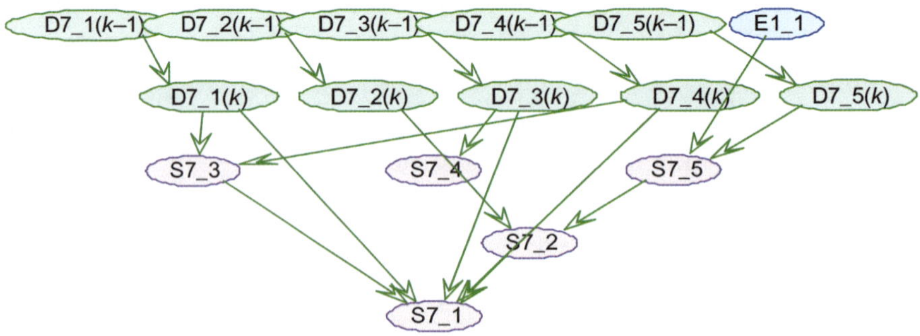

Fig. 6.11. Dynamic network structure of lubricating oil system.

that lubricating oil tubes and oil pumps have lowest reliabilities and need certain repair.

(2) The hazard reasons and consequences of gas turbine compressor system estimated by the assessment model are shown in Tables 6.10 and 6.11, from which the most possible hazard reasons are "smaller diameter of oil inlet throttle" with the

Table 6.10. Result of assessment model (possible hazard reasons with safety-related actions).

Possible hazard reasons	Probability of occurrence	Safety-related action required
1. Smaller diameter of oil inlet throttle	0.8035	1. Checking oil throttle orifice of bearing and increasing throttle diameter, making the flow of lubricating oil meet the operation requirement
2. Higher outlet water temperature	0.71424	2. Increasing the cooling circulating water
3. Fault in main oil pump	0.6829	3. Switching pump for inspection and repair
4. Insufficient cooling water	0.6351	4. Increasing the cooling circulating water
5. Oil pipe rupture or oil leakage at conjunctions	0.5122	5. Inspection or replacement of pipe section
6. Bearing clearance is on the small side, causing severe friction and a mass of heat	0.4763	6. Inspecting and appropriately adjusting (enlarging) the bearing clearance to decrease the heat caused by friction
7. Obstruction in oil filter or oil passage way, which causes oil pressure loss	0.4	7. Disassembly checking and cleaning the lubricating oil filter
8. Deterioration of lubricant	0.3969	8. Releasing the gas from lubricating oil system or changing lubricating oil
9. Bearing fault	0.3616	9. Checking and repairing or replacing the bearing
10. Lubricating oil leakage at pipeline and valves	0.2561	10. Changing conjunction gasket at leak location, tightening connecting bolt, and making no leakage occurrence

Table 6.11. Result of assessment model (possible hazard consequences).

Possible hazard consequences	Probability of occurrence
1. Oil film is difficult to sustain; bearing will be burnt; seal will be damaged or impeller will be destroyed	0.9132
2. Shutdown caused by low oil pressure	0.6034
3. Carbonization of lubricating oil or decreasing lubrication performance	0.3886

probability of 0.8035, and "fault in main oil pump" with the probability of 0.6829. The most possible hazard consequence is "Oil film is difficult to sustain; bearing will be burnt; seal will be damaged or impeller will be destroyed" with the largest probability of 0.9132. Therefore, the optimal safety-related actions advised by the ISPM are given as "checking oil throttle orifice of bearing and increasing throttle diameter" and "switching pump for inspection and repair".

6.5.2. *The results of risk evaluation*

From October 2007 to May 2008, there were 117 fault accidents, where 37 fault accidents were related to "High lubricating oil temperature", 18 fault accidents were related to "High axial displacement of bearing in gas generator", and three times of more than two fault accidents simultaneously. The severity of each fault consequence was investigated based on its influence on the customer, environment, and economy, for example, the duration of gas supply disruption, economic loss, interference on surrounding residents' lives, and maintenance cost, etc. Expert rating method and statistic analysis are adopted for the above research purpose in this case study.

Based on the results from the assessment model in ISPM, the risk evaluation model extracts the possible fault propagation paths from the DBN model in ISPM, as shown in Fig. 6.12, the legend in which is as follows:

R1–R40 indicate fault causes in the lubrication system, and are shown in Table 6.12.

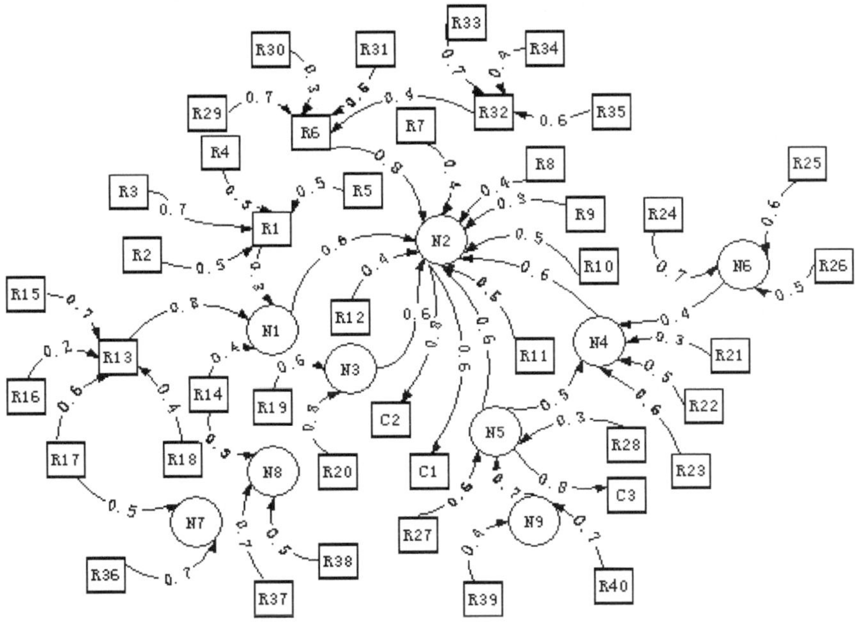

Fig. 6.12. Fault propagation paths in lubricating oil system.

N1–N9 indicate fault deviations in the lubrication system: "the pressure of lubrication filter is on the high side" (N1), "the pressure of lubricating oil is on the low side" (N2), "the level of lubricating oil tank is on the low side" (N3), "the temperature of lubricating oil flowing from bearing is on the high side" (N4), "lubricating oil temperature in lubrication pipeline is on the high side" (N5), "lubricating oil pressure in lubrication pipeline is on the low side" (N6), "lubricating oil pressure in lubrication pipeline is on the high side" (N7), "the pressure difference of lubricating oil tank is on the high side" (N8), and "the temperature of lubricating oil heater is on the high side" (N9).

C1–C3 indicate possible fault consequences in the lubrication system: "oil film is difficult to sustain; bearing will be burnt; seal will be damaged or impeller will be destroyed" (C1), "shutdown caused by low oil pressure" (C2), and "carbonization of lubricating oil or decreasing lubrication performance" (C3). The severity of each fault consequence is 0.9, 0.7, and 0.4, respectively.

Table 6.12. Possible fault causes in lubrication system.

Notation	Possible fault causes	Notation	Possible fault causes
R1	Differential pressure transmitter fault in the oil filter	R11	Oil control valve failure or bypass valve is opened
R2	Transmitter wire is damaged	R12	Discharge valve is opened
R3	Transmitter wire is broken	R13	Oil filter is blocked
R4	Surge protector fault	R14	Connecting pipe of differential pressure transmitter is blocked
R5	Differential pressure transmitter fault	R15	Lubrication system equipment, valves and pipes are cleared incompletely, which blocks the oil filter due to its internal residues with welding slag and other solid particles flowing into the filter with oil
R6	Lubrication pump fault	R16	Pollutants produced by poor management block the oil filter
R7	Bearing failure, and excessive oil consumption	R17	Lubricating oil system is not pre-heated when starting up, which causes low oil temperature and viscosity thus blocking the filter
R8	Lubricating oil pipe is broken, or there is flange leakage	R18	System equipment, valves and pipeline produce corrosion products, which flow into oil filter with lubricating oil and block the filter
R9	Malfunction of pressure self-control equipment or pressure transmitter	R19	Lubricating oil is insufficient
R10	Pump suction pipe leakage	R20	Leakage causes excessive oil consumption

(*Continued*)

Table 6.12. (*Continued*)

Notation	Possible fault causes	Notation	Possible fault causes
R21	Temperature detector failure (wire damage, corrosion, fracture, or RTD fault)	R31	Oil pump motor fault
R22	Space between the rotor and bearing oil tube is blocked or contaminated	R32	Pressure switch on the outlet pipeline fails
R23	Bearing seat wears down	R33	Wire damage or corrosion
R24	Speed of lubrication pump is on the low side	R34	Pressure switch wire is broken
R25	Setting value of relief valve VR902 on the outlet pipe of lubrication pump is too low	R35	Pressure switch failure
R26	Inlet filter is blocked	R36	Speed of lubrication pump is on the high side
R27	Lubricating oil heater fault	R37	Oil separator FSA901 is blocked, and the efficiency decreases
R28	Temperature transmitter of oil tank fault	R38	Differential pressure transmitter TPD324 is connected with the external environment, which causes the filter of connecting pipe blocked
R29	Pump control box failure	R39	Lubrication temperature transmitter fault
R30	Power line from control box to oil pump fault	R40	MCC Oil heating control module faults

In the risk evaluation model, the parameters of ant colony algorithms are set as $m = 10$, $\alpha = 1$, $\beta = 2$, $\rho = 0.7$, $Q = 50$, and the quantitative risk evaluations are performed by combining hazard probability and consequence severity of each node along the fault propagation path using Eqs. (6.7)–(6.10). When the ant colony algorithm tends to be stabilized in the iterative loop, the overall risk evaluation results are obtained, in which the overall risk evaluation values of C1 and C2 are 0.476 and 0.635, respectively. All the fault propagation paths with their occurrence probabilities and risk values are shown in Table 6.13, where the fault propagation paths with

Table 6.13. Fault propagation paths and responding risk evaluation values.

No.	Propagation paths	Occurrence probability	Risk evaluation	No.	Propagation paths	Occurrence probability	Risk evaluation
1	R6 → N2 → C1	0.3048	0.2743	14	R12 → N2 → C2	0.1018	0.0713
2	R6 → N2 → C2	0.4064	0.2845	15	R21 → N4 → N2 → C1	0.0366	0.0329
3	R7 → N2 → C1	0.0763	0.0687	16	R21 → N4 → N2 → C2	0.0488	0.0342
4	R7 → N2 → C2	0.1018	0.0713	17	R22 → N4 → N2 → C1	0.1019	0.0917
5	R8 → N2 → C1	0.0763	0.0687	18	R22 → N4 → N2 → C2	0.1358	0.0951
6	R8 → N2 → C2	0.1018	0.0713	19	R23 → N4 → N2 → C1	0.1467	0.1320
7	R9 → N2 → C1	0.1142	0.1028	20	R23 → N4 → N2 → C2	0.1956	0.1369
8	R9 → N2 → C2	0.1523	0.1066	21	R24 → N6 → N4 → N2 → C1	0.0422	0.0380
9	R10 → N2 → C1	0.1191	0.1072	22	R24 → N6 → N4 → N2 → C2	0.0563	0.0394
10	R10 → N2 → C2	0.1588	0.1112	23	R25 → N6 → N4 → N2 → C1	0.0310	0.0279
11	R11 → N2 → C1	0.1191	0.1072	24	R25 → N6 → N4 → N2 → C2	0.0414	0.0290
12	R11 → N2 → C2	0.1588	0.1112	25	R26 → N6 → N4 → N2 → C1	0.0215	0.0194
13	R12 → N2 → C1	0.0763	0.0687	26	R26 → N6 → N4 → N2 → C2	0.0287	0.0201

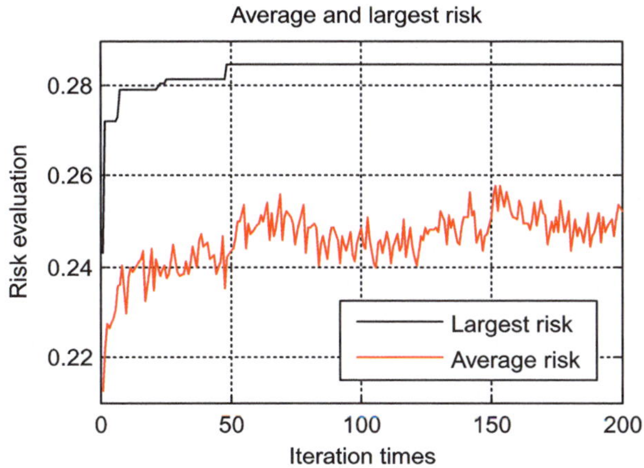

Fig. 6.13. Average and largest risk of all fault propagation paths.

largest probabilities are: R6 → N2 → C1 and R6 → N2 → C2. The average and largest risk values (on pathway R6 → N2 → C2) of all fault propagation paths determined by the ant colony algorithm in risk evaluation model are shown in Fig. 6.13.

Therefore, the responding safety contingent control strategies are made as follows: (1) to open and check lubrication system and lubricating oil pump; (2) to further monitor the temperature, pressure, and flow of lubrication oil flowing through bearings. The related regulation is needed if the observable parameter is out of normal range; otherwise, emergent shutdown of gas turbine compressor system is needed if the regulation is invalid.

The field safety engineer checked the gas turbine compressor system according to the advised safety-related actions (provided by the assessment model) and contingent control strategies (provided by the risk evaluation model), and found that the lubrication pump working state was unstable and oil supply was insufficient. Further inspection report indicated that the skeleton oil seal in the pump was aging, and the bearing was partially worn. After replacing the oil seal, washer, and bearing in the oil pump, all monitoring parameters of gas turbine compressor system fell in the safety range, and no fault alarm was given by ISPM. It is proved that the results given by the

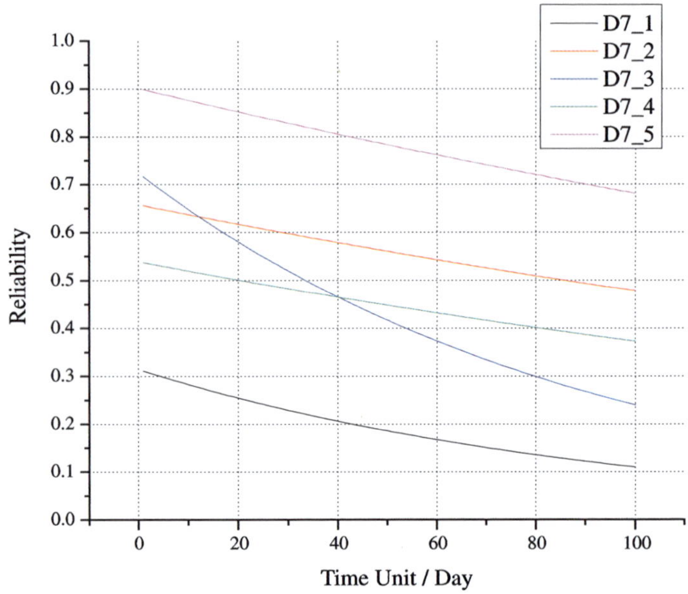

Fig. 6.14. Future reliability trends of degraded lubricating subsystem.

ISMP system proposed in the chapter fitted well with the actual situation, and the fault propagation paths as a pre-warning alarm are conducive for the safety engineer to make contingency plans to avoid accidents in advance.

6.5.3. *The results of safety prediction*

According to the algorithm of the prediction model, future reliability trends of lubricating subsystems beginning with the present degraded states are predicted, and shown in Fig. 6.14, which indicates that D7_1 (Oil Tube Degradation) and D7_4 (Oil Pump Degradation) with least reliabilities will deteriorate gradually and their reliabilities will be under 0.5 in 30 days. Although D7_3 (Oil Filter Degradation) has higher reliability at present, it will deteriorate severely, which will affect other components. The influence on the bearing degradation process is shown in Fig. 6.15, which indicates that in the situation of higher lubricating oil temperature and lower lubricating oil pressure, the degradation trends of bearings (solid lines in Fig. 6.15) are more

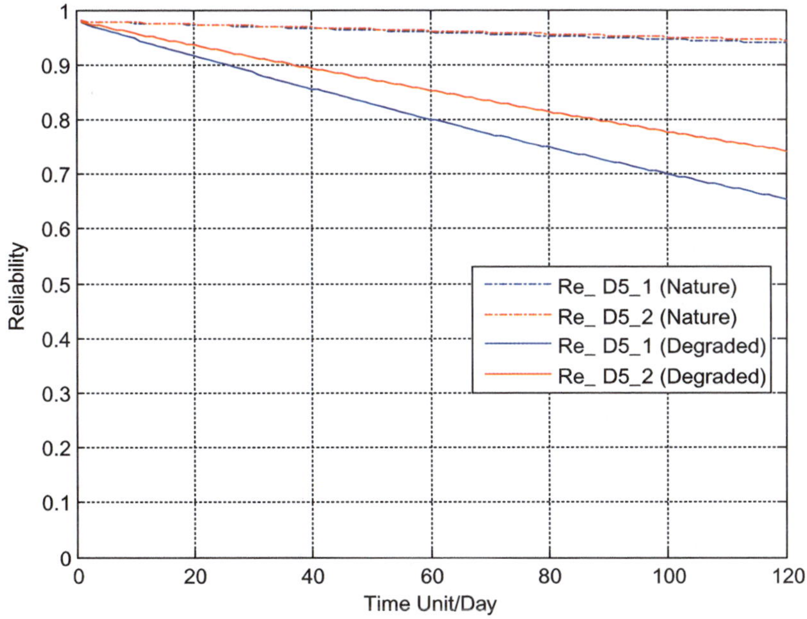

Fig. 6.15. Bearing degradation trends in the situation of normal and degraded lubricating system.

severe than the natural degradation process (dash lines in Fig. 6.15). In other words, the degraded states of the lubricating system shorten the RUL of bearings.

Considering the results of safety assessment, risk evaluation, and prediction by ISPM, proactive maintenance decisions are made as follows:

(1) replacing oil tube, which is considered as perfect maintenance, so the state of D7_1 can be set as [1 0 0 0] after repair;
(2) cleaning oil filter, which is considered as unperfected maintenance, so the state of D7_3 can be set as [0.9 0.1 0] after repair;
(3) repairing oil pump, which is also considered as unperfected maintenance, so the state of D7_4 can be set as [0.85 0.15 0] after repair;

Figure 6.16 shows the future reliability trends of lubricating subsystems beginning with the states after maintenance. The reliabilities

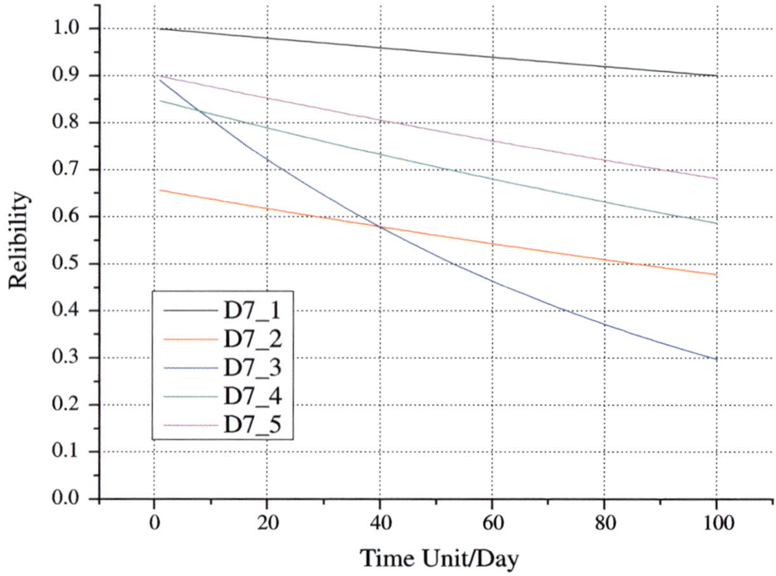

Fig. 6.16. Future reliability trends of repaired lubricating subsystem.

of the repaired subsystem are all above 0.5, which are acceptable in industrial practice, and except D7_3, the other four subsystems will run functionally well in the next four months. Due to the degradation mechanism of the oil filter, its reliability deteriorates more severely than other subsystems, so the oil filter should be cleaned (or replaced) every two or three months to ensure the whole system works normally.

6.6. Conclusion

(1) Safety prognosis plays a fundamental role in industrial safety management. Considering the randomness, complexity, and uncertainty of fault propagation, an integrated safety prognosis model (ISPM) is proposed in this chapter using DBNs and ant colony algorithm, integrating safety assessment, risk evaluation, and prediction for safety prognosis in a unified framework. The interaction and dependency among entities in complex systems are considered and then modeled into ISPM. By integrating the HAZOP model,

degradation model, DBN model, monitoring model, assessment model, risk evaluation model, and prediction model, the ISPM is developed to reduce fault occurrence probability effectively by relying longer on the assumptions of failure independence.

(2) Since degradation generally occurs before failures, predicting the trend of system degradation allows the degraded behavior or hazards to be corrected before they cause failure and system breakdowns. Ant colony algorithm is used on the basis of the DBN model in the ISPM to search the most reliable fault propagation path with estimated risk value, which helps to make safety-related actions in time to control fault influence range and reduce the loss of fault consequence.

(3) The effectiveness and accuracy of the ISPM are demonstrated through a real case study, and the application results of gas turbine compressor systems are discussed in detail, which indicate that the ISPM is able to improve the accuracy and efficiency of safety management for multicomponent and multihazard complex system, providing adequate advise on safety-related actions, contingent plans and proactive maintenance plans.

(4) Future researches will focus on applying the proposed ISPM to other complex industrial systems, and introducing proper intelligent optimization method to accelerate the calculation process.

References

[1] M. Dong, D. He, "Hidden semi-Markov model-based methodology for multisensor equipment health diagnosis and prognosis," *European Journal of Operational Research*, vol. 178, no. 3, pp. 858–878, 2007.

[2] H. B. Duan, *Ant Colony Algorithms: Theory and Applications*, Science Press, Beijing, 2005 (in Chinese).

[3] K. Ghorbanian, M. Gholamrezaei, "An artificial neural network approach to compressor performance prediction," *Applied Energy*, vol. 86, no. 7–8, pp. 1210–1221, 2009.

[4] X. B. Gu, *Safety Analysis Method and Its Application in Petrochemical Industry*. Chemical Industry Publishing House, Beijing, 2001 (in Chinese).

[5] A. Heng, A. C. C. Tan, J. Mathew, N. Montgomery, D. Banjevic, A. K. S. Jardine, "Intelligent condition-based prediction of machinery reliability," *Mechanical Systems and Signal Processing*, vol. 23, no. 5, pp. 1600–1614, 2009.

[6] A. Heng, S. Zhang, A. C. C. Tan, J. Mathew, "Rotating machinery prognostics: State of the art, challenges and opportunities," *Mechanical Systems and Signal Processing*, vol. 23, no. 3, pp. 724–739, 2008.

[7] M. A. Herzog, T. Marwala, P. S. Heyns, "Machine and component residual life estimation through the application of neural networks," *Reliability Engineering & System Safety*, vol. 94, no. 2, pp. 479–489, 2009.

[8] J. Q. Hu, L. B. Zhang, Z. H. Wang, W. Liang, "The application of integrated diagnosis database technology in safety management of oil pipeline and transferring pump units," *Journal of Loss Prevention in the Process Industries*, vol. 22, no. 6, pp. 1025–1033, 2009.

[9] R. Huang, L. Xi, X. Li, C. R. Liu, H. Qiu, J. Lee, "Residual life predictions for ball bearings based on self-organizing map and back propagation neural network methods," *Mechanical Systems and Signal Processing*, vol. 21, no. 1, pp. 193–207, 2007.

[10] B. Iung, M. Monnin, A. Voisin, P. Cocheteux, E. Levrat,"Degradation state model-based prognosis for proactively maintaining product performance," *CIRP Annals-Manufacturing Technology*, vol. 57, no. 1, pp. 49–52, 2008.

[11] A. K. S. Jardine, D. Lin, D. Banjevic, "A review on machinery diagnostics and prognostics implementing condition-based maintenance," *Mechanical Systems and Signal Processing*, vol. 20, no. 7, pp. 1483–1510, 2006.

[12] R. Kothamasu, S. H. Huang, W. H. VerDuin, "System health monitoring and prognostics-a review of current paradigms and practices," *International Journal of Advanced Manufacturing Technology*, vol. 28, pp. 1012–1024, 2006.

[13] B. S Lee, H. S. Chung, K. T. Kim, F. P. Ford, P. L. Andersen, "Remaining life prediction methods using operating data and knowledge on mechanisms," *Nuclear Engineering and Design*, vol. 191, no. 2, pp. 157–165, 1999.

[14] J. B. Liu, D. Djurdjanovic, J. Ni, N. Casoetto, J. Lee, "Similarity based method for manufacturing process performance prediction and diagnosis," *Computers in Industry*, vol. 58, no. 6, pp. 558–566, 2007.

[15] F. Louzada-Neto, "Extended hazard regression model for reliability and survival analysis," *Lifetime Data Analysis*, vol. 3, no. 4, pp. 367–381, 1997.

[16] D. Lugtigheid, D. Banjevic, A. K. S. Jardine, "System repairs: When to perform and what to do. *Reliability Engineering & System Safety*, vol. 93, no. 4, pp. 604–615, 2008.

[17] A. Muller, M. C. Suhner, B. Iung, "Probabilistic vs. dynamical prognosis process-based e-maintenance system," In: *Proceedings of IFACINCOM' 04, Information Control in Manufacturing*, Salvador, 2004.

[18] A. Muller, M. C. Suhner, B. Iung, "Formalisation of a new prognosis model for supporting proactive maintenance implementation on industrial system," *Reliability Engineering and System Safety*, vol. 93, no. 2, pp. 234–253, 2008.

[19] K. Murphy, *Dynamic Bayesian Networks: Representation, Inference and Learning*, Thesis of the University of California, Berkley, 2002.

[20] M. Samrout, E. Chatelet, R. Kouta, N. Chebbo, "Optimization of maintenance policy using the proportional hazard model," *Reliability Engineering & System Safety*, vol. 94, no. 1, pp. 44–52, 2009.

[21] Y. Sun, L. Ma, J. Mathew, W. Wang, S. Zhang, "Mechanical systems hazard estimation using condition monitoring," *Mechanical Systems and Signal Processing*, vol. 20, no. 5, pp. 1189–1201, 2006.

[22] H. Sutherland, T. Repoff, M. House, G. Flickinger, "Prognostics, a new look at statistical life prediction for condition-based maintenance," In: *Proceedings of the 2003 IEEE Aerospace Conference,* pp. 3131–3136, 2003.

[23] V. T. Tran, B. S. Yang, A. C. C. Tan, "Multi-step ahead direct prediction for the machine condition prognosis using regression trees and neuron-fuzzy systems," *Expert Systems with Applications*, vol. 36, no. 5, pp. 9378–9387, 2009.

[24] P. J. Vlok, M. Wnek, M. Zygmunt, "Utilising statistical residual life estimates of bearings to quantify the influence of preventive maintenance actions," *Mechanical Systems and Signal Processing*, vol. 18, no. 4, pp. 833–847, 2004.

[25] W. Q. Wang, M. F. Golnaraghi, F. Ismail, "Prognosis of machine health condition using neuro-fuzzy systems," *Mechanical Systems and Signal Processing*, vol. 18, no. 4, pp. 813–831, 2004.

[26] P. Weber, L. Jouffe, "Complex system reliability modelling with dynamic object oriented Bayesian networks (DOOBN)," *Reliability Engineering & System Safety*, vol. 91, no. 2, pp. 149–162, 2006.

[27] L. B. Zhang, J. Q. Hu, W. Liang, Z. H. Wang, "Quantitative HAZOP analysis of compressor units based on fuzzy information fusion," In: *2nd World Conference on Safety of Oil and Gas Industry*, Texas, 2008.

Appendix

Table A.1. Information of dynamic nodes in DBN model.

Subsystem	Dynamic nodes	Node ID	State space	Parent nodes	Children nodes
1. Air system	1. Air filter degradation (AFD)	D1_1	{Normal, Fouling, Blocked}	∅	{S1_1, S1_2, S1_5}
	2. Ventilation system degradation (VSD)	D1_2	{Normal, Degraded, Failure}	∅	{S1_2, S1_3}
2. Axial compressor system	1. Compressor blade fouling degradation (CBFD)	D2_1	{Normal, Fouling, Failure}	{S1_3}	{S2_1, S2_2}
3. Fuel system	1. Fuel adjustor degradation (FAD)	D3_1	{Normal, Degraded, Failure}	∅	{S3_1, S3_2}
	2. Fuel tube degradation (FTD)	D3_2	{Normal, Slightly leakage, Severe leakage}	∅	{S3_1}
4. Combustion system	1. Hot path degradation (HPD)	D4_1	{Normal, Fouling, Corrosion, Thermal distortion, Fouling & corrosion & distortion, Failure}	{S1_3}	{S4_1}
	2. Air cooling tube degradation (ACTD)	D4_2	{Normal, Leakage, Blocked, Failure}	{S1_3}	{S4_1}

5. Turbine system_1 (gas turbine subsystem)	1. GT 1#2#3# bearing degradation (GTBD)	D5_1	{Normal, Degraded, Failure}	{S7_1, S7_2, S5_1, S5_3}	{S7_3, S5_3, S5_4}
	2. GT thrust bearing degradation (GTTBD)	D5_2	{Normal, Degraded, Failure}	{S5_1, S5_2, S7_1, S7_2}	{S7_3, S5_5, S5_2}
	3. GT rotor degradation (GTRD)	D5_3	{Normal, Unbalanced, Unaligned, Failure}	{S5_1}	{S5_3}
	5. GT blade degradation (GTBD)	D5_4	{Normal, Thermal distortion, Corrosion, Failure}	{S4_1}	{S5_1}
6. Turbine system_2 (power turbine subsystem)	1. PT bearing degradation (PTBD)	D6_1	{Normal, Degraded, Failure}	{S7_1, S7_2, S6_3, S6_5}	{S7_3, S6_5, S6_6}
	2. PT thrust bearing degradation (PTTBD)	D6_2	{Normal, Degraded, Failure}	{S7_1, S7_2, S6_3, S6_4}	{S6_4, S7_3, S6_7}
	3. PT rotor degradation (PTRD)	D6_3	{Normal, Unbalanced, Unaligned, Failure}	{S6_3}	{S6_5}
	4. PT blade degradation (PTBD)	D6_4	{Normal, Thermal distortion, Corrosion, Failure}	{S4_1}	{S6_3}

(*Continued*)

Table A.1. (*Continued*)

Subsystem	Dynamic nodes	Node ID	State space	Parent nodes	Children nodes
7. Lubricating oil system	1. Oil tube degradation (OTD)	D7_1	{Normal, Fouling, Leakage, Failure}	{S7_1}	{S7_1, S7_3}
	2. Oil cooler degradation (OCD)	D7_2	{Normal, Degraded, Failure}	∅	{S7_2,}
	3. Oil filter degradation (OFD)	D7_3	{Normal, Fouling, Blocked}	{S7_4}	{S7_1, S7_4}
	4. Oil pump degradation (OPD)	D7_4	{Normal, Degraded, Failure}	∅	{S7_3, S7_1}
	5. Oil heater degradation (OHD)	D7_5	{Normal, Degraded, Failure}	∅	{S7_5}

Table A.2. Information of static nodes in DBN model.

Subsystem	Static nodes	Node ID	State space	Normal interval	Classification critiria	Parent nodes	Children nodes
1. Air system	1. Differential pressure of inlet filter (kPaD)	S1_1	{Normal; On the high side; Superhigh}	< 0.75	$\{< 0.75; [0.75,1.5); \geq 1.5\}$	{D1_1}	{S2_1, S2_2}
	2. Cabinet pressure (kPa)	S1_2	{Normal; On the low side; Ultra low}	> 0.06227	$\{> 0.06227; (0.03736, 0.06227); \leq 0.03736\}$	{D1_1, D1_2}	∅
	3. Cabinet temperature (°C)	S1_3	{Normal; On the high side}	< 75	$\{< 75; \geq 75\}$	{E1_1, D1_2}	{D2_1, D5_1, D4_1, D4_2, D6_1}
	4. T1 temperature (°C)	S1_4	{Normal; On the low side}	> 5	$\{> 5; \leq 5\}$	{E1_1}	{S6_2}
	5. Inlet differential pressure (kPa)	S1_5	{Normal; On the high side}	< 1	$\{< 1; \geq 1\}$	{D1_1}	{S2_2}
2. Axial compressor system	1. PCD (kPa)	S2_1	{Normal; On the low side}	> 551	$\{> 551; \leq 551\}$	{S1_1, D2_1, S2_2}	{S4_1, S5_1, S6_2, S6_3}

(*Continued*)

Table A.2. (*Continued*)

Subsystem	Static nodes	Node ID	State space	Normal interval	Classification critiria	Parent nodes	Children nodes
	2. Serge (Bool)	S2_2	{Normal, Surge}	False	{False; True}	{D2_1, S1_1, S1_5}	{S2_1}
3. Fuel system	1. Fuel flow (kNm^3/h)	S3_1	{Normal; On the low side; On the high side}	[1900, 2400]	{[1900, 2400]; $<$ 1900; $>$ 2400}	{E2_3, D3_1, D3_2}	{S4_1}
	2. Fuel pressure (kPa)	S3_2	{Normal; On the low side; On the high side}	[2000, 3450]	{[2000, 3450]; $<$ 2000; $>$ 3450}	{D3_1, E2_2}	∅
	3. Fuel temperature (°C)	S3_3	{Normal; On the low side; On the high side; Superhigh}	[38, 85]	{[38, 85]; $<$ 38; (85, 96); $\geq$ 96}	{E2_1}	{S6_2}
4. Combustion system	1. T5 temperature (°C)	S4_1	{Normal; On the low side; On the high side}	(500, 760)	{(500, 760); $\leq$ 500; $\geq$ 760}	{S2_1, S3_1, D4_1, D4_2}	{S5_1, D5_4, S6_2, S6_3}
5. Turbine system_1 (gas turbine subsystem)	1. NGP (%)	S5_1	{Normal; On the low side; On the high side; Superhigh}	(80, 100) (100% = 11220 rpm)	{(80, 100); $\leq$ 80; [100, 102.5); $\geq$ 102.5}	{S2_1, S4_1, D5_4}	{S5_2, S5_3, S6_1, D5_1, D5_2, D5_3}

2. Axial displacement (mm)	S5_2	{Normal; On the high side; Superhigh}	(−0.508, 0.102)	{(−0.508, 0.102); [0.102, 0.178) or (−0.584, −0.508]; ≥ 0.178 or ≤ 0.584}	{S5_1, D5_2}	{D5_2}
3. Radial vibration (μm)	S5_3	{Normal; On the high side; Superhigh}	< 63.5	{< 63.5; [63.5, 101.6); ≥ 101.6}	{D5_1, D5_3, S5_1}	{D5_1}
4. 1#2#3# bearing oil return temperature (°C)	S5_4	{Normal; On the high side; Superhigh}	1#< 75 2#, 3# < 111	1#: {< 75; [75, 85); ≥ 85} 2#, 3#: {< 111; [111, 121); ≥ 121}	{D5_1, S7_2}	∅
5. Thrust bearing oil Return temperature (°C)	S5_5	{Normal; On the high side; Superhigh}	< 110	{< 110; [110, 121); ≥ 121}	{D5_2, S7_2}	∅

(Continued)

Table A.2. (*Continued*)

Subsystem	Static nodes	Node ID	State space	Normal interval	Classification critiria	Parent nodes	Children nodes
6. Turbine system_2 (power turbine subsystem)	1. Output power (kW)	S6_1	{Normal; On the low side}	> 5000	{> 5000; ≤ 5000}	{S6_3, S5_1}	∅
	2. Efficiency (%)	S6_2	{Normal; On the low side}	> 0.24	{> 0.24; ≤ 0.24}	{S1_4, S2_1, S3_3, S4_1}	∅
	3. NPT (%)	S6_3	{Normal; On the low side; On the high side; Superhigh}	(80, 93.7) (100% = 8856 rpm)	{(80, 93.7); ≤ 80; [93.7, 98.4); ≥ 98.4}	{S2_1, S4_1, D6_4}	{D6_1, D6_2, D6_3, S6_1, S6_4, S6_5}
	4. Axial displacement (mm)	S6_4	{Normal; On the high side; Superhigh}	(−0.508, 0.102)	{(−0.508, 0.102); [0.102, 0.178) or (−0.584, −0.508]; ≥ 0.178 or ≤ 0.584}	{D6_2, S6_3}	{D6_2}
	5. Radial vibration (μm)	S6_5	{Normal; On the high side; Superhigh}	< 63.5	{< 63.5; [63.5, 101.6); ≥ 101.6}	{D6_1, D6_3, S6_3}	{D6_1}

	6. 4#, 5# bearing oil return temperature (°C)	S6_6	{Normal; On the high side; Superhigh}	< 75	{< 75; [75, 85); $\geq$ 85}	{D6_1, S7_2}	∅
	7. Thrust bearing oil return temperature (°C)	S6_7	{Normal; On the high side; Superhigh}	< 110	{< 110; [110, 121); $\geq$ 121}	{D6_2, S7_2}	∅
7. Lubricating oil system	1. Oil supply pressure (kPa)	S7_1	{Normal; On the low side; Ultra low; On the high side}	(210, 449)	{(210, 449); (173, 210]; $\leq$ 173; $\geq$ 449}	{D7_1, D7_3, D7_4, S7_2, S7_3}	{D7_1, D5_1, D5_2, D6_1, D6_2}
	2. Oil supply temperature (°C)	S7_2	{Normal; On the low side; On the high side; Superhigh}	(35, 55)	{(35, 55); $\leq$ 35; [55, 74); $\geq$ 74}	{D7_2, S7_5}	{S7_1, S5_5, S5_4, S6_6, S6_7, D6_1, D6_2, D5_1, D5_2}

(*Continued*)

Table A.2. (*Continued*)

Subsystem	Static nodes	Node ID	State space	Normal interval	Classification critiria	Parent nodes	Children nodes
	3. Oil level (cm)	S7_3	{Normal; On the low side; Ultra low; On the high side}	(48.3, 55.9)	{(48.3, 55.9); (40.6, 48.3]; $\leq$ 40.6; $\geq$ 55.9}	{D5_1, D5_2, D6_1, D6_2, D7_1, D7_4}	{S7_1}
	4. Differential pressure of oil filter (kPa)	S7_4	{Normal; On the high side}	< 207	{< 207; $\geq$ 207}	{D7_3}	{S7_3}
	5. Oil tank temperature (°C)	S7_5	{Normal; On the low side; On the high side; Superhigh}	(21, 68)	{(21, 68); $\leq$ 21; [68, 74); $\geq$ 74}	{D7_5, E1_1}	{S7_2}
8. Exogenous environment (or external system)	Environment temperature (°C)	E1_1	{Normal; On the low side; On the high side}	(0, 40)	{(0, 30); $\leq$ 0; $\geq$ 40}	∅	{S7_5}

Fuel supply temperature (°C)	E2_1	{Normal; On the low side; On the high side}	Adjustable	According to manual setting based on supply and demand	∅	{S3_3}
Fuel supply pressure (kPa)	E2_2	{Normal; On the low side; On the high side}	Adjustable	According to manual setting based on supply and demand	∅	{S3_2}
Fuel supply flow (kNm^3/h)	E2_3	{Normal; On the low side; On the high side}	Adjustable	According to manual setting based on supply and demand	∅	{S3_1}

Chapter 7

An Intelligent Fault Diagnosis System for Process Plant Using a Functional HAZOP and DBN Integrated Methodology

Integration of a functional hazard operability (HAZOP) approach with dynamic Bayesian network (DBN) reasoning is presented in this chapter. The presented methodology can unveil early deviations in the fault causal chain online. A functional HAZOP study is carried out firstly where a functional plant model (i.e., multilevel flow modeling [MFM]) assisted in the goal-oriented decomposition of the plant's purpose into the means of achieving that purpose. DBN model is then developed based on the functional HAZOP results to provide a probability-based knowledge representation which is appropriate for the modeling of causal processes with uncertainty. An intelligent fault diagnosis system (IFDS) is proposed based on the whole integrated framework, and investigated in a case study of process plants at a petrochemical corporation. The study shows that the IFDS provides a very efficient paradigm for facilitating HAZOP studies and for enabling the reasoning to reveal potential causes and/or consequences far away from the site of the deviation online.

7.1. Introduction

Modern technological advances are creating a rapidly increasing number of complex engineering systems, processes, and products. It is their scale, nonlinearities, interconnectedness, and interactions with humans and the environment that can make these complex process plant systems fragile, when the cumulative effects of multiple abnormalities can propagate in numerous ways to cause systemic failures. One of the main reasons behind accidents is that it is often too late to correct the problems by the time they are detected. Given the size, scope, and complexity of the systems and interactions, it is becoming difficult for plant personnel to anticipate, diagnose, and control serious abnormal events in a timely manner. In a large process plant, there may be as many as 1500 process variables observed every few seconds leading to information overload. Furthermore, the measurements may be insufficient, incomplete, and/or unreliable due to a variety of causes such as sensor biases or failures [28]. Usually monitoring systems such as distributed control systems (DCS) have no "understanding" of the actions required for changes in the process state, nor of actions that an operator takes to correct the state. This often leads to alarms that many cases are inappropriate, and require interpretation from the operator.

Hazard studies provide a systematic methodology for identification, evaluation, and mitigation of potential process hazards which can cause severe human, environmental, and economic losses. However, there exist nonlinear interactions among a large number of interdependent components and the environment. The nonlinear interactions can be further compounded by human errors, equipment failures, and dysfunctional interactions among components and subsystems, that make accident scenarios diversified, random, and can also lead to "emergent" behavior [18].

There exist considerable incentives in developing appropriate diagnostic methodologies for monitoring, analyzing, interpreting, and controlling such abnormal events in complex process plant systems. Effective diagnosis of the fault causes and prediction of their consequence can reduce the investigation time of abnormal events and improve the effectiveness of accident prevention. Diagnosis methods for process system can be mainly divided into two categories: model-based diagnosis methods [25, 26] and historical data-based diagnosis methods [27, 29]. Data-based methods are usually used to detect abnormal events and set off alarms, but they are unable to reveal the underlying causes which is of capital importance for field operators. Whereas in order to present the cause–consequence relationship in a complex process plant, various models have been put forward to identify potential hazard sometimes far away from the alarming position, such as signed directed graph (SDG) [4, 9, 19], Petri network [2], Fault semantic network (FSN) [6] and a variety of methods are integrated for some typical complex systems [13, 32].

The main advantages of model-based approaches consist in the causal models which capture more deep-level knowledge than a data-based method [14]. In general, models for the analysis of processes have been derived from expert or operator knowledge of the process or from known model equations that define the behavior of the system. The cornerstone of the above modeling is fault propagation analysis [7]. Yuan [31] indicated that fault propagation and its cause–effect relationship in the system were of priority for the fault diagnosis. Among qualitative reasoning methods, hazard and operability (HAZOP) analysis is the preferred approach in the chemical process industry. HAZOP is a structured and systematic examination of a process operation so as to identify and evaluate the existing or impending problems [3]. A typical HAZOP provides an identification of accidental events (top events, TEs) and operability problems by using logical sequences of cause–deviation–consequence of process parameters. Such method is usually used offline but can be helpful for the design of online FDI algorithms by identifying critical components to be monitored. Therefore, the integration of model-based approaches and HAZOP analysis is of great interest as an interesting solution for fault diagnosis [22, 24].

Unfortunately, in spite of the abundant representation of specialized knowledge and expertise during the HAZOP study, it is not possible to develop a systematic way to fully study all the fault propagation behavior. Some of the weaknesses that were addressed relate to the coupling of vulnerabilities of the method with the human limitations of practitioners; causes of deviations and the identification of initiating events [3]. Rodriguez presented the use of D-higraphs to perform HAZOP studies for fault propagation analysis [20]. Rossing [21] presented a HAZOP methodology where a functional plant model assisted in the goal-oriented decomposition of the plant's purpose into the means of achieving that purpose. This approach led to nodes with simple functions from which the selection of process and deviation variables followed directly. The method provided a good way for implementation into a computer-aided reasoning tool to perform root cause and consequence analysis. However, the rule-based reasoning in functional model may also be combined with case-based reasoning techniques [23]. Another reason that limits the functional model in online diagnosis for a real industrial plant lies in its qualitative reasoning capability rather than in the quantitative way. That is, it does not lend itself to quantitative analysis, to rank the effects of failures, and to study the relative effectiveness of the proposed corrective actions [8].

Therefore, some disadvantages of above qualitative reasoning consist in the poor capability to handle uncertainties in the cause–effect structure, limited representation of observable node states, and only diagnosis of single faults is possible [17]. Probabilistic graphical models are highly advantageous for analyzing the cause–effect relationship with uncertainty. The probabilistic graphical model consists of a graphical structure and a probabilistic description of the relationships among random variables under system uncertainty. Bayesian network (Br) is one of the major classes of graphical models and has been applied to various fields. BNs have been employed to identify the root cause of process variations and give a probabilistic confidence level of the diagnosis [1, 10, 15, 30]. Nevertheless, for BN-based process monitoring techniques, potential root causes need to be specified and added to hidden nodes in advance. The biggest problem with the application of BN-based methods is that they require the in-depth process knowledge to design the network structure for well-performed process diagnosis. In addition, it can be time-consuming to build precise a graphical model for complex processes, and it is also challenging to check the accuracy of the inferred structure.

It is well known that, in general, no single method is sufficient for a wide range of problem-solving tasks. This chapter presents a functional approach integrated with HAZOP study for hazard analysis and develops a functional model as a basis for dynamic Bayesian network (DBN) reasoning on causes and consequences of deviations monitored by the condition monitoring system. In this chapter, the use of multilevel flow modeling (MFM) is proposed as a technique to obtain a representation of technical processes suitable for reasoning about goal-oriented actions in complex and heterogeneous processes. While reasoning based on a DBN model representing hazard cause–effect relationship to handle uncertainty should enable the construction of intelligent aids for the operator, it can function as a better assistant for presently used DCS-based monitoring systems. The resulting tool is called intelligent fault diagnosis system (IFDS). By using the reasoning system implemented with the inherent DBN reasoning scheme, the most possible initial reason(s) when observable deviations are detected by the condition monitoring system can be found out accurately, and also the future possible consequences can be predicted in a timely manner for proactive maintenance or emergency decision making.

Section 7.2 presents a functional HAZOP study based on the qualitative MFM model of the process system with a few examples applied on an FCCU plant. Section 7.3 presents an IFDS based on the whole functional HAZOP and DBN integrated framework. In this section, after a brief presentation of basic DBN theory and its interest for quantitative causal reasoning, we introduce how the functional HAZOP results are transformed to a DBN model for abnormal event identification and fault cause online diagnosis. The developed methodology is applied in Sec. 7.4 for online diagnosis of an FCCU process. Finally, Sec. 7.5 concludes the work.

7.2. MFM Modeling and Functional HAZOP Study

In this chapter, multilevel flow modeling (MFM) as one of the main functional modeling methods is used to represent the knowledge of plant functions. MFM combines the means-end dimension with the whole-part dimension, to describe the functions of the process under study and enable modeling at different abstraction levels. MFM is a modeling methodology which has been developed to support functional modeling of process plants involving interactions between material, energy, and information flows [12]. Along the means-end dimension, MFM represents a system in terms of goals, objectives, functions, and components each of which can be described at different levels of part-whole decomposition (see Fig. 7.1). This means

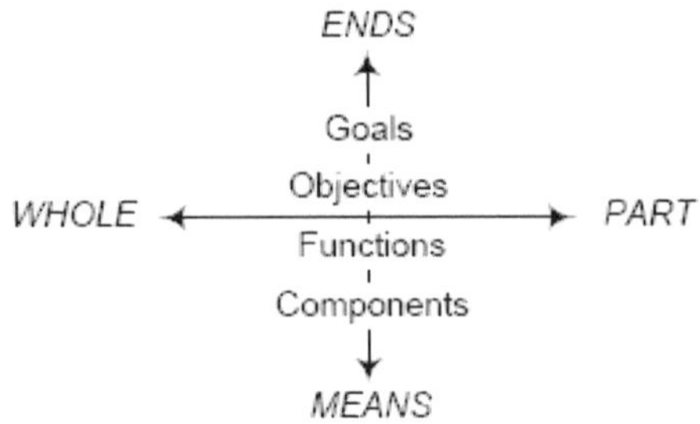

Fig. 7.1. Means-ends and part-whole dimensions in MFM.

that an MFM model consists of chunks of interrelated means-end structures, each associated with goals belonging to different levels of decomposition. Top-level goals typically reflect external requirements for the overall system, whereas subgoals express requirements for proper function of a specific system part, which has to be met by the function of another system part (not necessarily at a lower level). Because of the relatively detailed functional description of MFM, such dependencies can be explicitly represented by specific relation types. Such relations tie together the individual chunks of means-end knowledge. The level of function is represented by mass and energy flow structures. The flow structures consist of interconnected flow functions (Fig. 7.2), which represent the purposeful behavior of specific components in view of an overall flow structure achieving specific objectives. Therefore, functions are here represented by elementary flow functions interconnected to form flow structures representing a particular goal-oriented view of the system.

The use of flow concepts provides a uniform representation of the intended system behavior at multiple levels of plant aggregation, abstracted from the actual physical implementation. MFM is accordingly a plant model which can serve as the central core of an intelligent system for HAZOP analysis [21]. It provides a plant representation which is more generic and less dependent on knowledge about specific situations or incidents compared to, e.g., fault trees. The effort in acquisition of plant knowledge for the HAZOP analysis is therefore greatly reduced compared to methods based on fault trees. MFM provides a formalization of means-end

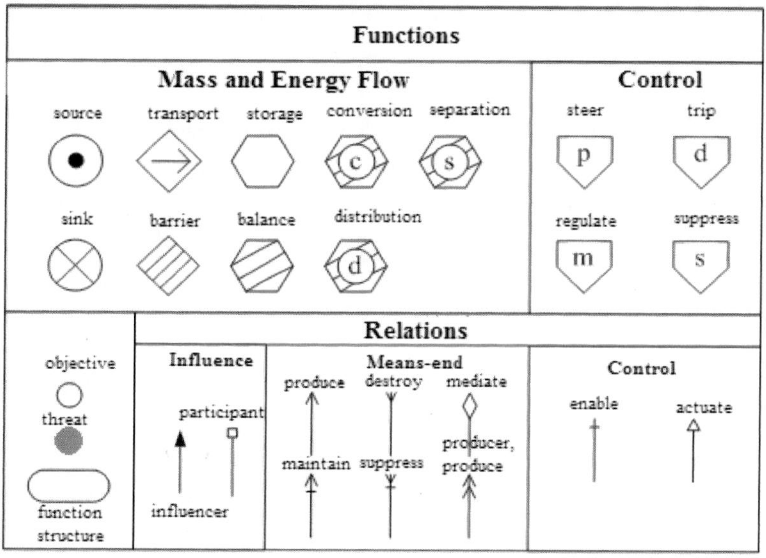

Fig. 7.2. The basic MFM symbols.

concepts which play a fundamental role in HAZOP when reasoning about causes, consequences, and counteraction plans. The approach reduces the work involved in the HAZOP of a plant by dividing the plant along functional lines and analyzing nodes with the same function once only.

7.2.1. *Traditional HAZOP study*

Hazard and operability (HAZOP) analysis is widely accepted as the method for conducting the Preliminary Hazard Analysis (PHA) analysis in the process industry It is used to identify design defects and the hazard issues as well which prevent efficient operation. The P&ID is divided into sections or nodes and then each section is studied applying an algorithm. Usually, nodes are equipment items. Once a node is chosen, each line of the node is analyzed applying certain deviations. These deviations result from the combination of a "guide word" with a "property" of the line, i.e., "GuideWord + Property = Deviation". GuideWords are described in IEC61882, which include but not limited to NONE, MORE OF, LESS OF, PART OF, MORE

THAN, OTHER THAN, REVERSE, etc. Properties or parameters are flow, temperature, composition, etc.

HAZOP analysis is also a thorough bidirectional analysis method. In each analysis step, bidirectional analysis is used to identify hazard scenarios (accident chains) consisting of each deviation's causes and consequences. Reasons which could cause the abnormal event are identified by reverse analysis and the negative consequences are also predicted by forward analysis along accident chains according to each deviation of process parameters.

Although HAZOP studies are easy to learn, reusable, and systematic, it is a procedure that consumes a lot of time and effort, and the results may have a lot of redundant or incomplete items which make the report lack accuracy.

7.2.2. *Functional HAZOP study*

Rossing [21] proposed a functional HAZOP assistant which was divided into three phases corresponding to the traditional HAZOP. The first phase corresponds to the pre-meeting phase, the second phase to the meeting phase, and the third phase to the post-meeting phase. Here, in this chapter, only Phase 1 is used for fault propagation analysis, while the diagnosis on the MFM workbench as Phase 2 in Rossing's work will be replaced by the DBN model for quantitative reasoning and online diagnosis which will be mentioned in Sec. 5.3.

Thus, the functional HAZOP for intelligent fault diagnosis involves the following steps (also illustrated in Fig. 7.3):

- **Phase 1: MFM model development**
  (1) Comprehensively analyze the production process and fully know the system prototype. Fully understand the physical system and if needed, collect supporting documents from the design plan (e.g., P&ID diagram, operation manual) or organize a pre-meeting made up of 3–4 experts to understand the process.
  (2) Problem definition and goal decomposition. Define the modeling system scope and identify the abstraction level. Decompose the overall goal into a series of subgoals.

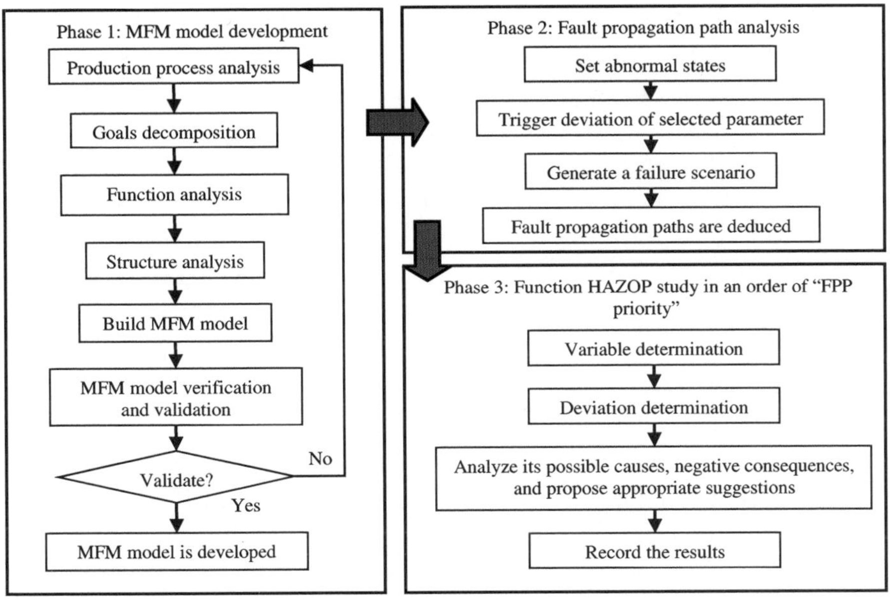

Fig. 7.3. Procedure for functional HAZOP study.

(3) Analyze and study the function characteristics of each part of the system which aim at realizing the goals and list all conditions and limitations of function realization. Select the function symbols in Fig. 7.2 to abstract the function flows represented by mass flow, energy flow, and control flow.

(4) Find equipment components and its corresponding functional mapping relation. In order to prevent unexisting function representation errors, check whether there is a corresponding structure to support the required function. In this step, find out the mapping relationship between function and structure.

(5) Establish the multilevel flow models based on relation among goals, functions, and components. Connect different abstraction mass or energy flow structures by means-end relations selected from Fig. 7.2 and label means-end relations with corresponding main function names in each flow structure. Connect the control flow structure and mass or energy flow structure by control relations selected from Fig. 7.2 and label

the relations with corresponding flow function names that are controlled or manipulated variables.

(6) Model verification and validation.

- **Phase 2: Fault propagation path analysis**
Since there are strong correlations among process units, deviation of a certain parameter from normal operating condition may further spread to affect other units, or evolve to reach a more serious negative state which will eventually lead to accidents. Due to the propagation characteristics of faults in complex process system, the final accident may happen far away from the initial cause, which makes the fault diagnosis quite difficult. Set abnormal state of any function to trigger deviation of the selected parameter so as to generate a failure scenario. A schematic diagram of the fault propagation path (FPP) can be described as root cause (initial event) → indirect cause → direct cause → variable deviations → deviation spread → alarming → accidents.
MFM is used to abstract the process system into various flow structures, in order to simplify the complex process. Therefore, with the help of MFM, the propagation path of a series of fault in a hazard scenario can be analyzed clearly (see Fig. 7.8). The FPPs developed and used for HAZOP have the following advantages:

(1) HAZOP study will be carried out along each FPP, so the nodes on FPP will be analyzed to reveal corresponding deviations, and their possible reasons and consequences with suggested appropriate safety measures. In this way, the previous nodes' possible consequences may have relation to the subsequent nodes' possible reasons. Therefore, along the FPP, hazard scenarios can be developed successfully, which are very important for the work of the fault cause reasoning and consequence prediction in the IFDS. At the same time, the redundancy of the traditional HAZOP results can be reduced.
(2) The proposed functional HAZOP, which focuses on the fault propagation concentrating main manpower and economic resources to solve the critical safety problems, can imply a

reduction in financial costs and reduce the probability of man-made errors or omissions.

(3) The systematic approach is able to find causes far from the node of the current deviation. Hence, it provides a convenient way to build a DBN model for root cause reasoning and consequence prediction which will be elaborated in Sec. 5.3. It also provides great help for the development of a computer-aided intelligent diagnosis system including both hardware and software.

- **Phase 3: function HAZOP study in an order of 'FPP priority'**

1. For each type of the node on an FPP in a verified MFM model, i.e., each physical or chemical phenomenon, describe the process variable(s), which identifies the design intent or normal operation. For a node with the function, "gas transport", normal operation could be described by flow rate, temperature, pressure, and number of phases.
2. Select one guide word to combine with the parameter as a deviation. For each process variable, specify the relevant deviations. For flow rates, typical deviations are qualitative, e.g., more, less, and reverse.
3. Systematically question every node on the FPP to discover how deviations from the design intention can occur. Decide as to whether these deviations can give rise to hazards. To those meaningful deviations, analyze their possible causes, negative consequences after their occurrence, and propose appropriate suggestions.
4. Repeat step 3, till all the possible deviations of this parameter have been analyzed.
5. Select the next parameter. Repeat steps 2–4, till all the possible deviations of all parameters of this node have occurred.
6. Repeat steps 2–5, till all the nodes of the FPP have been analyzed.
7. Develop hazard scenarios. Along the selected FPP, the previous nodes' possible consequences should be linked to the subsequent nodes' possible reasons. Analyzing node by

node along the FPP, the hazard scenarios can be developed successfully, which will be used to develop a DBN model for fault diagnosis reasoning.

8. Select the next FPP in the MFM of the underlining plant unit, repeat steps 1–7, till all the FPPs in the MFM of the given process have been analyzed.
9. Carryout risk evaluation according to each hazard scenario (Optional choice). Check whether the risk exceeds the threshold value. Define measures to mitigate the risk.
10. Repeat above step until all hazards are mitigated to a risk value below the threshold.
11. Record the results. Follow up action items.

Using MFM to model the plant all that is needed is a basic understanding of chemical unit operations, their purposes, and the fundamentals on which these purposes are built, i.e., transport phenomena, thermodynamics, and kinetics. This means that functional HAZOP study may be efficiently performed by less experienced personnel [21]. Meanwhile, the functional HAZOP enables a thorough analysis not only within the single nodes but also between nodes and sections and thereby facilitates the revealation of more complex causes of deviations than possible using the traditional approach. It may also be utilized for consequence analysis. Therefore, the benefit of the functional HAZOP analysis is that it provides in-depth process knowledge to design the DBN structure for well-performed process diagnosis. In addition, the challenge to check the accuracy of the inferred structure can be guaranteed since it is based on MFM which is developed by a systematic and scientific way and is also verified and validated.

The following sections take fluidized catalytic cracking unit (FCCU) as an example, which is widely used in the petrochemical process, and illustrate the implementation of the above steps of the functional HAZOP study.

7.2.3. *Phase 1: MFM modeling of FCCU*

FCCU is a typical refinery system, including reaction regeneration unit, fractionation unit, and absorption–stabilization unit. The MFM model of the reaction regeneration unit is illustrated as an example, which will be further used in the case study of the proposed IFDS. According to the method of MFM modeling, the MFM models of the typical parts in FCCU are developed and shown in Table 7.1.

7.2.3.1. *Analysis of the reaction–regeneration process*

From a system point of view, FCCU reactor–regenerator can be considered as an interconnected and complex system. The main process of the reaction–regeneration system is shown in Fig. 7.4. The preheated hot feedstock from atmospheric distillation unit is mixed with recycled oil from the fractionator bottom. The mixture is vaporized once in contact with the hot regenerated catalyst from the regenerator and injected into the bottom of the riser reactor. Catalytic cracking reactions occur in the lower area of the riser. Spent catalyst is discharged into the bottom of the reactor after separation from oil vapor, then enters the stripping section to be further stripped from product vapors by hot steam. The flow of catalyst from reactor to the regenerator is controlled by a slide valve presented in the standpipe connecting the regenerator and the reactor. This slide valve is used to control the catalyst bed level in the reactor. Regenerated catalyst flows to the riser through a slide valve located in the standpipe connecting the regenerator and the riser. This valve is used to control the heat supply to the riser by regulating the catalyst flow rate to maintain the riser temperature at a desired level and therefore the product composition.

7.2.3.2. *Target decomposition of the reaction–regeneration unit*

The general objective of the unit is to complete the catalytic cracking reaction of the heavy oil. It can be decomposed as two subgoals.

Subgoal I: Oil and gas reaction. It is accomplished by heating furnace, reactor, settlement, etc.

Subgoal II: Ensure the supply of catalyst. It is mainly completed by the regenerator and its attachment.

Table 7.1. The MFM models of some typical parts in FCCU.

Typical parts	MFM representation	Explanation	Typical parts	MFM representation	Explanation
Furnace	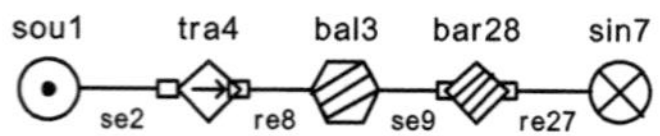	Heating energy flow	Separator	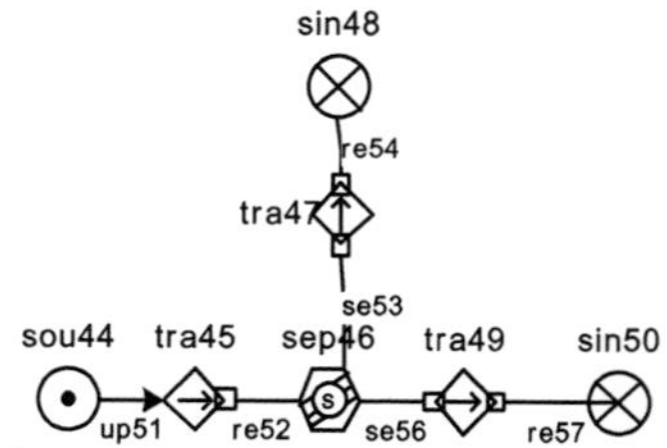	Separation of different mass or energy flows
Pump	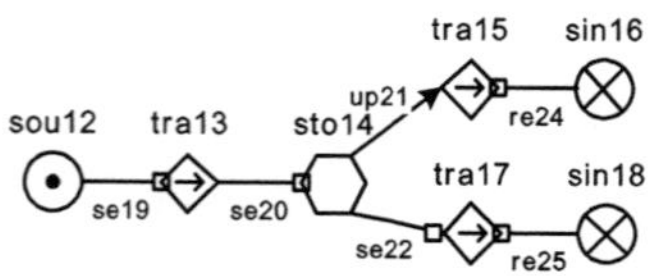	Pump power energy flow	Reactor/ Tower	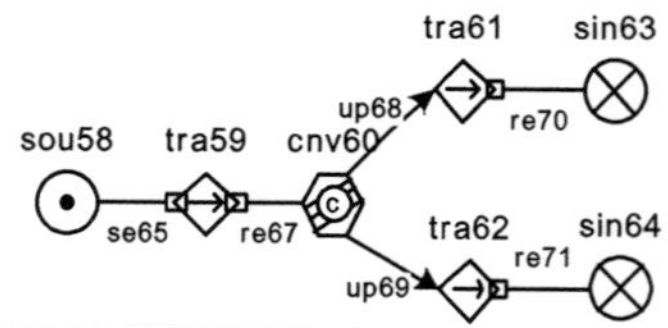	Convert more than one form of mass or energy flow
Pipeline	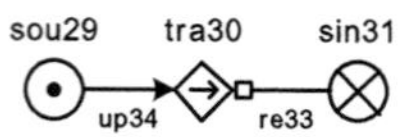	Transmission energy or mass flow	Vessel/ Boiler	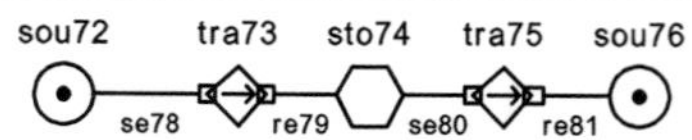	Accumulation of mass or energy flows
Valve	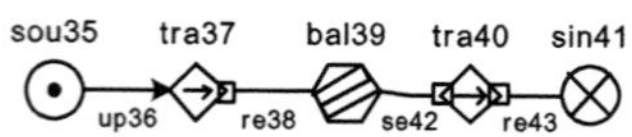	Keep the balance of flow			

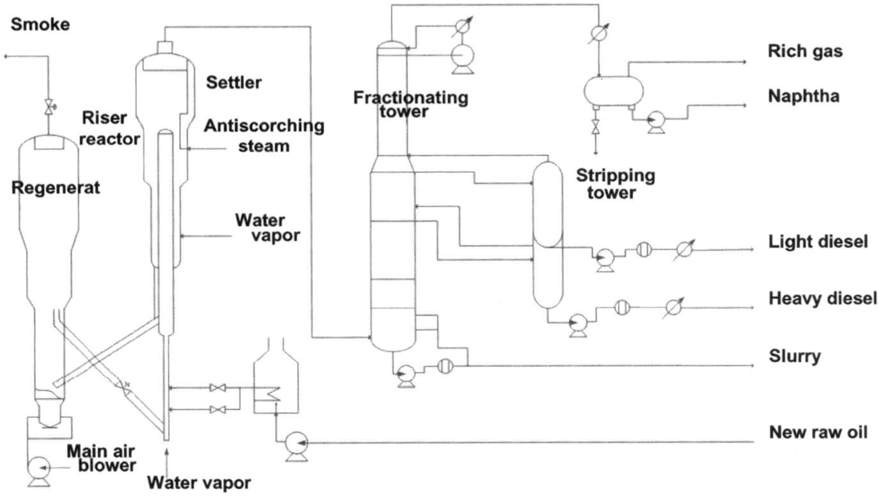

Fig. 7.4. Detailed process of the reaction–regeneration system and fractionation unit of FCCU.

Subgoal II is the condition of Subgoal I, i.e., in order to ensure the oil and gas reaction normally, the normal circulation of catalyst should be guaranteed (Fig. 7.5).

7.2.3.3. *Analysis of the main components and functions of the regeneration–reaction*

Reactor: The riser reactor provides enough contact space for the catalyst and the raw oil. The fast separator and the cyclone separator are built in the settlement, both of which are used for the reaction of oil and gas and the gas–solid separation of the catalyst. The riser reactor can be regarded as the transfer function in the MFM model, and the settlement as well as its internal structures can be regarded as the balance function. The MFM model of the reaction regeneration unit is developed as shown in Fig. 7.4, and the meaning of the involved symbols is shown in Table 7.2.

In the same way, the MFM models of the fractionation unit and the absorption–stabilization unit are developed and shown in Figs. 7.6 and 7.7, respectively. The meaning of the involved symbols in each model is shown in Tables 7.3 and 7.4.

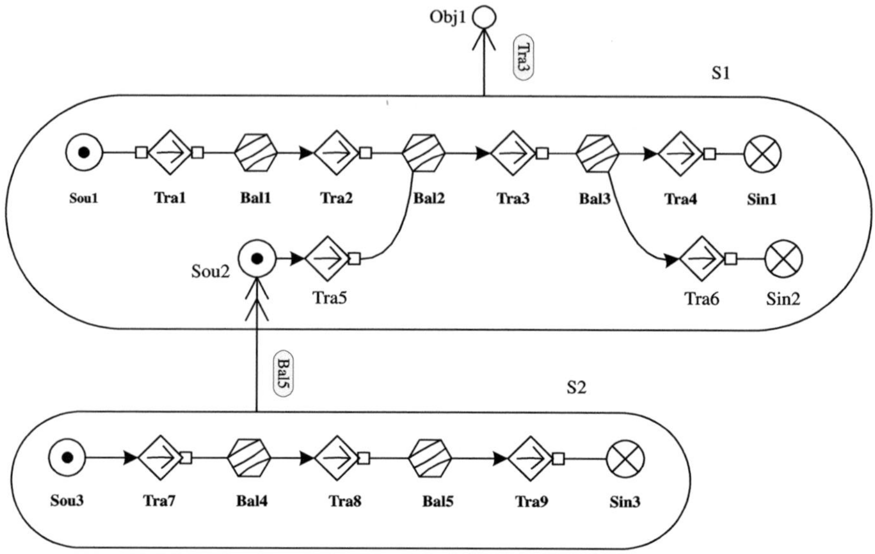

Fig. 7.5. The MFM model of the reaction–regeneration unit.

7.2.4. *Phase 2: MFM-based FPP analysis*

According to the MFM model of the reaction–regeneration unit (see Fig. 7.4), taking the functional node "Tra5" losing its normal transfer function as an example (i.e., regenerative inclined tube may be plugged), the FPP can be analyzed as in Fig. 7.8, where the solid line represents the forward propagation path, and the dashed line represents the reverse path. Therefore, the overall propagation path from the initial deviation or event to the final accident can be revealed as in Fig. 7.9.

7.2.5. *Phase 3: Functional HAZOP study results of FCCU*

According to the MFM model (Fig. 7.4), the FPP (as an example in Fig. 7.8) is further analyzed in this section for the final functional HAZOP study, in order to reveal the fault propagation mechanisms. In this case study, the riser reactor is selected as an example to demonstrate the process and result of the functional HAZOP analysis.

Table 7.2. Meaning of the symbols in the MFM model of the reaction regeneration unit.

Symbol	Meaning	Symbol	Meaning	Symbol	Meaning
S1	Flow structure of the oil and gas reaction	Tra1	Raw oil pump	Bal1	Heating furnace
S2	Flow structure of the catalyst circulation	Tra2	Feeding tube	Bal2	Riser Intersection
Obj1	Oil and gas reaction	Tra3	Riser reactor	Bal3	Settler
Sou1	Raw oil input	Tra4	Pipe at the top of the settler	Bal4	Premixed tube
Sou2	Catalyst input	Tra5	Inclined tube for regenerated catalyst	Bal5	Regenerator
Sou3	Spent catalyst input	Tra6	Inclined tube for spent catalyst		
Sin1	Oil and gas flow to the fractionating tower	Tra7	Inclined tube for spent catalyst		
Sin2	Catalyst output	Tra8	Charring tank		
Sin3	Catalyst flows to reaction	Tra9	Inclined tube for regenerated catalyst		

(1) Determine the nodes in HAZOP analysis: The analytical nodes of the reaction regeneration unit are determined including raw oil feed pipeline (node 1), riser reactor (node 2), depressor (node 3), regenerator (node 4), air feed pipeline for regenerator (node 5), regenerated flue gas output pipeline (node 6), prestripping pipeline (node 7).

(2) According to the MFM and FPP of the reaction regeneration unit, the process parameters and corresponding guidewords are selected. Then possible causes of the deviation and possible consequences are analyzed, and the corresponding safety measures are put forward as shown in Table 7.5.

The parameter "regenerator reserve" is selected as an example, and the GuideWord "MORE" is used to match the parameter,

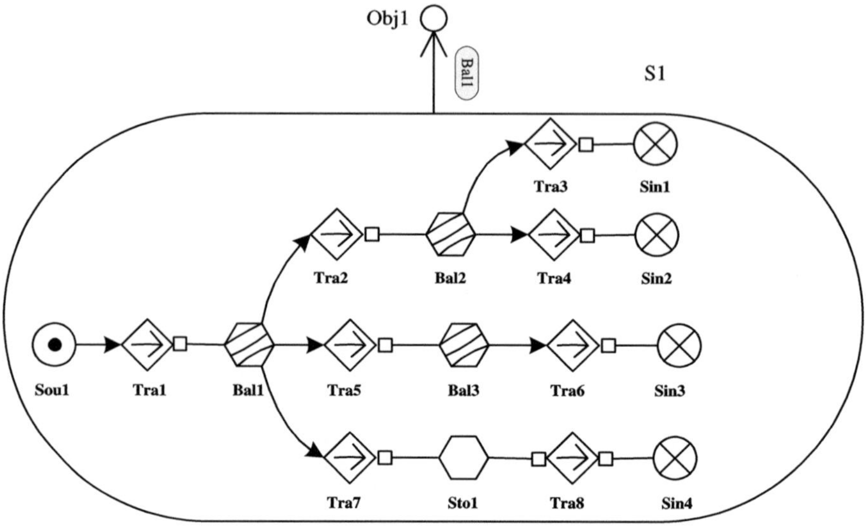

Fig. 7.6. The MFM model of the fractionation unit.

which constitute the deviation as "high regenerator reservoir". According to the developed FPP (Fig. 7.8), the reasons which may cause the deviation are analyzed as follows:

The flow of the inclined tube I (i.e., the inclined tube which is connected to the input of the regenerator) may get much higher than normal, which may be caused by the valve fault of the inclined tube I (i.e., overlarge opening degree) or the cyclic catalyst flow in the system may get higher than normal.

The flow of the inclined tube II (i.e., the inclined tube which is connected to the output of the regenerator) may get much lower than normal, which may be caused by the valve fault of the inclined tube II (i.e., undersize opening degree), or the inclined tube II blocking.

Some possible consequences which may be caused by the current deviation are analyzed as follows:

The flow of the inclined tube II will get much higher than normal, which will further cause the flow of the riser reactor, input flow of the depressor, and the flow of the inclined tube I gets much higher.

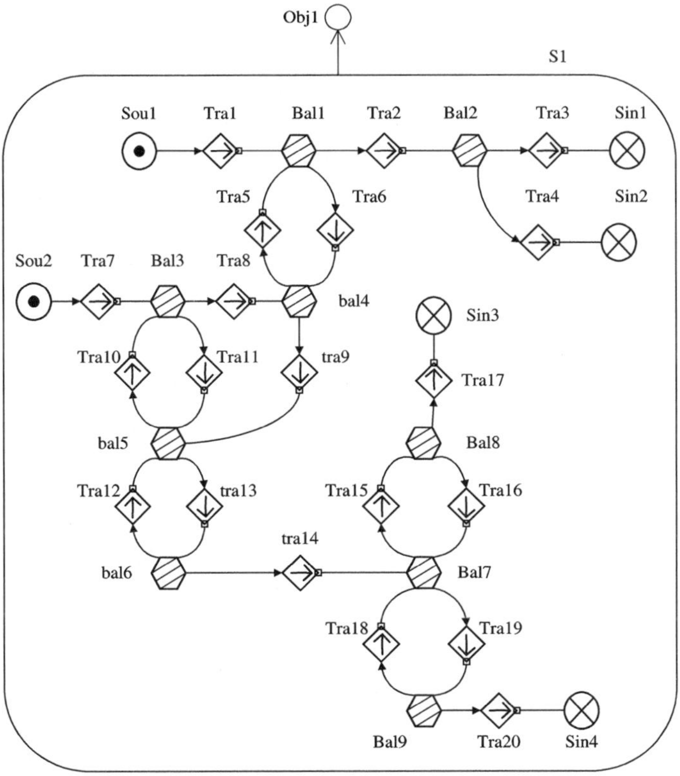

Fig. 7.7. The MFM model of the absorption–stabilization unit.

If the main booster fan is unable to meet the requirements of the catalyst regeneration, the effect of the catalyst regeneration will be poor.

Safety measures are further determined for maintenance, including adjusting the open degree of the valve, reducing the cyclic catalyst flow, etc.

In the same way, the deviations corresponding to the nodes upstream and downstream of the FPP of the regenerator can be further analyzed, such as the flow of the inclined tube I and the flow of the inclined tube II, etc. The reasons and consequences of the above deviations as well as corresponding safety measures are summarized and shown in Table 7.5 (only partial results are demonstrated as examples).

Table 7.3. Meaning of the symbol in the MFM model of the fractionation unit.

Symbol	Meaning	Symbol	Meaning	Symbol	Meaning
S1	Oil–gas separation	Tra1	Fractionator feed pipe	Bal1	Fractionating tower
Obj1	Implementation of oil and gas separation	Tra2	Vent pipe	Bal2	Oil and gas separator on the top of the tower
Sou1	Oil and gas for reaction input	Tra3	Rich gas output pipe	Bal3	Stripper
Sin1	Rich gas flowing to absorb stable device	Tra4	Naphtha pump	Sto1	Heat exchanger
Sin2	Crude gasoline output	Tra5	First middle discharging pipeline in the fractionating tower		
Sin3	Light diesel oil output	Tra6	Light diesel oil discharging pump		
Sin4	Heavy diesel oil output	Tra7	Second middle discharging pipeline in the fractionating tower		
		Tra8	Heavy diesel oil output pump		

(3) The development of hazard scenarios: According to above results, the possible reasons and consequences of each node can be linked and developed into several hazard scenarios. The hazard scenarios of the event "Regenerator reserve + MORE" are demonstrated as examples in Fig. 7.10.

7.3. Intelligent Fault Diagnosis System

7.3.1. *Dynamic Bayesian network*

After the functional HAZOP study, the next step is to develop a quantitative reasoning procedure online to reveal fault causes and

Table 7.4. Meaning of the symbol in the MFM model of the absorption–stabilization unit.

Symbol	Meaning	Symbol	Meaning
S1	absorption-stabilization	Tra1	Naphtha pipeline
Obj1	Separation of rich gas and naphtha	Tra2	Gas pipeline connected to the top of the absorption tower
Sou1	Naphtha (from fractionation)	Tra3	Dry gas pipeline connected to the top of the reabsorber
Sou2	Rich gas (from fractionation)	Tra4	Rich oil pipeline connected to the bottom of the reabsorber
Sin1	Dry gas desulfurization tank	Tra5	Airflow in absorption tower
Sin2	Rich oil storage tank from absorption	Tra6	Liquid flow in absorption tower
Sin3	Liquid hydrocarbon storage tank	Tra7	Rich oil pipeline
Sin4	Stabilized gasoline storage tank	Tra8	Non-condensable gas pipe
Bal1	Top of the absorption tower	Tra9	Pipe and pump connected to the bottom of the absorption tower
Bal2	Reabsorber	Tra10	Feed pipe and pump of the desorber
Bal3	Condensation oil settling tank	Tra11	Lean gas pipe connected to the top of the desorber
Bal4	The bottom of the absorption tower	Tra12	Boiling gas reflux pipe connected to the bottom of the desorber
Bal5	Desorber	Tra13	Outlet conduit connected to the bottom of the desorber
Bal6	Reboiler at the bottom of the desorber	Tra14	Inlet pump of the stabilizer
Bal7	Stabilizer	Tra15	Gas pipeline connected to the top of the stabilizer
Bal8	Reflux tank connected to the top of the stabilizer	Tra16	Reflux pump connected to the top of the stabilizer
Bal9	Reflux tank connected to the bottom of the stabilizer	Tra17	Liquid hydrocarbon output pipe
		Tra18	Outlet pipe of the reboiler at the bottom of the stabilizer
		Tra19	Outlet pipe of the bottom of the stabilizer
		Tra20	Stabilized gasoline pipeline

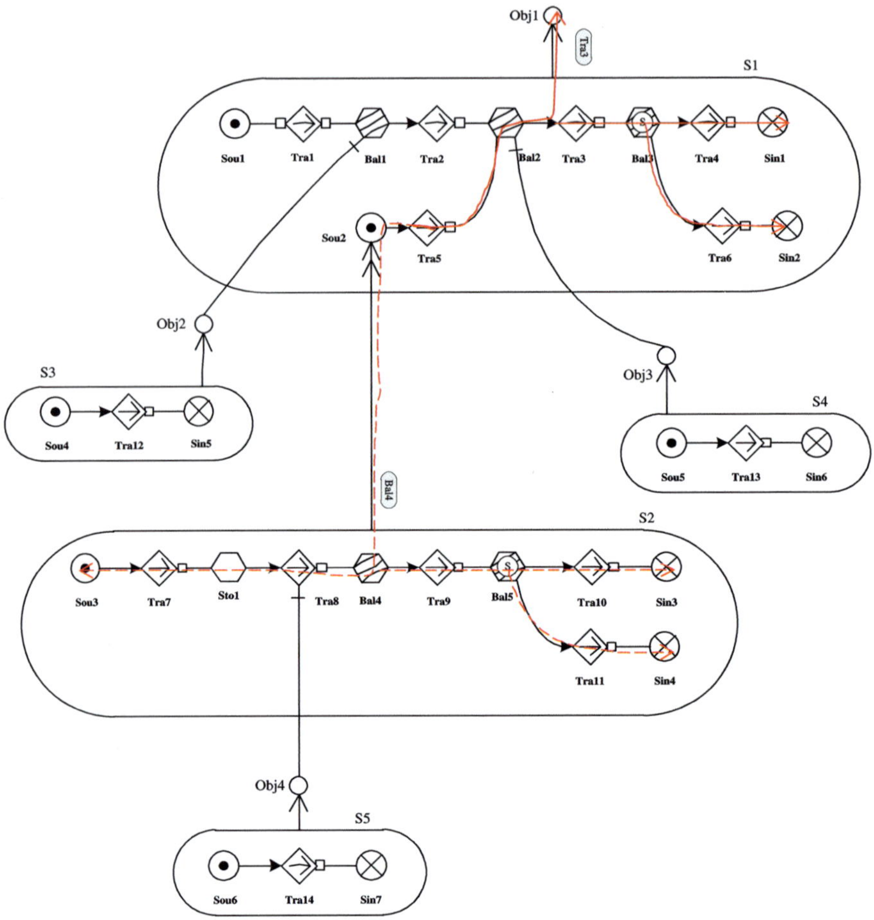

Fig. 7.8. FPP derived from MFM (Tra5 malfunction).

consequences on each level along with the input of the condition monitoring data. In this way, when deviations (or abnormal events) are detected by the condition monitoring system (e.g., DCS, MES, etc.), the reasons why deviations occur will be analyzed online automatically and the results provided to operators to find out safety countermeasures as soon as possible. Dynamic Bayesian network (DBN) is introduced to develop such quantitative reasoning procedure for fault diagnosis.

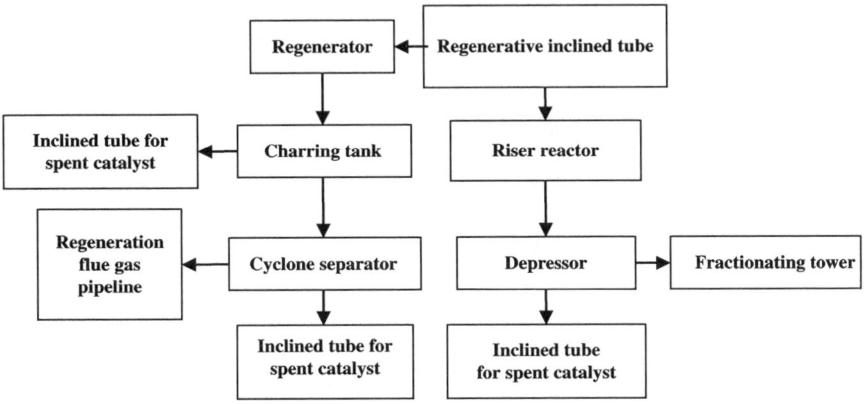

Fig. 7.9. The overall propagation path from the initial deviation to the final accident of the reaction regeneration unit (Tra5).

A Bayesian network as a probability-based knowledge representation method is appropriate for the modeling of causal processes with uncertainty. A BN is a directed acyclic graph (DAG) whose nodes represent random variables and the links define probabilistic dependences between variables. These relationships are quantified by associating a conditional probability table (CPT) with each node, given any possible configuration of values for its parents.

The static Bayesian network (SBN) can be extended to a DBN model by introducing relevant temporal dependencies that capture the dynamic behaviors of the domain variables between representations of the static network at different times. Two types of dependencies can be distinguished in a DBN: contemporaneous dependencies and non-contemporaneous dependencies. Contemporaneous dependencies refer to arcs between nodes that represent variables within the same time period. Non-contemporaneous dependencies refer to arcs between nodes that represent variables at different times.

Therefore, a DBN is a way to model probability distributions over semi-infinite collections of random variables $\{Z_1, Z_2, \ldots\}$. A DBN is defined to be a pair, $(B_1, B_{\rightarrow})$, where B_1 is a Bayesian network which defines the prior $P(Z_1)$, and $B_{\rightarrow}$ is a two-slice temporal Bayesian

Table 7.5. Functional HAZOP studies of the reaction regeneration unit with MFM function "Bal5", "Tra5" and "Tra6" (partial).

MFM functions	Deviation	Reasons	Consequence	Safety measures
Bal5	Regenerator reserve + "*MORE*"	1. The slide valve of the regenerator is malfunctionally closed 2. The catalyst is added too fast and too much 3. The catalyst losses are very high	1. The effect of the regeneration is poor, even causing the catalyst deactivation	1. Reduce the feed rate 2. Adjust the opening degree of the regeneration valve 3. Increasing the circulation of the external coolers 4. Shut down when the catalyst losses are very high
Tra5	The flow of the inclined tube I + "*MORE*"	1. The flow of the inclined tube II increased 2. The time of the spent catalyst staying in the premixer becomes shorter 3. The opening degree of the valve of the inclined tube I gets larger	1. The catalyst stacks, which makes the regeneration effect poor 2. The supply of the catalyst for reaction is not adequate, which makes the reaction target difficult to realize	1. Close the opening degree of the valve of the inclined tube I

Tra5	The flow of the inclined tube I + "*LESS*"	1. The feed of the spent catalyst is insufficient 2. The inclined tube I leaks 3. The pressure of the regenerator gets higher 4. The inclined tube I is blocked	1. The feed of the regenerated catalyst is insufficient 2. The amount of material in the reactor is relatively high, which will render the output not up to the requirements	1. Increase the opening degree of the valve of the inclined tube I 2. Repair the inclined tube I
Tra6	The flow of the inclined tube II + "*MORE*"	1. The flow of the inclined tube I gets larger 2. The pressure of the regenerator gets higher 3. The inclined tube II leaks	1. Excess catalyst is provided, which will make the reaction out of control	1. Close the opening degree of the valve of the inclined tube I 2. Check if the inclined tube I leaks 3. Inspect the pressure gauge of the regenerator
Tra6	The flow of the inclined tube II + "*LESS*"	1. The regenerator leaks 2. The flow of the inclined tube I is insufficient 3. The provision of the catalyst is insufficient	1. The regenerated catalyst is insufficient, making the reaction target difficult to realize	1. Open the catalyst supply pipeline with fresh catalyst instead of the regenerated catalyst 2. Increase the opening degree of the valve of the inclined tube I

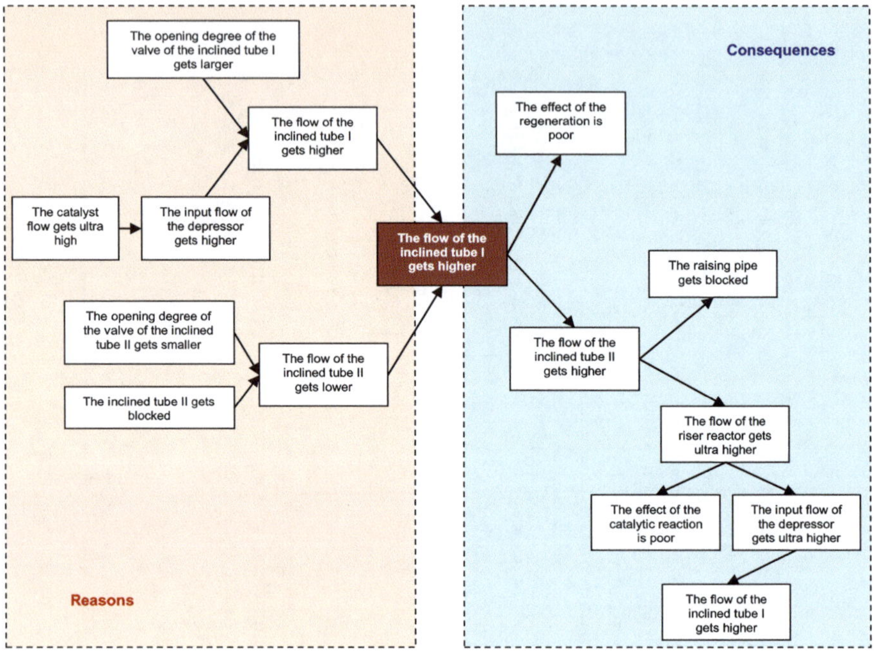

Fig. 7.10. Hazard scenarios of "Regenerator reserve + MORE".

network (2TBN) which defines $P(Z_t|Z_{t-1})$ by means of a DAG as follows:

$$P(Z_t|Z_{t-1}) = \prod_{t=1}^{N} P(Z_t^i|\mathrm{Pa}(Z_t^i)), \tag{7.1}$$

where Z_t^i is the ith node at time t and $Pa(Z_t^i)$ are the parents of Z_t^i in the graph.

The nodes in the first slice of a 2TBN do not have any parameters associated with them, but each node in the second slice of the 2TBN has an associated conditional probability distribution (CPD) for continuous variables or conditional probability table (CPT) for discrete variables, which defines $P(Z_t^i|\mathrm{Pa}(Z_t^i))$ for all $t > 1$. The parents of a node $Pa(Z_t^i)$ can either be in the same time slice or in the previous time slice. The arcs between slices are from left to right, reflecting the causal flow of time. If there is an arc from

Z^i_{t-1} to Z^i_t, this node is called persistent. The arcs within a slice are arbitrary. Directed arcs within a slice represent "instantaneous" causation. In this chapter, the parameters of the CPTs used by the proposed model are assumed time-invariant i.e., the model is time-homogeneous.

The semantics of a DBN can be defined by "unrolling" the 2TBN until there are T time-slices. The resulting joint distribution is then given by

$$P(Z_{1:T}) = \prod_{t=1}^{T} \prod_{i=1}^{N} P(Z^i_t \mathrm{Pa}(Z^i_t)). \tag{7.2}$$

Several inference methods for a discrete-state DBN can be used, i.e., forward–backward algorithm, unrolled junction tree, and the frontier algorithm. In this chapter, the forward–backward method is used for Bayesian inference in the DBN reasoning stage. For a more detailed presentation of the DBN model construction and learning, see, e.g., [11, 16].

7.3.2. *Integrated methodology procedure*

In this section, the integrated methodology procedure (shown in Fig. 7.11) employed by the presented framework is described and explained in detail. The main steps of the integrated methodology procedure include the following:

- **Stage I: DBN modeling**
  - (1) Functional HAZOP study is implemented to provide the in-depth process and hazard knowledge. The outcomes from this functional HAZOP study are FPPs or cause–consequence paths.
  - (2) Development of a DBN quantitative reasoning model based on functional HAZOP study.
    - (a) Determination of the model nodes. Nodes in DBN are determined corresponding to the nodes analyzed in the functional HAZOP study. Among all the parameters in HAZOP nodes, observable variables are considered as

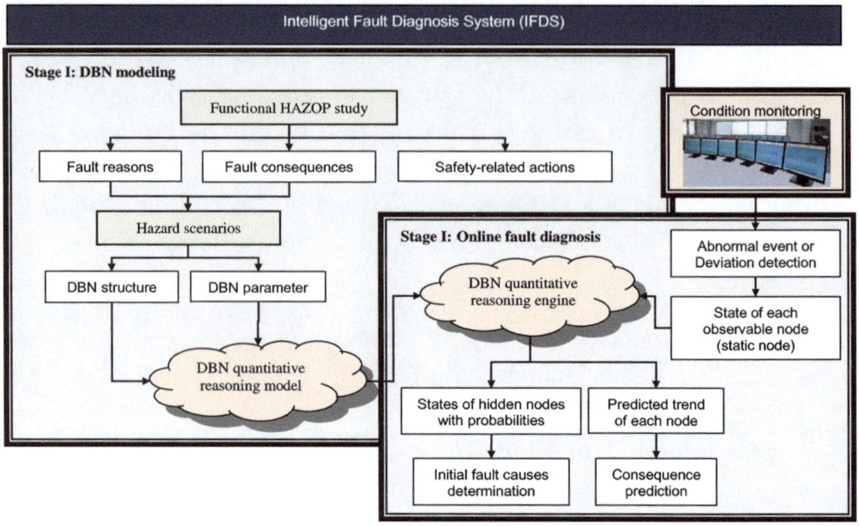

Fig. 7.11. The integrated methodology procedure for IFDs.

static nodes in DBN, whose values can be monitored by programmable logic controller (PLC) or distributed control systems (DCS). Their states can be described by HAZOP guide words.

(b) Another kind of nodes in DBN, named as dynamic nodes, are related to the hidden states of the system, representing various fault or hazard statuses in the HAZOP analysis (such as the fault condition of oil pump, corrosion condition of equipment, etc.). The status of these hidden nodes should be calculated by the reasoning algorithm according to the states of static nodes.

(3) Determination of the model structure. Model structure representing the casual relationship between nodes is determined based on the FPP from the functional HAZOP study. The development of the nodes and the structure of the model comply with the principle of DBN. The model structure can be optimized with K2 algorithm, and more details can be found in [5].

(4) Determination of conditional probability. CPT is selected to present the causal relationship between nodes. CPTs can be either obtained by analyzing historical data or determined based on experts' experience. Later, as new data are available and accumulated, it may be possible to facilitate this step further to update these CPTs.

- **Stage II: online fault diagnosis**

(1) Condition monitoring of deviation by DCS so as to generate a fault scenario.
(2) Run the DBN reasoning scheme to implement cause analysis and consequence analysis for the deviation.

The key issue in fault diagnosis by DBN is to obtain the probability $P(X^T|Y^T)$, where Y^T is the observation set of variables within a limited time-series, and X^T is the hidden variable set, which should be estimated based on the observed value of Y^T. In this chapter, FB algorithm is used for forward and backward inference calculation to search the fault root cause and consequences. It includes two steps as follows:

(a) *Fault initial cause inference*:
The forward recursive calculation is applied to compute $\alpha_t(i)$ as follows:

$$\begin{aligned}\alpha_t(i) &= P(X_t = i|y_{1:t}) \\ &= P(y_t|X_t)\left[\sum x_{t-1}P(X_t(x_{t-1})P(x_{t-1}|Y_{1:t-1})\right],\end{aligned} \tag{7.3}$$

where, $\alpha_t(i)$ indicates the state probability, $X_t = i$ at moment t, and it meets the given observable sequence $(y_1, y_2, \ldots, y_t)$ before moment t.

In this way, the state of hidden variables in DBN can be revealed. Alarms which show the most likely initial causes will be set off. Proper safety suggestions are also provided simultaneously based on functional HAZOP study results.

(b) *Fault consequence inference*:
The backward recursive calculation is applied to compute $\beta_t(i)$ as follows:

$$\beta_t(i) = P(y_{t+1}|Y_0^t) = \frac{\sum_{x_{t+1}} \alpha_{t+1}(x_{t+1})}{\sum_{x_t} \alpha_t(x_t)} = \frac{\sum_{x_{t+1}} P(X_{t+1} = i|y_{1:t+1})}{\sum_{x_t} P(X_t = i|y_{1:t})} \quad (7.4)$$

where $\beta_t(i)$ represents the probability of the observable sequence at moment $t+1$, and, by recursive computation, the probability that the observable sequence is in keeping with $(y_{t+1}, y_{t+2}, \ldots, y_T)$ after moment t can be calculated.

(3) Existing safeguards for each consequence are provided and required mitigating action(s) are proposed.

7.4. Case Study

7.4.1. *Stage I: DBN modeling*

Riser reactor is one of the important units in FCCU. A systematic functional HAZOP analysis is carried out in the first stage (the MFM model is shown in Fig. 7.4). Table 7.5 shows the results of the functional HAZOP analysis of MFM function "Tra3" (only a part of the functional HAZOP results is listed as an example).

According to the functional HAZOP results, parameters which can reflect the operating status of the plant are selected for the development of DBN. Information of the selected static and dynamic nodes are shown in Tables 7.6–7.8.

According to FPP and hazard scenarios (Figs. 7.9 and 7.10), the causal impact relationships between static and dynamic nodes are analyzed, and the DBN model can be built according to the procedure in Sec. 7.3.2. Figure 7.12 shows the optimal DBN model of the reactor.

Table 7.6. HAZOP results of the reactor section (partial).

MFM function	Deviation	Probable causes	Consequences	Safety precautions
Tra3	Riser reactor temperature + "*MORE*"	1. Large opening degree of regeneration slide valve 2. High regeneration temperature 3. High feed temperature of raw oil 4. Large feed amount of raw oil 5. Light quality of raw oil 6. Large flow of pre-lifting steam	1. Pressure increase of disengager 2. Depth increase of reaction, coking	1. Adjust opening degree of regeneration slide valve, control circulation of catalyzer 2. Reduce feed temperature of raw oil 3. Adjust handling capacity in time according to raw material composition 4. Reduce the flow of pre-lifting steam
	Riser reactor temperature + "*LOW*"	1. Small opening degree of regeneration slide valve 2. Small opening degree of feed control valve 3. Low feed temperature of raw oil 4. Little feed amount of raw oil 5. Heavy quality of raw oil 6. Water bearing in pre-lifting steam	1. Rapid decrease in pressure of disengager	1. Change regeneration slide valve and feed control valve into manual operation 2. Strengthen the raw oil heat exchanger, increase feed temperature 3. Increase feed amount 4. Adjust handling capacity in time according to raw material composition 5. Strengthen dewatering of pre-lifting steam

(*Continued*)

Table 7.6. (*Continued*)

MFM function	Deviation	Probable causes	Consequences	Safety precautions
	Riser reactor temperature + "*Fluctuation*"	1. Abnormal fluidization 2. Fluctuation of reaction pressure 3. Fluctuation of regeneration pressure 4. Fault indication of the reaction temperature instrument 5. water bearing in crude oil, water bearing in crude steam 6. large variation of temperature of the regenerator, leakage of external heat, or superheated steam coil	1. Likely to result in fire, explosion, serious leakage of material, etc. hazard	1. Reduce fluctuation of pressure and storage 2. Change to be manual or control by subline when instrument failure, switch to stand-by pump when there is pump failure 3. Strengthen dewatering of raw material and steam 4. Shut down external heat

Table 7.7. The dynamic nodes in the DBN Model of the reactor.

Subsystem	Dynamic nodes	ID	State set
Reactor	1. Raw oil pump	D2_1	{1:normal, 2:fault}
	2. Raw oil feeding pipeline valve	D2_2	{1:normal, 2:excessive, 3:small}
	3. Raw oil atomization steam pipeline valve	D2_3	{1:normal, 2:excessive, 3:small}
	4. Steam system	D2_4	{1:normal, 2:fault}
	5. Pre-lifting steam pipeline valve	D2_5	{1:normal, 2:excessive, 3:small}
	6. Regeneration slide valve	D2_6	{1:normal, 2:excessive, 3:small}

Table 7.8. The static nodes in the DBN Model of the reactor.

Subsystem	Static nodes	ID	State set
Reactor	1. Feed amount of raw oil	S2_1	{1:normal, 2:high, 3:low}
	2. Temperature of reaction zone	S2_2	{1:normal, 2:high, 3:low}
	3. Overall pressure drop of riser	S2_3	{1:normal, 2:high, 3:low}
	4. Temperature of pre-stripper	S2_4	{1:normal, 2:high, 3:low}
	5. Steam amount of pre-stripper	S2_5	{1:normal, 2:high, 3:low}
	6. Atomizing steam flow	S2_6	{1:normal, 2:high, 3:low}

7.4.2. *Stage II: online fault diagnosis*

In the stage of online fault diagnosis, OPC technology is applied in the condition monitoring system to obtain the online process parameters from DCS (see Fig. 7.13). The real-time data is then compared with the safety threshold to reveal if there is any abnormal event. As shown in Fig. 7.14, if the observed parameter overpasses the upper or lower threshold, it should be considered as an abnormal event which will cause an alarm. By the way, in the field workshop, the thresholds are usually determined as a static value according to expertise. However, we have a project of safety threshold optimization, in which the thresholds need to be dynamically calculated and the application performance seems to be satisfying. Related results will be published soon.

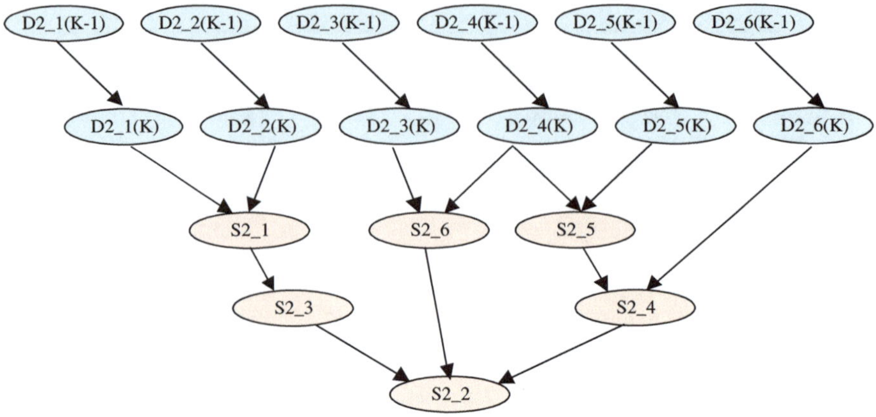

Fig. 7.12. The optimal DBN model of the reactor.

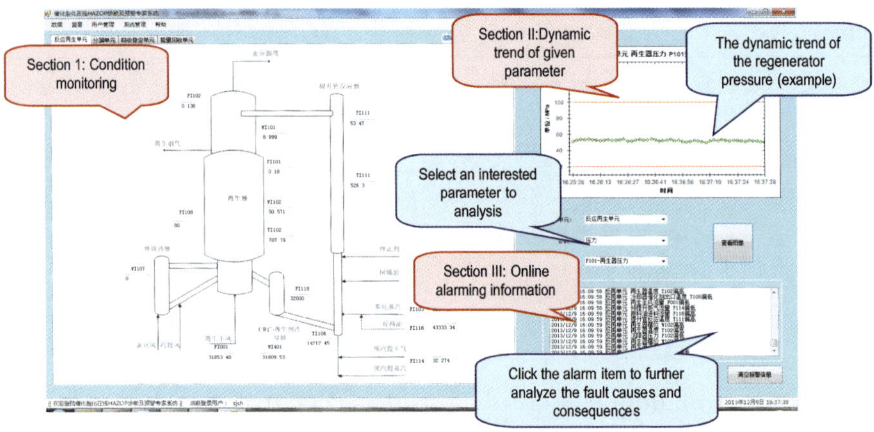

Fig. 7.13. OPC-based condition monitoring system.

When the plant was operated normally, all parameters of the FCCU reactor were normal, and the states of the hidden nodes were calculated as "normal" state. Figure 7.15 shows the reasoning results of the hidden nodes (the meaning of the index "1", "2", "3" in the figure indicates the different hidden states explained in Table 7.7).

At a certain moment, the "pre-stripping section temperature" low alarm occurred, followed by low alarm of "reaction zone temperature", while other parameters were in normal states. By the

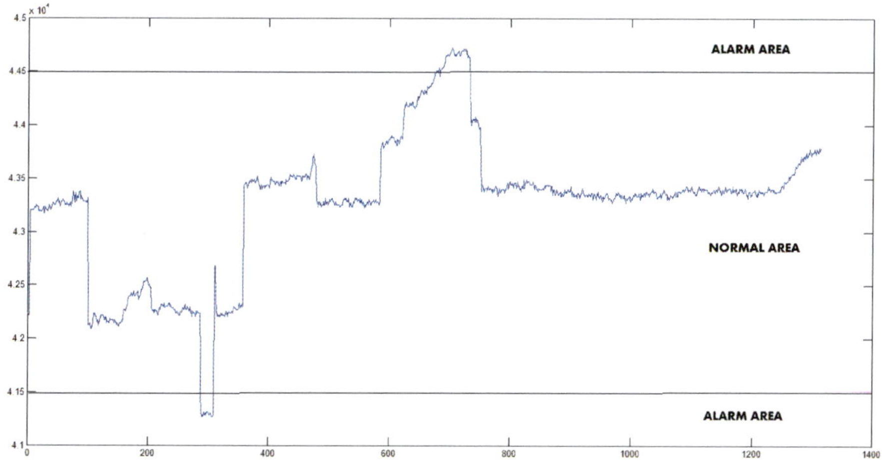

Fig. 7.14. Real-time process data with its safety thresholds (as an example).

inference of the DBN model, it was known that the state of node D2_6 "regeneration slide valve" had changed at this moment. Its probability of being in a "over opening" state was calculated as 87.5%, while the other nodes' states did not change. Figure 7.16 shows the reasoning results of each hidden node's state.

The fault cause reasoning results showed "over opening of the regeneration slide valve" which further resulted in an increase in the catalyst flow in the stripper reactor, while "stripping vapor convection" was in normal, therefore it then led to a decrease in the temperature of the pre-stripping section and a low temperature alarm happened. In a word, the direct cause was analyzed as "increase in the catalyst flow in the stripper reactor" and the initial cause was "over opening of the regeneration slide valve". Safety countermeasures were provided to "adjust opening degree of regeneration slide valve, control circulation of catalyzer".

Consequence analysis indicated that it would further affect the subsequent reaction stage, and a low temperature alarm of the stripping reactor would also happen later. The state of node S2_4 (temperature of pre stripper) would be in "low" state, and it would further cause the state of S2_2 (temperature of reaction zone) be in "low" state with probability as 84.7%.

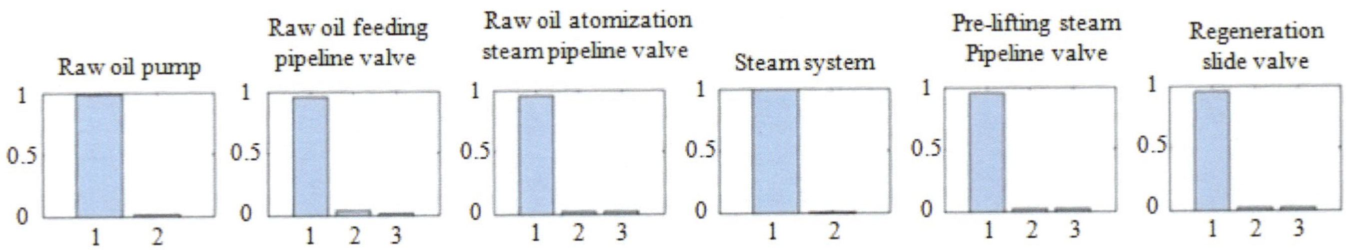

Fig. 7.15. Reasoning results of the states of hidden nodes (normal condition).

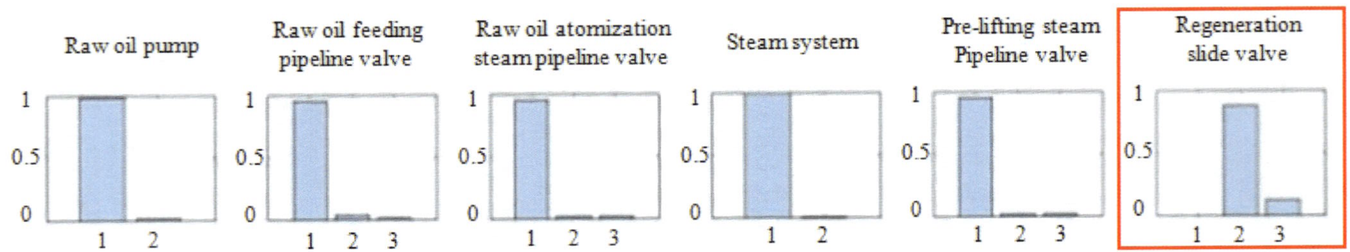

Fig. 7.16. Reasoning results of the states of hidden nodes (abnormal event).

Figure 7.17 shows the historical data with respect to the whole process from alarming to adjustment. Related alarming concerned with three process parameters: "reaction temperature", "pre-lift section temperature", and "riser section total pressure drop". The pre-lift zone temperature exceeded the threshold value firstly and a low alarm occurred, followed by a low alarm in reaction zone temperature. It is also noticed that there was a sharp increase of riser overall pressure drop before the low alarm of the two parameters occurred. After the increase, pre-riser section temperature kept decreasing, till a low alarm occurred.

The above information and field control records provided a validation of the reasoning results. It was the increase in the handling capacity of reactor that led to a low temperature alarm in the reaction zone. The direct cause of the abnormal event was an increase in the catalyst flow running into the reactor, which brought about a drop-down in the pre-lift section temperature. Through the case study, the proposed IFDS was validated with adequate consistency with the reality of the process plant operation.

7.4.3. *Traditional versus IFDS*

7.4.3.1. *Traditional HAZOP versus functional HAZOP study*

This research demonstrated a promising potential and feasibility of functional knowledge-based tool to assist some of the tasks involved in HAZOP and as such contributed to better use of resources and time for a HAZOP team. The functional HAZOP study proposed in this chaper allows the HAZOP meeting to change focus from identifying possible causes of deviations to being concerned with FPP and hazard scenarios and revealing the initial causes. The availability of such a scientifically based systematic approach for performing the HAZOP helps the HAZOP study to thoroughly cover possible hazards in a plant.

7.4.3.2. *DCS versus IFDS*

In domestic refineries, DCS is still a widely used system for condition monitoring and abnormal event handling. We make a comparison

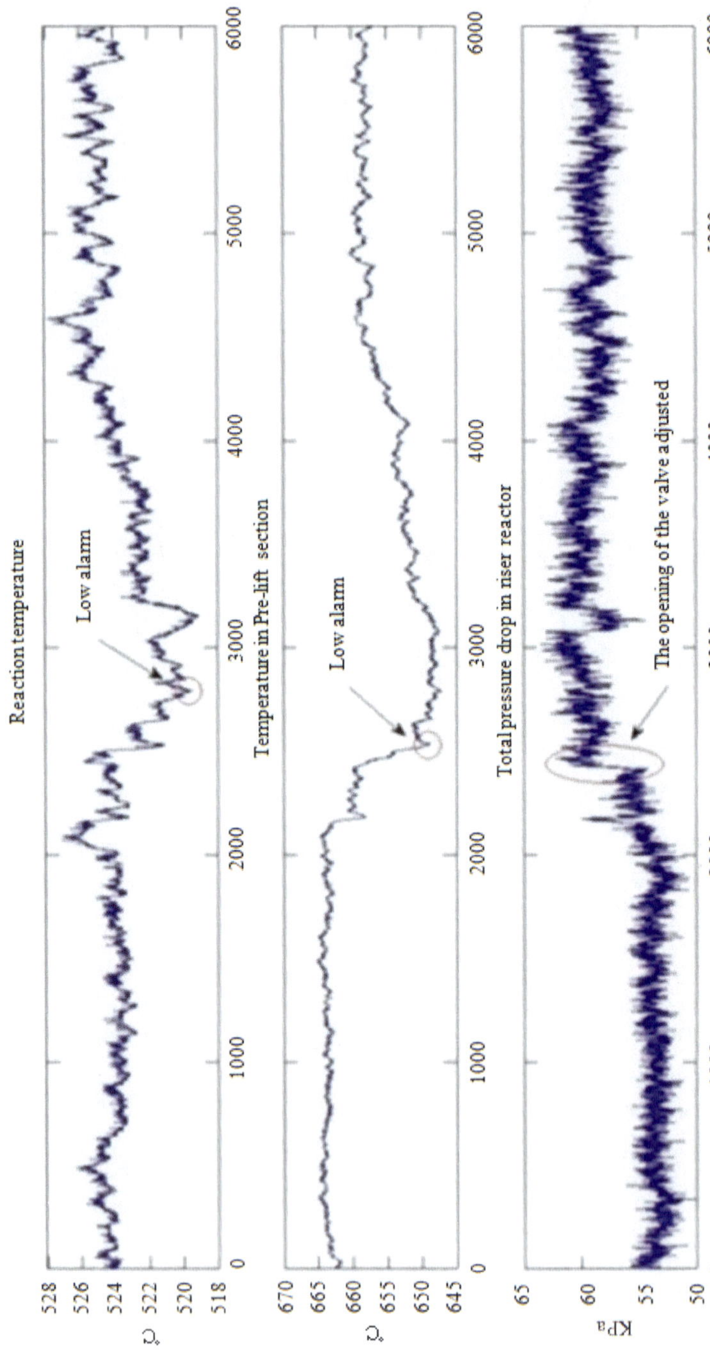

Fig. 7.17. Historical data of related process parameters.

Table 7.9. DCS versus IFDS.

Item	Function	DCS system	Proposed IFDS system
1	Condition monitoring	√	√
2	Process control (chain and sequence control)	√	×
3	Abnormal observable value alarming	√	√
4	Abnormal event risk assessment	×	√
5	Fault online diagnosis	×	√
6	Initial cause reasoning	×	√
7	Providing safety countermeasures	×	√
8	Consequence prediction	×	√
9	Computer-assisted HAZOP platform	×	√
10	HAZOP repository	×	√

between the existing DCS system and the proposed IFDS system (shown in Table 7.9) to illustrate their different functions and application focuses. Actually, both are essential and neither is superior to the other; they are not in conflict but complementary, each with its own sphere of competence.

7.4.3.3. *Existing diagnosis methods versus IFDS*

An exact comparison is not possible, since the modeling of FCCU by one or two other methods for the case study is quite a large project. Traditional and widely used methods such as signed directed graph (SDG), Petri network (PN), ANN, and a variety of methods integrated for some typical complex systems, have the capacity of identifying potential hazard, intelligent diagnosis, etc. However, SDG is still considered as the representation method that requires robust mechanism to construct fault propagation models and to identify faults, causes, and consequences. There still exist other weaknesses such as qualitative reasoning rather than the quantitative way, limited state representation, and weak uncertainty information processing. The comparison of strengths and weaknesses of each typical method is shown in Table 7.10, which help the counselor in safety engineering area to choose an appropriate method for a specific problem.

Table 7.10. Typical methods versus IFDS.

Method / Function	Online monitoring	System-level diagnosis	Qualitative reasoning	Quantitative reasoning	Node state representation	Safety measures	Risk assessment	HAZOP analysis
SDG-based method	√	√	√	×	**Limited (usually three states)**	√	√	√
ANN-based method	√	×	×	√	**Limited (usually appointed one state)**	×	×	×
PN-based method	×	√	√	×	**Limited (usually appointed one state)**	×	×	×
IFDS-based method	√	√	√	√	**Unlimited**	√	√	√

7.5. Conclusion

According to the strong interdependencies between various plants and components in the complex process system, which usually cause cascading faults or accidents, an integration of a functional HAZOP approach with DBN reasoning is presented and illustrated in detail in this chapter.

In order to build the model which is able to systematically, accurately, and entirely reflect the interdependency between parameters of the process plant, MFM is used at first to represent a system in terms of goals, objectives, functions, and components, each of which can be described at different levels of the part-whole decomposition. Based on the MFM concept, functional HAZOP analysis is introduced, by which all the possible deviations and their corresponding potential fault causes and consequences are analyzed carefully according to different FPPs, and finally hazard scenarios are developed. This approach significantly reduces the effort involved in the traditional HAZOP method.

DBN is further used to quantitatively represent the fault causal relationships which represent the fault interdependencies in the complex process system. With its powerful quantitative inference mechanism of handling uncertainty information, the most possible initial reason(s) that could happen in the abnormal event can be found out accurately and also the future possible consequences can be predicted for proactive maintenance or emergency decision making.

In each step of the IFDS implementation, particular examples that can be applied to the FCCU are presented and explained in detail to guide people on how to carry out the technological modeling and analyzing process. Finally, by a real case study of FCCU in a particular petrochemical company in China, the IFDS system was practically used in the field for pilot application. The effectiveness and accuracy were validated. Future research work will focus on the optimization for low-probability events in DBNs and inference algorithm to handle the missing data situation.

References

[1] A. Alaeddini, I. Dogan, "Using Bayesian networks for root cause analysis in statistical process control," *Expert Systems with Applications*, vol. 38, pp. 11230–11243, 2011.

[2] S. Babaie, A. Khosrohosseini, A. Khadem-Zadeh, "A new self-diagnosing approach based on petri nets and correlation graphs for fault management in wireless sensor networks," *Journal of Systems Architecture*, vol. 59, no. 8, pp. 582–600, 2013.

[3] P. Baybutt, "A critique of the hazard and operability (HAZOP) study," *Journal of Loss Prevention in the Process Industries*, vol. 33, pp. 52–58, 2015.

[4] C. Chang, C. Chen,"Fault diagnosis with automata generated languages," *Computers & Chemical Engineering*, vol. 35, no. 2, pp. 329–341, 2011.

[5] G. Cooper, E. Hersovits, "A Bayesian method for the induction of probabilistic networks from data," *Machine Learning*, vol. 9, pp. 309–347, 1992.

[6] H. Gabbar, S. Hussain, A. Hosseini, "Simulation-based fault propagation analysis — Application on hydrogen production plant," *Process Safety and Environmental Protection*, DOI: 10.1016/j.psep.2013.12.006, 2014.

[7] H. Gabbar, "Improved qualitative fault propagation analysis," *Journal of Loss Prevention in the Process Industries*, vol. 20, no. 3, pp. 260–270, 2007.

[8] M. Giardina, M. Morale, "Safety study of an LNG regasification plant using an FMECA and HAZOP integrated methodology," *Journal of Loss Prevention in the Process Industries*, vol. 35, pp. 35–45, 2015.

[9] B. He, T. Chen, X. Yang, "Root cause analysis in multivariate statistical process monitoring: Integrating reconstruction-based multivariate contribution analysis with fuzzy-signed directed graphs," *Computers & Chemical Engineering*, vol. 64, pp. 167–177, 2014.

[10] J. Hu *et al.*, "Fault propagation behavior study and root cause reasoning with dynamic Bayesian network based framework. *Process Safety and Environmental Protection*, DOI: 10.1016/j.psep.2015.02.003, 2015.

[11] N. Khakzad, F. Khan, P. Amyotte, "Dynamic safety analysis of process systems by mapping bow-tie into Bayesian network," *Process Safety and Environmental Protection*, vol. 91, no. 1–2, pp. 46–53, 2013.

[12] M. Lind, "Modeling goals and functions of complex industrial plant," *Applied Artificial Intelligence*, vol. 8, no. 2, pp. 259–283, 1994.

[13] M. R. Maurya, R. Rengaswamy, V. Venkatasubramanian,"A signed directed graph-based systematic framework for steady-state malfunction diagnosis inside control loops," *Computers and Chemical Engineering*, vol. 61, pp. 1790–1810, 2006.

[14] M. R. Maurya, R. Rengaswamy, V. Venkatasubramanian, "A signed directed graph and qualitative trend analysis-based framework for incipient fault diagnosis," *Chemical Engineering Research and Design*, vol. 85, no. 10, pp. 1407–1422, 2007.

[15] J. Mori, V. Mahalec, J. Yu, "Identification of probabilistic graphical network model for root-cause diagnosis in industrial processes," *Computers & Chemical Engineering*, vol. 71, pp. 171–209, 2014.

[16] K. Murphy, *Dynamic Bayesian Networks: Representation, Inference and Learning*, Ph.D. thesis, University of California, Berkeley, 2002.

[17] B. Ould-Bouamama *et al.*, "Bond graphs for the diagnosis of chemical processes," *Computers & Chemical Engineering*, vol. 36, no. 10, pp. 301–324, 2012.

[18] H. J. Pasman, B. Knegtering, W. J. Rogers, "A holistic approach to control process safety risks: Possible ways forward," *Reliability Engineering & System Safety*, vol. 117, pp. 21–29, 2013.

[19] M. Ram, R. Rengaswamy, V. Venkatasubramanian, "Application of signed digraphs-based analysis for fault diagnosis of chemical process flowsheets," *Engineering Applications of Artificial Intelligence*, vol. 17, no. 5, pp. 501–518, 2004.

[20] M. Rodríguez, J. L. de la Mata, "Automating HAZOP studies using D-higraphs," *Computers & Chemical Engineering*, vol. 45, pp. 102–113, 2012.

[21] N. L. Rossing *et al.*, "A functional HAZOP methodology," *Computers & Chemical Engineering*, vol. 34, no. 2, pp. 244–253, 2010.

[22] R. Diego *et al.*, "On-line fault diagnosis system support for reactive scheduling in multipurpose batch chemical plant," *Computers & Chemical Engineering*, vol. 25, no. 4–6, pp. 829–837, 2001.

[23] M. M. van Paassen, P. Wieringa,"Reasoning with multilevel flow model," *Reliability Engineering & System Safety*, vol. 64, no. 2, pp. 151–165, 1999.

[24] V. Venkatasubramanian, J. Zhao, S. Viswanathan, "Intelligent systems for HAZOP analysis of complex process plants," *Computers & Chemical Engineering*, vol. 24, no. 9–10, pp. 2291–2302, 2000.

[25] V. Venkatasubramanian *et al.*, "A review of process fault detection and diagnosis: Part I: Quantitative model-based methods," *Computers & Chemical Engineering*, vol. 27, no. 3, pp. 293–311, 2003.

[26] V. Venkatasubramanian, R. Rengaswamy, S. N. Kavuri, "A review of process fault detection and diagnosis: Part II: Qualitative models and search strategies," *Computers & Chemical Engineering*, vol. 27, no. 3, pp. 313–326, 2003.

[27] V. Venkatasubramanian *et al.*, "A review of process fault detection and diagnosis: Part III: Process history based methods," *Computers & Chemical Engineering*, vol. 27, no. 3, pp. 327–346, 2003.

[28] V. Venkatasubramanian, "Prognostic and diagnostic monitoring of complex systems for product lifecycle management: Challenges and opportunities," *Computers & Chemical Engineering*, vol. 29, no. 6, pp. 1253–1263, 2005.

[29] C. Wang, W. Zhang, C. Wu,"The theory and application and development trend of the fault diagnosis technology," *Automation Petro-chemical Industry*, vol. 6, pp. 7–13, 2008.

[30] G. Weidl, A. L. Madsen, S. Israelson, "Applications of object-oriented Bayesian networks for condition monitoring, root cause analysis and decision

support on operation of complex continuous processes," *Computers & Chemical Engineering*, vol. 29, pp. 1996–2009, 2005.

[31] H. Yuan, "Network topology for the application research of electrical control system fault propagation," *Procedia Engineering*, vol. 15, pp. 1748–1752, 2011.

[32] Z. Zhang *et al.*, "SDG multiple fault diagnosis by real-time inverse inference," *Reliability Engineering & System Safety*, vol. 87, no. 2, pp. 173–189, 2005.

Chapter 8

DBN-Based Failure Prognosis Method Considering the Response of Protective Layers for Complex Industrial Systems

In complex industrial systems, operating, regulating, maintenance activities, and external incidents occur dynamically and multiple entities in the same or different subsystems interact in a complex manner. Most of the single faults have multiple propagation paths. Any local slight deviation is able to propagate, spread, accumulate, and increase through system fault causal chains. It will finally result in system failure and unplanned outages or even catastrophic accidents. The key issues focus on both how to reduce the probability of fault occurrence and decrease the loss of fault consequence. The implementation of such requirements can be studied in terms of the determination of the fault root causes, prediction of the possible consequence, and also estimation of the risk and timing of various maintenance activities, which are considered in a failure prognosis scheme.

This study proposes a DBN-based failure prognosis method for complex systems. Not only the interaction between components, but also the influence of the layers of protection in the system is considered when the dynamic failure scenarios are analyzed. Therefore, the proposed method considers multiple factors including degradation mechanism, parameter deviation, the response of the layers of protection, and also the external environment. With this model, the dynamic influence diagram of the components' degradation trends can be calculated and used to evaluate the different effects of the layers of protection quantitatively. Some key problems are also explained such as how to determine the new nodes in DBN representing the behavior of protective layers and how to update the conditional probability table in the extended model.

In the case study, the proposed method is tested on the flue gas energy recovery system (FGERS) which is widely used in the petrochemical industry to demonstrate its effectiveness. It is of great help for early warning and optimization of the layers of protection in complex industrial systems.

8.1. Introduction

Prognostics is a technology used to monitor degradation in engineering systems, predict when failures may occur, improve reliability, and provide a cost effective strategy for scheduled maintenance [30]. Prognostics of engineering systems or products has become very important as degradations of the individual parts may cause a severe (and irreversible) damage to the entire system, environment, and users. Ultimately, it may lead to failures and will result in significantly costly repairs, which could otherwise have been avoided. In complex industrial systems, equipment and systems are anticipated to perform their function safely and reliably to meet the production requirements. Operating, regulating, maintenance activities, and external incidents occur dynamically and multiple entities in the same or different subsystems interact in a complex manner. The science of implementing into failure prognosis procedures requires a broad understanding of the complex interaction of industrial process, mechanical and process design, process safety protection and control systems. This information can be utilized to formulate detailed operational and maintenance procedures, as well as being fed forward to improve the safety and reliability of subsequent designs.

There are three main approaches to prognostics: (i) data driven; (ii) model driven; and (iii) fusion approach. Fusion approach is a combination of both (i) and (ii) methodologies [16, 30].

Data-driven approach can be further classified into statistical and machine learning techniques. Statistical techniques can be either parametric or non-parametric. Machine learning techniques can be either supervised learning, where test data is available, or unsupervised learning, where test data is not available. The principal disadvantage of the data-driven approach is that the confidence level in the predictions depends on the available historical and empirical data. Data-driven strategies to prognostics have been applied in a number of engineering applications [1, 3, 5, 11, 23, 26, 38].

Model-driven approach is based on the physics of failure (PoF) or system models [8, 15, 28, 31]. PoF models are based on the underlying physical phenomena of failures which require detailed FMEA study. System model relates the system's output to its input, and it can be derived from the first principles or test data. It requires knowledge of the failure mechanisms, geometry of the system, material properties, and the external loads that are applied to the system.

Fusion approach entails a combination of data- and model-driven approaches which incorporates the benefits and eliminates the drawbacks from both approaches. Therefore, the accuracy of the fusion approach should be higher than both model- and data-driven approaches when used individually [9, 18, 25, 29, 32, 35, 37]. Al-Dahidi [2] proposed an approach for RUL estimation from heterogeneous fleet data that consists of building a homogeneous discrete-time finite-state semi-Markov model, whose states are the degradation levels that the equipment can experience throughout life and that are identified by resorting to an unsupervised ensemble clustering approach. Zaidan [36, 37] selected a Bayesian hierarchical model to utilize fleet data from multiple assets to perform a probabilistic estimation of remaining useful life (RUL) for civil aerospace gas turbine engines. The hierarchical formulation allows Bayesian updates of an individual predictive model to be made based on data received asynchronously from a fleet of assets with different in-service lives and for the entry of new assets into the fleet.

Therefore, as a fusion approach, dynamic Bayesian networks (DBNs) have been gaining popularity gradually for reliability and risk evaluation as a robust and viable alternative to most traditional methods [12, 13, 17]. DBN is appropriate for monitoring and predicting values of random variables, and capable of representing the system states at any time [20, 21, 34]. Liu [19] integrated Parallel Monte Carlo simulation and recursive Bayesian method in the proposed modeling framework to estimate the degradation state of a system composed of dependent degradation components whose conditions are monitored (even without knowing the initial system degradation state) and to dynamically assess the system risk and RUL.

However, considering the complexity of modern engineering systems, most DBN-based prognostics are limited to finding degradation paths between predefined components and especially fail to express the effect of the protective layers and their capacity. Protective layers (also known as safety barrier systems) may be classified according to the schematization introduced by Center of Chemical Process Safety (CCPS) [4] as follows: (i) inherently safer design; (ii) passive protection systems; (iii) active protection systems; (iv) procedural and emergency safeguards. The protective layers should be considered in the failure prognosis study, the reasons of which are the failure scenarios of a complex industrial system which usually include the driving process of multiple failure factors, the degradation process of components, the effect of the protective layers (PLs), and the variation of the environment. These failure scenarios must be analyzed properly to identify the root cause(s), according to which the corrective actions (i.e., preventive replacement (PR), corrective replacement (CR), etc.) can be implemented. Ramzali *et al.* [27] employed the event tree analysis (ETA), fault tree analysis (FTA), and reliability block diagram (RBD) methods to build a safety barrier system and quantify barriers FP (failure probability). Duijm [7] described the syntax and principles for constructing consistent and valid safety-barrier diagrams with the focus on deliberately inserted safety systems that supported the management and maintenance of the systems.

However, several questions still need to be noted: (1) desired optimal protective system tends to involve a multiplicity of process variables, while by each measure, in terms of a procedural or technical barrier, only a single one can be controlled, i.e., what type of variable with what characteristic should be selected?; (2) how will the barriers interact under various process conditions and states?; (3) how is the functioning of barriers controlled over time? [24]. Hence, it is important to develop a method that has the ability to quantitatively describe the protective system's behaviors and its time-dependent effects along with the component's degradation.

This study proposes a DBN-based failure prognosis method for complex systems considering the response of the protective layers.

Not only the interaction between components, but also the influence of the protective layers in the system should be considered when the dynamic failure scenarios are analyzed. Therefore, the proposed method considers multiple factors including degradation mechanism, parameter deviation, the response of the protective layers, and also the external environment. With this model, the dynamic influence diagram of the components' degradation trends can be calculated and be used to evaluate the different effects of the protective layers quantitatively. In the case study, the proposed method is tested on the flue gas energy recovery system (FGERS) which is widely used in the petrochemical industry to demonstrate its effectiveness.

The remainder of the chapter is organized as follows. Section 8.2 presents an introduction to the basic theory of the DBN-based failure prognosis framework. Section 8.3 presents the proposed method and describes especially how to consider the behavior and characteristics of PLs and how to integrate into the basic DBN model. Section 8.4 provides an instantiation of the DBN-based dynamic degradation model with the example of FGERS. Case studies to demonstrate the effectiveness of the failure prognosis by the proposed method are illustrated in Sec. 8.5. The conclusions are drawn in Section 8.6.

8.2. DBN-Based Root Cause Analysis and Failure Prognosis Framework

A Bayesian network (BN) as a probability-based knowledge representation method is appropriate for the modeling of causal processes with uncertainty. A BN is a directed acyclic graph (DAG) whose nodes represent random variables and links define probabilistic dependencies between variables. These relationships are quantified by associating a conditional probability table (CPT) with each node, given any possible configuration of values for its parents [13]. The set of nodes of BN represents the system variables (which can be discrete or continuous), and the set of directed arcs represents the dependencies or influence among the variables. In discrete BNs, variables are defined over a set of mutually exclusive states, and a probability is associated to each state.

So, the fault propagation behaviors can be studied by BN, in which the different states of nodes represent various abnormal conditions of components or units of the complex system; while directed arcs indicate the system functional relationship and also fault interdependencies. The quantification of probabilities in discrete BNs consists of assigning prior probabilities to the nodes without parents, and defining a CPT for the nodes with parents. The CPT specifies the probability that the node is in a particular state given any combination of parent states. Prior knowledge can be acquired from previous safety analysis, such as HAZOP study, Failure mode and Effect Analysis (FMEA), FTAs, etc.

BNs use probability theory for handling uncertainty. Values of the variables are expressed as probability distributions. As information accumulates (e.g., observations of variables from a condition monitoring system in the field), the posterior probability distribution of variables of interest (such as the unknown root causes of a certain abnormal event) can be computed conditioned on the variables that have been observed, and knowledge of the true value of the variables usually increases. Hence, the uncertainty of the value is reduced, and the probability distribution becomes less spread. So, BN has the advantage both of model-based and data-based methods for modeling and predicting the dynamic fault propagation and degradation behavior. Prior knowledge can be used first to develop a BN model, then as more data are accumulated by the condition monitoring system the parameters and structure of BN model can be updated and more and more close to its physical truth.

The static Bayesian network (SBN) can be extended to a dynamic Bayesian network (DBN) model by introducing relevant temporal dependencies that capture the dynamic behaviors of the domain variables between representations of the static network at different times. DBNs consist of a sequence of time slices, and each slice comprises a static BN describing the system in the corresponding time step. Temporal links between variables in different time slices represent a temporal probabilistic dependence between the variables. Therefore, two types of dependencies can be distinguished in a DBN: contemporaneous dependencies and non-contemporaneous dependencies. Contemporaneous dependencies refer to arcs between nodes that represent variables within the same time period. Non-contemporaneous dependencies refer to arcs between nodes that represent variables at different times [14].

Therefore, a DBN is a way to model probability distributions over semi-infinite collections of random variables $\{Z_1, Z_2, \ldots\}$. A DBN is defined to be a pair, $(B_1, B_{\rightarrow})$, where B_1 is a BN which defines the prior $P(Z_1)$, and $B_{\rightarrow}$ is a two-slice temporal Bayesian network (2TBN) which defines $P(Z_t|Z_{t-1})$ by means of a DAG as follows:

$$P(Z_t \mid Z_{t-1}) = \prod_{t=1}^{N} P(Z_t^i \mid \mathrm{Pa}(Z_t^i)), \tag{8.1}$$

where Z_t^i is the ith node at time t and $\mathrm{Pa}(Z_t^i)$ are the parents of Z_t^i in the graph.

The nodes in the first slice of a 2TBN do not have any parameters associated with them, but each node in the second slice of the 2TBN has an associated CPT for discrete variables, which defines $P(Z_t^i|\mathrm{Pa}(Z_t^i))$ for all $t > 1$. The parents of a node $\mathrm{Pa}(Z_t^i)$ can either be in the same time slice or in the previous time slice. The arcs between slices are from left to right, reflecting the causal flow of time. If there is an arc from Z_{t-1}^i to Z_t^i, this node is called persistent. The arcs within a slice are arbitrary. Directed arcs within a slice represent "instantaneous" causation. In this chapter, the parameters of the CPTs used by the proposed model are assumed time-invariant, i.e., the model is time-homogeneous. The semantics of a DBN can be defined by "unrolling" the 2TBN until there are T time slices. The resulting joint distribution is then given by

$$P(Z_{1:T}) = \prod_{t=1}^{T} \prod_{i=1}^{N} P(Z_t^i \mathrm{Pa}(Z_t^i)). \tag{8.2}$$

DBNs are excellent tools for many types of probabilistic inference, such as fault propagation study and failure prognosis, since they encode all relevant qualitative and quantitative information contained in a full probabilistic model. There are three important types of inferences used in relation to dynamic models: smoothing, filtering, and prediction. Here for root cause reasoning, the filtering and smoothing are used. By filtering, the unknown state of unobservable nodes (e.g. current fault mode of a certain component) at the current time slice can be calculated by given related observations; while by smoothing, the root causes can be deduced at one or several previous time slices given the current observable and unobservable variables. Several inference methods for a discrete-state DBN can be used [17, 22], i.e., forward–backward algorithm, unrolled junction tree, and the frontier algorithm. In this chapter, the forward–backward method is used for Bayesian inference in the DBN reasoning stage.

A DBN-based framework for failure prognosis is proposed, since the failure or degradation behavior should be studied and modeled in a scientific and systematic way. The overall workflow is shown in Fig. 8.1, from which there are three key stages: Stage I: Hazard scenarios development; Stage II: DBN development; Stage III: DBN-based root cause and consequence reasoning. For a more detailed presentation of the method, see for example, [11].

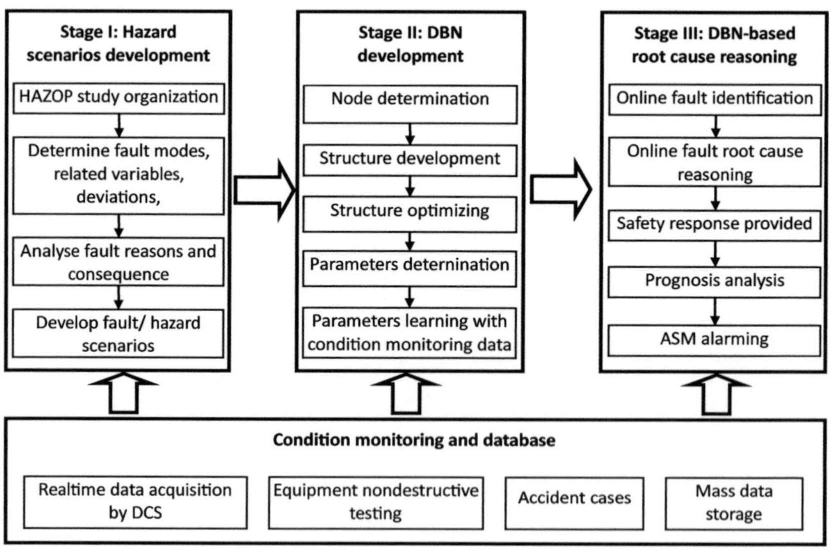

Fig. 8.1. DBN-based framework for root cause reasoning and failure prognosis.

8.3. DBN-Based Failure Prognosis Method Considering the Effect of Protective Layers

8.3.1. *Functional analysis of the protective layers*

Four main categories of protective layers (or safety barriers) are defined as follows [6]:

Passive barriers: These barriers are always functioning (permanent), without a need for human actions, energy sources, or information sources. Passive barriers may be physical barriers (retention bund, wall, etc.), permanent barriers (corrosion prevention systems), or inherently safe design.

Activated barriers: These barriers set up preconditions that need to be met before the action can be carried out. So, these barriers must be automated or activated manually to work or these barriers can be mechanical barriers that require an activation (hardware) to achieve their function.

Human actions: The effectiveness of these barriers relies on the knowledge of the operator in order to reach the purpose. Human actions are to be interpreted broadly, including observations by all senses, communication, thinking, physical activity and also rules, guidelines, safety principles, etc. Human actions may be a part of a detection–diagnosis–action sequence. Human actions contain on-site inspections, emergency plan for unexpected condition, etc. Inspectors often check devices based on noise, temperature, and vibration to find abnormality such as temperature drift and excessive vibration.

Symbolic barriers: These barriers need an interpretation by a person in order to achieve their purpose. The typical example can be passive warnings (like keeping out of prohibited areas, opening labelled pipes, refraining from smoking, etc.)

In this chapter, activated barriers in abnormal conditions and human actions are mainly considered in the DBN-based failure prognosis modeling.

Taking FGERS as an example, its main protective layers are analyzed but not limited to as follows.

Monitoring and control system of lubricating oil: The monitoring and control system of the lubricating oil (LO) is used to monitor oil pressure, liquid level in an oil tank, and differential pressure of the oil filter. It is also called a "temperature control system" since it is conducive to control the temperature of other important components such as bearings by adjusting the oil inflow.

The lubricating oil monitoring system is composed of three components: main lubricating oil pump, oil cooler, and several duplex strainers. There is another spare pump in the case of oil pressure falling below the alarming threshold. If an abnormal event such as "low pressure" is detected, the standby pump will get started automatically. In addition, the signal of "low pressure" will be detected by the oil pressure transmitter located on the downstream of oil dampers if the lube oil pressure falls down, i.e., the lubricating oil filter blocks. Then an alarm of "low pressure" will be issued. Shutdown for protection of the whole system will be taken into action automatically by the system, if the state of "low low pressure (LLP)" is detected by sensors embedded in the pressure switch located on the downstream of the lubricating oil pipe.

It can be known from the above working mechanism that the lubricating oil monitoring system works normally only when the three parts, monitoring systems of LO pressure, liquid level in the oil tank, and differential pressure of the oil filter, work successively.

Monitoring system of bearing temperature: It can be found that states of bearings are of great significance to equipment fault diagnosis, especially in the early stage of bearing degradation. Bearing temperature is a critical indicator which can be detected in real-time by the bearing temperature monitoring system.

Monitoring system of bearing vibration: Vibration amplitude is another critical indicator of condition monitoring. The incipient failure of a bearing can be detected by vibration monitoring. By studying the FFT spectrum of the bearing's vibration signal, the condition state and certain fault mode of the bearing can be identified. As it is difficult to measure the rotation amplitude of an impeller, vibration-based monitoring method is prone to be used to reflect the condition of the shaft.

Monitoring system of inlet flue gas: Flue gas turbine works in a rigorous condition with high temperature and catalyst particles. Inlet monitoring system is used to ensure gas temperature and catalyst particles conform to specific requirements strictly. An alarm will occur when the two indicators reach the warning thresholds.

Human inspection: Regular human inspection is a common way to prevent the failure of the alarm system. Inspectors check the situation of the whole plant in fixed intervals. The abnormality will usually be judged by observing the machine manually.

8.3.2. *Extended DBN model for failure prognosis considering PL effect*

In the basic DBN-based FP model, only failure and degradation factors are considered, while the effects of protective layers during the components' degradation process are usually neglected. In this chapter, how to model the PL effects and integrate them into the basic DBN analysis framework are studied.

Firstly, the PLs of each subsystem should be determined. In this chapter, activate protection and human action as PLs are considered as examples in Sec. 8.3.1. Secondly, the functional mechanism of each PL is analyzed by the FTA which will be transformed in the DBN model later. Quantitative assessment of the PL performance and CPT is required to be updated. The details of each step are presented as follows.

- ***Step 1: PL identification***

In this step, layer of protection analysis (LOPA) as a semi-quantitative method of risk analysis and evaluation can be used to identify the initial events and their corresponding PLs to assess the effectiveness of protective layers and the residual risk of an accident scenario [10].

- ***Step 2: Functional analysis of PL***

FTA is then applied to analyze the protective mechanism of PLs and determine new PL nodes and directed edge in the basic DBN model. Besides, FTA is required in the calculation of prior probability of PL nodes. Probability of basic events can be obtained by prior knowledge. Success and failure rates of PL can be calculated according to the computation rule of "gate" after the prior probabilities of

PL nodes are input into the basic DBN model. The PL nodes and corresponding directed edges will be added into the new extended DBN model with PL information by the FTA-DBN transformation method presented in step 3.

- ***Step 3: Construction of the extended DBN model***

The nodes in FTA are selected as dynamic nodes in the extended DBN model, and the related process variables are selected as static nodes. The identified PLs in Step 1 are transformed as PL nodes in the extended DBN model.

For example, to transform an FTA diagram to a DBN network, the problems to be solved include determination of nodes, directed edges, and conditional probability distributions. Since the purpose of the FTA analysis is to illustrate the mechanism of PLs and the interrelationship between PLs and component degradation behavior, the only problem to be analyzed is how the PL influences the parameter deviation. That is, the function of the PL is determined as the basic event and its corresponding parameter deviations are identified as the top events. Therefore, FTA events can be converted directly into the nodes of the DBN model. Figure 8.2 shows the transformation process from an FTA event to a DBN node and a directed edge.

The rest of the problem is the determination of CPTs for the PL nodes. Although the CPT conversion methodology for transforming an FTA to a DBN was given in literatures [33], it is not fully appropriate to deal with the PL nodes. When the PL fails, it has no

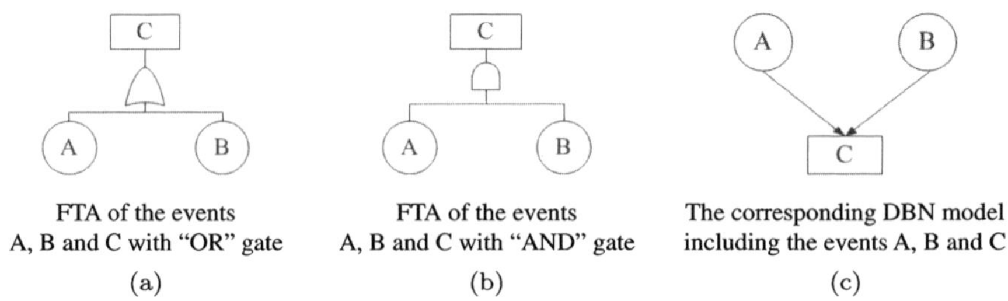

Fig. 8.2. Schematic diagram of the conversion from FTA to DBN.

effect on the parameters of the original CPT of the basic DBN model; however, when the safety barriers work successfully, the original CPT will be adjusted. It should be noted that, the probability of the "normal" state of the node may not be restored as "1", but the behavior of PL will increase this probability which will improve according to the actual effect of PL action. A quantitative evaluation is required to assess such protective effects.

Figure 8.3 defines several paths from the initial events to the top events, the difference of which are the probabilities of initial events and their effects. It is the illustration of the definition of "gate" used for the quantitative assessment of the protective effect. In Fig. 8.3, probability of failure on demand (PFD) indicates the FP of the protection layer and η indicates the protective efficiency, i.e., the adjusting degree of deviation (or the recovery capability of degradation) when the PL works. With the increase in η, the effect of the PL will improve, and the ability to restore the deviated parameter to the normal state is promoted. P_d indicates the probability of equipment failure, and M indicates the number of possible scenarios in the case of gate "c".

- ***Step 4: CPT updating***

In an abnormal situation, the process parameter will keep deviating from the normal threshold when PL fails; while when PL works, it will help correct the parameter deviation. In another words, when PL works, parameter deviations are more likely to return to the normal state instead of continuously deviating. The change of this probability is the efficiency of the protective layer.

To simplify the calculation process in the chapter, the probability of returning to "normal" state of observable node is set as 70% when PL works. That is, if PL works successfully, the state of the observable nodes will return to "normal" at the rate of 70%; while when the PL fails, its CPT will not be adjusted.

In Fig. 8.4, an observable variable is represented by node B, while node P indicates its related PL. In the original DBN model, the assumption that node $B(T+1)$ in "normal" state at the rate of 0.995

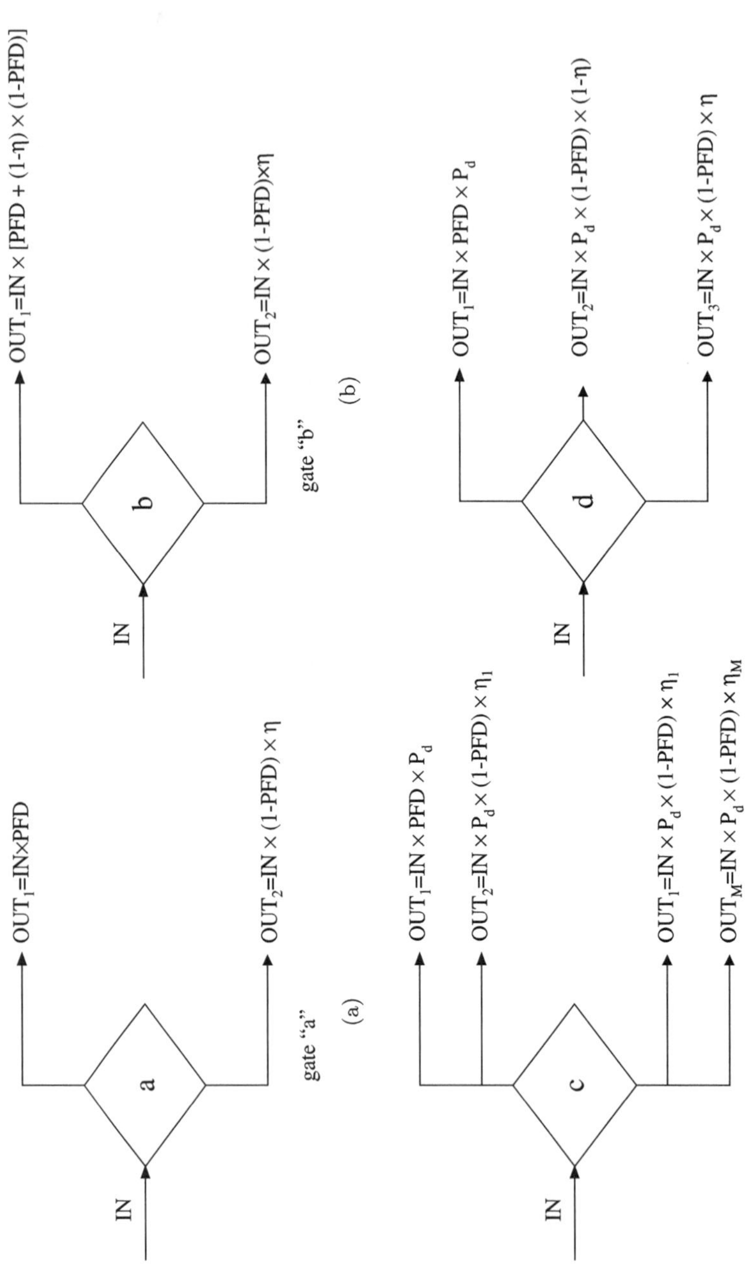

Fig. 8.3. Definition of "gates" for quantitative evaluation of PL.

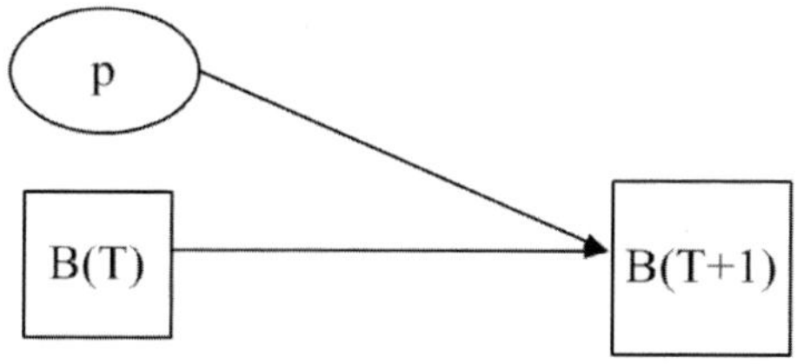

Fig. 8.4. Demonstration of the PL mechanism.

Table 8.1. CPT of node B in original DBN model.

$B(T+1)$	Normal	Deviated	$B(T+1)$	Normal	Deviated
$B(T)$ = Normal	0.995	0.005	B(T) = Deviated	0.005	0.995

Table 8.2. CPT of node B in the extended DBN model considering the PL effect.

$B(T+1)$		Normal	Deviated
$B(T)$ = Normal	P = Normal	0.995	0.005
$B(T)$ = Normal	P = Failed	0.995	0.005
$B(T)$ = Deviated	P = Normal	0.7	0.3
$B(T)$ = Deviated	P = Failed	0.005	0.995

and in "deviation" state at the rate of 0.005 is based on the premise that $B(T)$ is diagnosed as "normal" state. Node $B(T+1)$ in "normal" state at the rate of 0.005 and in "deviation" state at the rate of 0.995 is based on the premise that $B(T)$ is diagnosed as "deviation" state. Node P makes no contribution to the state of node $B(T+1)$ when node $B(T)$ is in "normal" state. Once deviation of node B(T) occurs, certain protective action of node P can be triggered. If node P works as expected, node $B(T+1)$ will be corrected to "normal" state at the rate of 70% as mentioned above. Tables 8.1 and 8.2 provide examples of the above description about the modification of CPT in the extended DBN model.

8.4. DBN-Based Failure Prognosis Modeling for FGERS

The proposed method is applied to the flue gas energy recovery system (FGERS) (shown in Fig. 8.5) which is one of the main systems in the typical catalytic cracking units. It consists of a flue gas turbine system, an axial flow compressor, a gear box, and an asynchronous motor which work together as a "coaxial" unit configuration. Its main function is to provide the oxygen for burning, ensure the catalyst of the regenerator is in burning state, and maintain the pressure balance.

The main air blower is started first by a steam turbine or an electric motor. When the flue gas turbine is put into operation, the steam turbine can supplement the work done by the flue gas turbine and drive the generator to generate electricity to the power grid. The flue gas discharged from the flue gas turbine is then transported to the waste heat boiler for heating water. Meanwhile, its production (i.e. the steam with medium-pressure) is used to promote the work done by the steam turbine. The flue gas with lower calorific value from the waste heat boiler is discharged into the atmosphere through the chimney (shown in Fig. 8.6). In this way, the flue gas with high temperature generated by the regenerator is used to carry out the expansion work for the flue gas turbine on the one hand, and on the other hand to heat the feedwater for the boiler. Therefore, the energy in the flue gas can be fully recycled and comprehensively utilized.

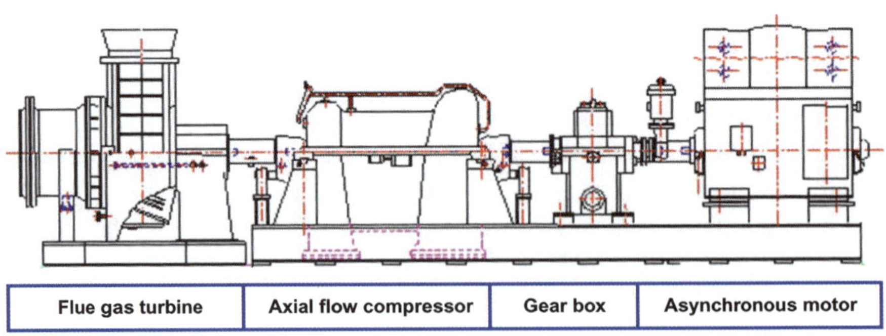

Fig. 8.5. Configuration of the integrated main air blower and flue gas turbine system.

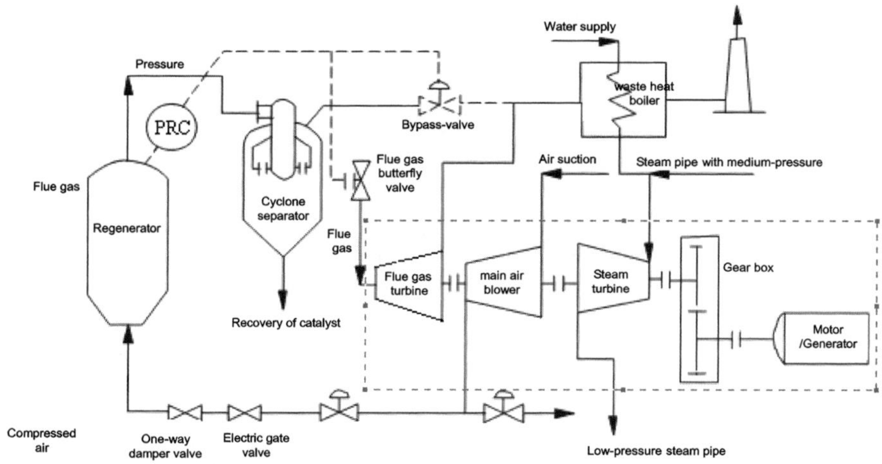

Fig. 8.6. Process flow of the FGERS in the catalytic cracking units.

Table 8.3. Identified PLs of the flue gas turbine system.

PL categories	PLs
Passive barriers	Monitoring system of lubricating oil pressure Monitoring system of lubricating oil tank liquid level Monitoring system of lubricating oil pressure Monitoring system of flue gas turbine bearing vibration Monitoring system of flue gas turbine bearing temperature Monitoring system of flue gas turbine inlet temperature
Human actions	Human response of operators in central control room Inspection

- ***Step* 1: *PL identification***

According to the principle presented in Sec. 8.3.2, the protective layers of the flue gas turbine system are identified and shown in Table 8.3.

- ***Step* 2: *Functional analysis of PL***

The results of FTA of the PLs in the flue gas turbine system are shown in Figs. 8.7–8.9. The detailed meanings of the symbols in Figs. 8.7–8.9 are listed in Table 8.4. Probabilities in Table 8.4 are calculated according to the principle in Sec. 8.3.2.

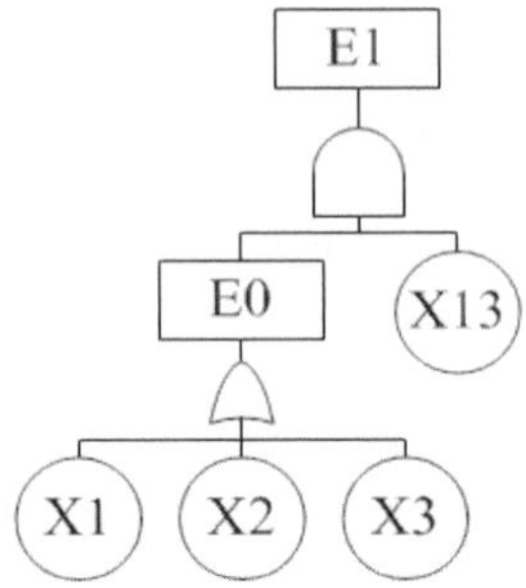

Fig. 8.7. FTA of monitoring system of lubricating.

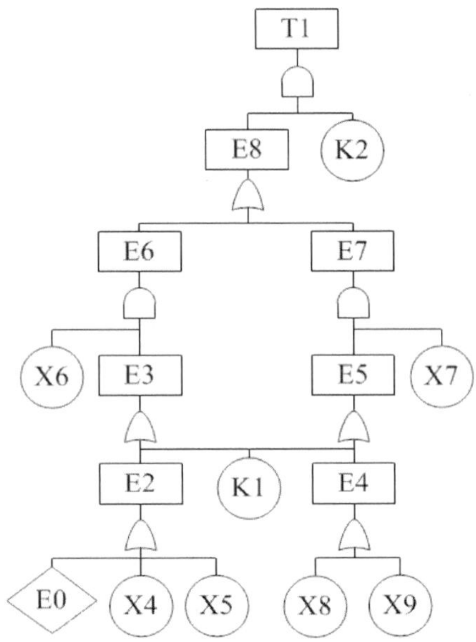

Fig. 8.8. FTA of flue gas turbine bearing.

- ***Step 3: Construction of the extended DBN model***

PL nodes are added in the original DBN model after the mechanism of the PLs is confirmed. In Fig. 8.11, PL nodes are shown in red color and represent the top-most protective layers in Table 8.3, in comparison with the original DBN model without consideration of the protective layers (shown in Fig. 8.10). Typical nodes

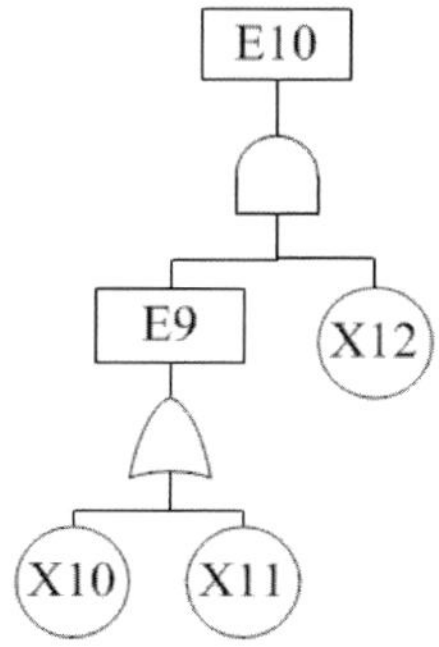

Fig. 8.9. FTA of monitoring system of flue gas turbine inlet gas.

representing lubricating oil monitoring, bearing temperature and vibration controlling and inspection are added compared with the original DBN model. The detailed information about the nodes in Figs. 8.10 and 8.11 is shown in Tables 8.5–8.7.

- ***Step 4: CPT updating***

Taking the node DA_l as an example, DA_l is a hidden node, and its current state is impacted by the corresponding node in its former time slice, i.e., DA_l (T) is the parent node of DA_l ($T + 1$). The original CPT of DA_l is shown in Table 8.8.

According to the principle in Sec. 8.3.2, the corresponding PLs of the node DA_l are triggered when the node DA_l lies in the state of "abnormal pressure or temperature". The prior probability of the PL node E0 was calculated in step 2 in Sec. 8.3.2, i.e., the probability of the state "success" is 0.4102 and "failure" is 0.5898.

If the PL node E0 works successfully, its influence on the node DA_l is as follows:

> The probability of the node DA_l being updated from the state of "abnormal temperature or pressure" to the state of "normal" is increased.
>
> The probability of the state of the node DA_l sustaining in the state of "abnormal temperature or pressure" is decreased.

Considering the effect of the PL node E0, the modified CPT of DA_l is shown in Table 8.9. In the same way, the modified

Table 8.4. The Detailed meaning of the symbols in Figs. 8.7–8.9.

Item	Event	Probability	Item	Event	Probability
X1	Failure of the lubricating oil pressure monitoring system	0.2570	E0	Failure of the lubricating oil supply monitor	0.5898
X2	Failure of the lubricating oil tank liquid level monitoring system	0.2570	E1	Failure of the temperature monitor	—
X3	Failure of the lubricating oil filter differential pressure monitoring system	0.2570	E2	Failure of the bearing temperature monitoring system	0.2570
X4	Failure of the lubricating oil temperature logic controller	—	E3	Failure of the bearing temperature controlling system	0.4056
X5	Failure of the bearing temperature sensor	—	E4	Failure of the bearing vibration monitoring system	0.2570
X6	High bearing temperature	—	E5	Failure of the bearing vibration controlling system	0.4056
X7	High bearing vibration amplitude	—	E6	Abnormal bearing temperature	—
X8	Failure of the bearing vibration sensor	—	E7	Abnormal bearing vibration	—
X9	Failure of the bearing vibration logic controller	—	E8	Bearing fault	—
X10	Failure of the flue gas detecting sensor	—	E9	Failure of the flue gas turbine inlet gas monitoring system	0.2570
X11	Failure of the flue gas logic controller sensor	—	E10	Deterioration of the flue gas abnormal state	—
X12	Abnormal state of flue gas	—	K1	Operation error of operators in the central control room	0.2
X13	Deviation of the parameter of lubricating oil	—	K2	Failure of inspection	0.2
		—	T1	Bearing failure	—

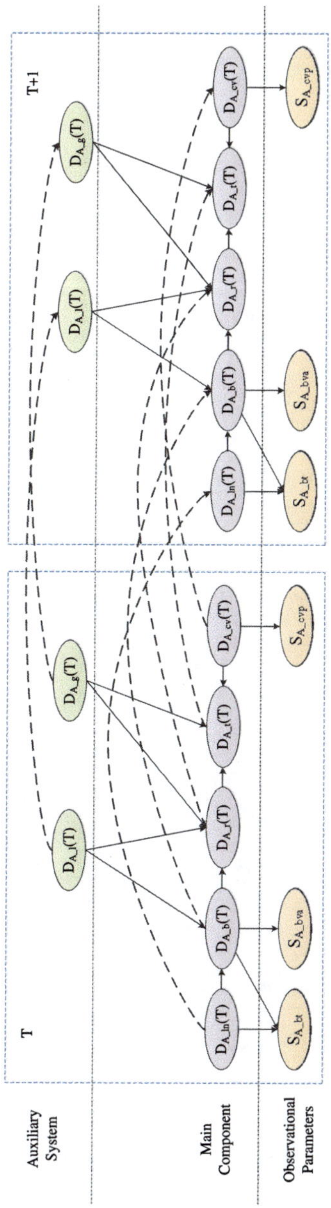

Fig. 8.10. Original DBN model without consideration of the protective layers.

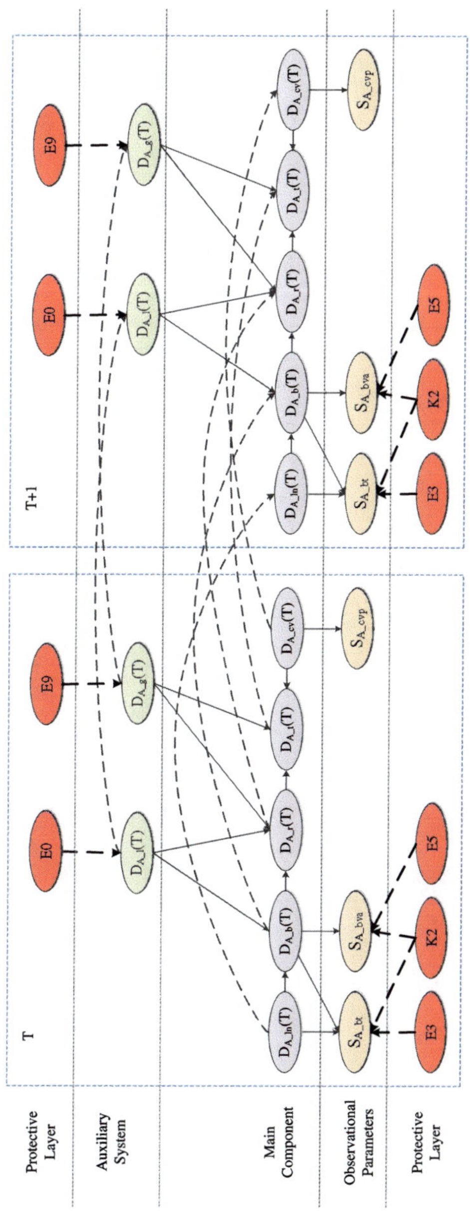

Fig. 8.11. New DBN model of the flue gas turbine subsystem considering the effect of PLs.

Table 8.5. Hidden nodes information for flue gas turbine.

Subsystem	No.	Dynamic nodes	Symbol	States distribution
Flue gas turbine	1	Lubricating oil	D_{A_l}	{1 Normal, 2 Abnormal temperature or pressure}
	2	Flue gas	D_{A_g}	{1 Normal, 2 Abnormal particle density or temperature or pressure}
	3	Lubricating oil nozzle	D_{A_ln}	{1 Normal, 2 Reduction in nozzle diameter, 3 Nozzle blockage}
	4	Bearing	D_{A_b}	{1 Normal, 2 Damaged}
	5	Rotor	D_{A_r}	{1 Normal, 2 Damaged}
	6	Turbine	D_{A_t}	{1 Normal, 2 Damaged}
	7	Cooling steam regulating valve	D_{A_cv}	{1 Normal, 2 Damaged}

Table 8.6. Observational nodes information for flue gas turbine.

Subsystem	No.	Observational nodes (Unit)	Symbol	States distribution	Normal range	Threshold
Flue gas turbine	8	Bearing temperature (°C)	S_{A_bt}	{1 Normal, 2 High, 3 Low}	(72, 95)	{(72, 95); $\geq$95; $\leq$72}
	9	Bearing vibration amplitude (mm)	S_{A_bva}	{1 Normal, 2 High}	(0, 0.05)	{(0, 0.05); $\geq$0.05}
	10	Cooling steam pressure (MPa)	S_{A_cvp}	{1 Normal, 2 High, 3 Low}	(0.6, 1.2)	{(0.6, 1.2); $\geq$1.2; $\leq$0.6}

Table 8.7. PL nodes information for flue gas turbine.

Subsystem	No.	Dynamic nodes	Symbol	States distribution
Flue gas turbine	11	Lubricating oil temperature monitoring system	E0	{1 Normal, 2 Damaged}
	12	Lubricating oil inlet gas monitoring system	E9	{1 Normal, 2 Damaged}
	13	Bearing temperature monitoring system	E3	{1 Normal, 2 Damaged}
	14	Human inspection	K2	{1 Normal, 2 Damaged}
	15	Bearing vibration monitoring system	E5	{1 Normal, 2 Damaged}

Table 8.8. The CPT of the node DA_l with no PL function.

$D_{A\_l}(T+1)$	Normal	Abnormal pressure or temperature
$D_{A\_l}(T)$ = Normal	0.99	0.01
$D_{A\_l}(T)$ = Abnormal pressure or temperature	0.01	0.99

Table 8.9. CPT of DA_l considering the PL protection.

	$D_{A_1}(T+1)$	Normal	Abnormal pressure or temperature
E0 = Normal	$D_{A\_l}(T)$ = Normal	0.99	0.01
E0 = Failed	$D_{A\_l}(T)$ = Normal	0.99	0.01
E0 = Normal	$D_{A\_l}(T)$ = Abnormal pressure or temperature	0.7	0.3
E0 = Failed	$D_{A\_l}(T)$ = Abnormal pressure or temperature	0.01	0.99

CPT of DA_g, SA_bt, SA_bva can also be calculated and shown in Tables 8.10, 8.12, 8.14 in comparison with their original CPT without consideration of the protective layers (shown in Tables 8.11 and 8.13).

Table 8.10. CPT of DA_g considering the PL protection.

	$\mathrm{D_{A\_g}}(T+1)$	Normal	Abnormal pressure or temperature
E9 = Normal	$\mathrm{D_{A\_g}}(T)$ = Normal	0.99	0.01
E9 = Failed	$\mathrm{D_{A\_g}}(T)$ = Normal	0.99	0.01
E9 = Normal	$\mathrm{D_{A\_g}}(T)$ = Abnormal pressure or temperature	0.7	0.3
E9 = Failed	$\mathrm{D_{A\_g}}(T)$ = Abnormal pressure or temperature	0.01	0.99

Table 8.11. CPT of SA_bt without the consideration of the PL protection.

$\mathrm{S_{A\_bt}}$		Normal	High	Low
$\mathrm{D_{A\_ln}}$ = Normal	$\mathrm{D_{A\_b}}$ = Normal	0.9940	0.0045	0.0015
$\mathrm{D_{A\_ln}}$ = Normal	$\mathrm{D_{A\_b}}$ = Failed	0.0075	0.9900	0.0025
$\mathrm{D_{A\_ln}}$ = Reduction in Nozzle diameter	$\mathrm{D_{A\_b}}$ = Normal	0.1500	0.8000	0.0500
$\mathrm{D_{A\_ln}}$ = Reduction in Nozzle diameter	$\mathrm{D_{A\_b}}$ = Failed	0.0045	0.9940	0.0015
$\mathrm{D_{A\_ln}}$ = Nozzle blockage	$\mathrm{D_{A\_b}}$ = Normal	0.0075	0.9900	0.0025
$\mathrm{D_{A\_ln}}$ = Nozzle blockage	$\mathrm{D_{A\_b}}$ = Failed	0.0015	0.9985	0

8.5. Case Study

The new extended DBN model considering the PL effect developed in Sec. 8.4 can be used for failure prognosis when an abnormal event happens. The results will play an important role both in the maintenance decision and the optimization of the safety barriers.

The protective effects of PL in time dimension are given in Sec. 8.5.1 while the effects in space dimension are given in Sec. 8.5.2. Reasoning results are drawn in bar graphs: the x-labels "1", "2", "3" correspond to the nodes information illustrated in Tables 8.5–8.7, representing the different states of each node, the y-axis represents the occurrence probability of each state.

If one or more nodes' states at a certain moment be obtained (by condition monitoring or inspection), the degradation state of other related nodes can also be calculated by the DBN reasoning

Table 8.12. CPT of SA_bt considering the PL protection.

S_{A_bt}				Normal	High	Low
D_{A_ln} = Normal	D_{A_b} = Normal	E3 = Normal	K2 = Normal	0.9940	0.0045	0.0015
D_{A_ln} = Normal	D_{A_b} = Normal	E3 = Normal	K2 = Failed	0.9940	0.0045	0.0015
D_{A_ln} = Normal	D_{A_b} = Normal	E3 = Failed	K2 = Normal	0.9940	0.0045	0.0015
D_{A_ln} = Normal	D_{A_b} = Normal	E3 = Failed	K2 = Failed	0.9940	0.0045	0.0015
D_{A_ln} = Normal	D_{A_b} = Failed	E3 = Normal	K2 = Normal	0.7	0.29	0.1
D_{A_ln} = Normal	D_{A_b} = Failed	E3 = Normal	K2 = Failed	0.7	0.29	0.1
D_{A_ln} = Normal	D_{A_b} = Failed	E3 = Failed	K2 = Normal	0.7	0.29	0.1
D_{A_ln} = Normal	D_{A_b} = Failed	E3 = Failed	K2 = Failed	0.0075	0.9900	0.0025
D_{A_ln} = Reduction in nozzle diameter	D_{A_b} = Normal	E3 = Normal	K2 = Normal	0.7	0.29	0.1
D_{A_ln} = Reduction in nozzle diameter	D_{A_b} = Normal	E3 = Normal	K2 = Failed	0.7	0.29	0.1
D_{A_ln} = Reduction in nozzle diameter	D_{A_b} = Normal	E3 = Failed	K2 = Normal	0.7	0.29	0.1
D_{A_ln} = Reduction in nozzle diameter	D_{A_b} = Normal	E3 = Failed	K2 = Failed	0.1500	0.8000	0.0500
D_{A_ln} = Reduction in nozzle diameter	D_{A_b} = Failed	E3 = Normal	K2 = Normal	0.7	0.29	0.1
D_{A_ln} = Reduction in nozzle diameter	D_{A_b} = Failed	E3 = Normal	K2 = Failed	0.7	0.29	0.1
D_{A_ln} = Reduction in nozzle diameter	D_{A_b} = Failed	E3 = Failed	K2 = Normal	0.7	0.29	0.1
D_{A_ln} = Reduction in nozzle diameter	D_{A_b} = Failed	E3 = Failed	K2 = Failed	0.0045	0.9940	0.0015
D_{A_ln} = Blockage	D_{A_b} = Normal	E3 = Normal	K2 = Normal	0.7	0.29	0.1
D_{A_ln} = Blockage	D_{A_b} = Normal	E3 = Normal	K2 = Failed	0.7	0.29	0.1
D_{A_ln} = Blockage	D_{A_b} = Normal	E3 = Failed	K2 = Normal	0.7	0.29	0.1
D_{A_ln} = Blockage	D_{A_b} = Normal	E3 = Failed	K2 = Failed	0.0075	0.9900	0.0025
D_{A_ln} = Blockage	D_{A_b} = Failed	E3 = Normal	K2 = Normal	0.7	0.29	0.1
D_{A_ln} = Blockage	D_{A_b} = Failed	E3 = Normal	K2 = Failed	0.7	0.29	0.1
D_{A_ln} = Blockage	D_{A_b} = Failed	E3 = Failed	K2 = Normal	0.7	0.29	0.1
D_{A_ln} = Blockage	D_{A_b} = Failed	E3 = Failed	K2 = Failed	0.0015	0.9985	0

Table 8.13. CPT of SA_bva without the consideration of the PL protection.

S_{A_bva}	Normal	High
D_{A_b} = Normal	0.9950	0.0050
D_{A_b} = Failed	0	1

Table 8.14. CPT of SA_bva considering the PL protection.

S_{A_bva}			Normal	High
D_{A_b} = Normal	K2 = Normal	E5 = Normal	0.9950	0.0050
D_{A_b} = Normal	K2 = Normal	E5 = Failed	0.9950	0.0050
D_{A_b} = Normal	K2 = Failed	E5 = Normal	0.9950	0.0050
D_{A_b} = Normal	K2 = Failed	E5 = Failed	0.9950	0.0050
D_{A_b} = Failed	K2 = Normal	E5 = Normal	0.7	0.3
D_{A_b} = Failed	K2 = Normal	E5 = Failed	0.7	0.3
D_{A_b} = Failed	K2 = Failed	E5 = Normal	0.7	0.3
D_{A_b} = Failed	K2 = Failed	E5 = Failed	0	1

scheme which is introduced in Sec. 8.2. The fault propagation paths can be deduced by reasoning which are beneficial to make proper maintenance plans or safety response programs. Furthermore, the variation trend of the probability of each node in each state can also be deduced in the time dimension. By comparing the effect of different PLs in the same degradation/failure scenario, the optimal safety barriers can be selected, and the best intervention time of each PL can be determined.

8.5.1. *Failure prognosis in time dimension*

By condition monitoring and fault diagnosis system, the bearing gets degraded to a certain degree, the probability of "normal" state is set as 0.7, while the probability of "normal" state of rotor is set as 0.5. Other dynamic nodes stay in the normal state with the probability as 100%. Degradation trend of dynamic nodes and deviation tendency in the following time slices can be calculated by the model developed in Sec. 8.4.

Reasoning results under with and without PL conditions are calculated, respectively. Deviation trend of bearing temperature

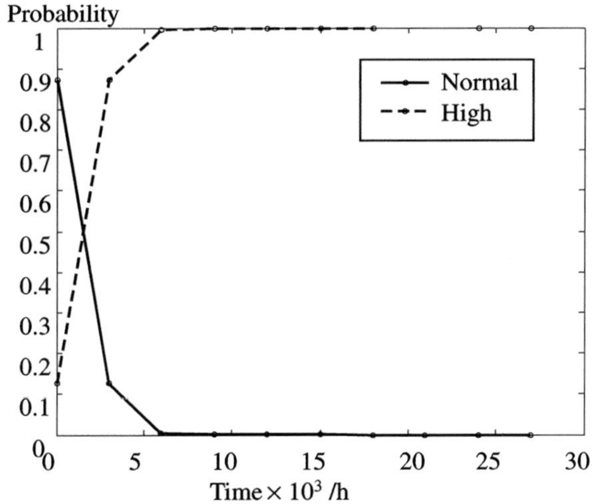

Fig. 8.12. Failure prognosis of the bearing temperature without PL protection.

without PLs is shown in Fig. 8.12. As time goes by, the bearing gets degraded gradually, with the increasing possibility of the state as "high bearing temperature" and the reducing possibility of the state as "normal". The variation trends of the bearing temperature considering the effect of PLs are shown in Fig. 8.13, from which the probabilities of "normal" and "high" states will be in the peak alternately. Since the original state of the bearing temperature is "normal" with the probability as 0.7 and the probability of the "normal" state of rotator is 0.5, the degraded rotator will accelerate the degradation process of bearing.

In the next time slice, the degradation state is deteriorated seriously, making the bearing temperature increase and the probability of the state as "high temperature" goes up. At this moment, the protective layer works, and the temperature is adjusted to normal. In the following time slices, as further degradation of the bearing occurs its temperature increases again, i.e., once the bearing temperature deviates from the normal state, it will be regulated in the next time slice. Therefore, the probability peaks of "high temperature" and "normal temperature" occur alternately.

Figure 8.14 shows the failure trend of the bearing in terms of the probability of the "normal" state of the bearing temperature by

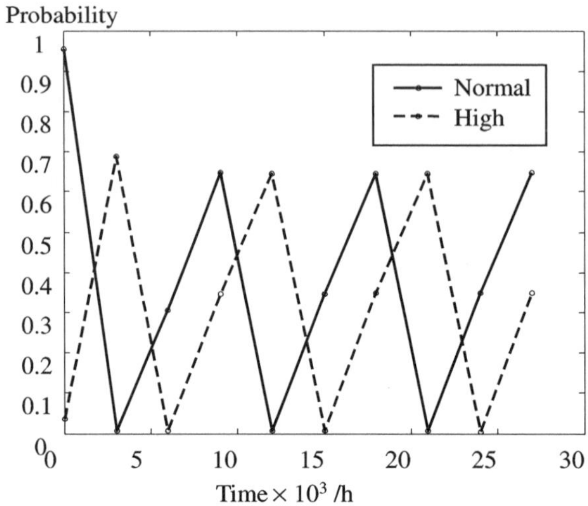

Fig. 8.13. Failure prognosis of bearing temperature with PL protection.

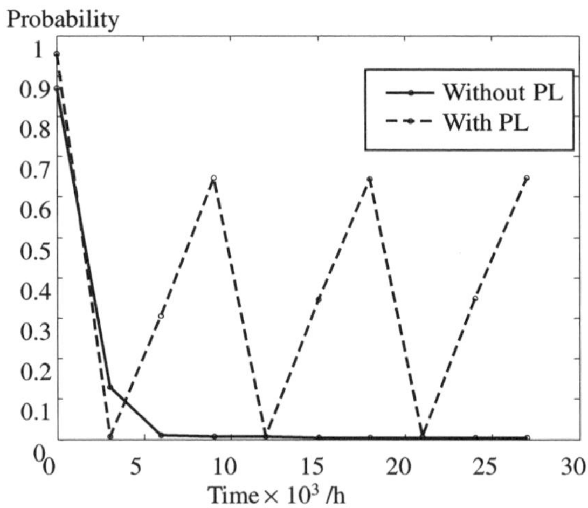

Fig. 8.14. Failure prognosis of the bearing temperature in "normal" state by comparison of the situation that with and without PL protection.

comparing the situations with and without PL protection. Despite being in the first 3000 hours, the probabilities of bearing temperature in the "normal" state in both the situations show a declining trend both with and without protective barrier, obviously after 3000 hours, the probability of the bearing temperature in the "normal" state remains at a low level without PL protection, while in the situation of the PL works, the temperature is continuously being adjusted to normal.

8.5.2. *Failure prognosis in space dimension*

When an abnormal event happened, the temperature of the bearing was 98°C, which was out of scope compared to the normal range [72, 95]°C, while other observable nodes were in the "normal" states. By the principle presented in Sec. 8.2, the probabilities of the degradation states (i.e. hidden nodes) of subsystems or components were calculated as shown in Fig. 8.15 and Table 8.15, according to which the root causes leading to the abnormal event could be deduced. By comparing the situations with and without

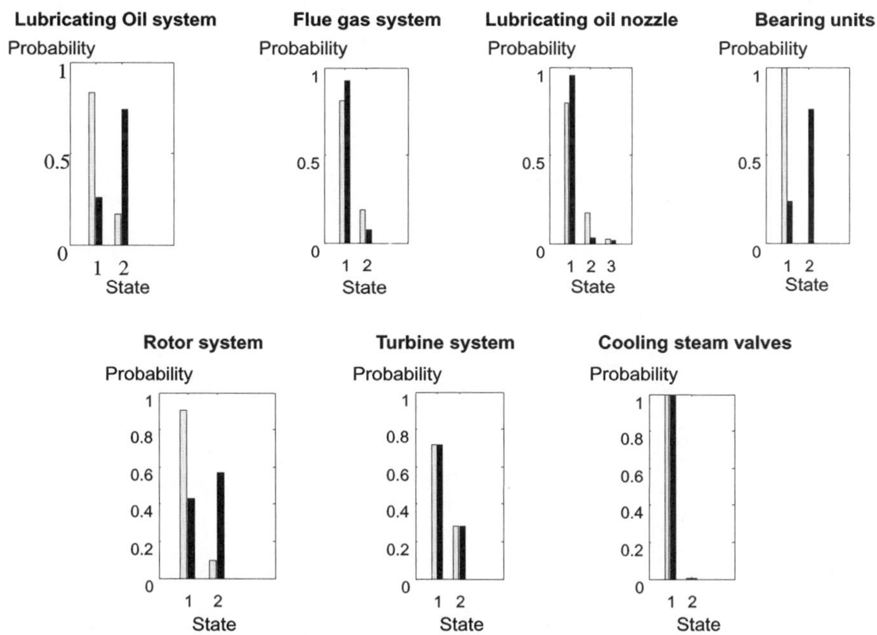

Fig. 8.15. Comparison of reasoning results for flue gas turbine under single node abnormality condition with and without PL protection.
(**Note*: The white bar on the left represents out of consideration of PL protection; while the black bar on the right represents considering the effect of PL protection).

Table 8.15. Backward reasoning results under single node abnormality of flue gas turbine.

Dynamic nodes	Symbol	Consideration of PL	State ID	State	Probability (%)
Lubricating oil system	D_{A_l}	True	1	Normal	26.09
			2	Abnormal temperature or pressure	73.91
		False	1	Normal	83.13
			2	Abnormal temperature or pressure	16.87
Flue gas system	D_{A_g}	True	1	Normal	92.51
			2	Abnormal temperature or pressure or particle density	7.49
		False	1	Normal	81.01
			2	Abnormal temperature or pressure or particle density	18.99
Cooling steam valves	D_{A_cv}	True	1	Normal	99.35
			2	Failed	0.65
		False	1	Normal	99.48
			2	Failed	0.52
Lubricating oil nozzle	D_{A_ln}	True	1	Normal	95.39
			2	Reduction in nozzle diameter	2.87
			3	Blockage	1.74
		False	1	Normal	79.79
			2	Reduction in nozzle diameter	17.53
			3	Blockage	2.68
Bearing units	D_{A_b}	True	1	Normal	23.42
			2	Damaged	76.58
		False	1	Normal	100
			2	Damaged	0

(*Continued*)

Table 8.15. (*Continued*)

Dynamic nodes	Symbol	Consideration of PL	State ID	State	Probability (%)
Rotor system	D_{A_r}	True	1	Normal	43.16
			2	Damaged	56.84
		False	1	Normal	90.87
			2	Damaged	9.13
Turbine system	D_{A_t}	True	1	Normal	71.79
			2	Damaged	28.21
		False	1	Normal	71.55
			2	Damaged	28.45

considering the effect of PL protection in Fig. 8.15 and Table 8.15, the degradation of the lubricating oil system, bearing units, and rotor system with higher probabilities of degradation states was revealed when considering the PL effects, while if the PL effects were not considered, the above hidden problems would be ignored, since their probabilities of being in the normal state were higher.

8.6. Conclusion

(1) This chapter proposed an extended DBN-based failure prognosis method for complex systems considering the effect of the protective layers. Not only the interaction between components, but also the influence of the layers of protection in the system is considered when the dynamic failure scenarios were analyzed. During the DBN modeling process, comprehensive information about the failure process was included, such as dynamic degradation mechanism, parameter deviation, the response of the protective layers, and also the external environment.

(2) Some key problems during modeling were discussed, such as how to determine the PL nodes, how to integrate the PL nodes into the original DBN model, and how to update the CPT after PL works.

(3) With the proposed model, the dynamic influence diagram of the components' degradation trends can be calculated and used

to evaluate the different effects of the layers of protection quantitatively. In the case study, the proposed method was tested on the FGERS which was widely used in the petrochemical industry to demonstrate its effectiveness. It is of great help for early warning and optimization of the layers of protection in the complex industrial system.

References

[1] D. Acuña, M. Orchard, "Particle-filtering-based failure prognosis via sigma-points: Application to lithium-ion battery state-of-charge monitoring," *Mechanical Systems and Signal Processing*, vol. 85, pp. 827–848, 2017.

[2] S. Al-Dahidi *et al.*, "Remaining useful life estimation in heterogeneous fleets working under variable operating conditions," *Reliability Engineering & System Safety*, vol. 156, pp. 109–124, 2016.

[3] D. Barraza-Barraza *et al.*, "An adaptive ARX model to estimate the RUL of aluminum plates based on its crack growth," *Mechanical Systems and Signal Processing*, vol. 82, pp. 519–536, 2017.

[4] CCPS — Center of Chemical Process Safety, *Layer of Protection Analysis: Simplified Process Risk Assessment*, American Institute of Chemical Engineers, Center of Chemical Process Safety, New York, 2001.

[5] J. Dai *et al.*, "Reliability risk mitigation of free air cooling through prognostics and health management," *Applied Energy*, vol. 111, pp. 104–112, 2013.

[6] V. Dianous, C. Fievez, "ARAMIS project: A more explicit demonstration of risk control through the use of bow-tie diagrams and the evaluation of safety barrier performance," *Journal of Hazardous Materials*, vol. 130, pp. 220–233, 2006.

[7] N. Duijm, "Safety-barrier diagrams as a safety management tool," *Reliability Engineering & System Safety*, vol. 94, pp. 332–341, 2009.

[8] O. Eker, F. Camci, I. Jennions, "Physics-based prognostic modelling of filter clogging phenomena," *Mechanical Systems and Signal Processing*, vol. 75, pp. 395–412, 2016.

[9] J. Fan, K. Yung, M. Pecht, "Predicting long-term lumen maintenance life of LED light sources using a particle filter-based prognostic approach," *Expert Systems with Applications*, vol. 42, no. 5, pp. 2411–2420, 2015.

[10] Y. Hong *et al.*, "A fuzzy logic and probabilistic hybrid approach to quantify the uncertainty in layer of protection analysis," *Journal of Loss Prevention in the Process Industries*, vol. 43, pp. 10–17, 2016.

[11] J. Hu *et al.*, "Fault propagation behavior study and root cause reasoning with dynamic Bayesian network based framework," *Process Safety and Environmental Protection*, vol. 97, pp. 25–36, 2015.

[12] J. Hu *et al.*, "An integrated method for safety pre-warning of complex system," *Safety Science*, vol. 48, pp. 580–597, 2010.

[13] J. Hu, L. Zhang, L. Ma, W. Liang, "An integrated safety prognosis model for complex system based on dynamic Bayesian network and ant colony algorithm," *Expert Systems with Applications*, vol. 38, pp. 1431–1446, 2011.

[14] J. Hu, L. Zhang, W. Liang, "Opportunistic predictive maintenance for complex multi-component systems based on DBN-HAZOP model," *Process Safety and Environmental Protection*, vol. 90, no. 5, pp. 376–388, 2012.

[15] M. Jouin *et al.*, "Degradations analysis and aging modeling for health assessment and prognostics of PEMFC," *Reliability Engineering & System Safety*, vol. 148, pp. 78–95, 2016.

[16] M. Kan, A. Tan, J. Mathew, "A review on prognostic techniques for non-stationary and non-linear rotating systems," *Mechanical Systems and Signal Processing*, vol. 62–63, pp. 1–20, 2015.

[17] N. Khakzad, F. Khan, P. Amyotte, "Risk-based design of process systems using discrete-time Bayesian networks," *Reliability Engineering & System Safety*, vol. 109, pp. 5–17, 2013.

[18] J. Liu *et al.*, "A data-model-fusion prognostic framework for dynamic system state forecasting," *Engineering Applications of Artificial Intelligence*, vol. 25, no. 4, pp. 814–823, 2012 (Special Section: Dependable System Modelling and Analysis).

[19] J. Liu, E. Zio, "System dynamic reliability assessment and failure prognostics," *Reliability Engineering & System Safety*, vol. 160, pp. 21–36, 2017.

[20] Z. Liu *et al.*, "Dynamic Bayesian network modeling of reliability of subsea blowout preventer stack in presence of common cause failures," *Journal of Loss Prevention in the Process Industries*, vol. 38, pp. 58–66, 2015.

[21] A. Muller, M. Suhner, B. Iung, "Formalisation of a new prognosis model for supporting proactive maintenance implementation on industrial system," *Reliability Engineering & System Safety*, vol. 93, pp. 234–253, 2008.

[22] K. Murphy, *Dynamic Bayesian Networks: Representation, Inference and Learning*, Ph.D. thesis, University of California, Berkeley, 2002.

[23] V. Nistane, S. Harsha, "Failure evaluation of ball bearing for prognostics," *Procedia Technology*, vol. 23, pp. 179–186, 2013.

[24] H. J. Pasman, B. Knegtering, W. J. Rogers,"A holistic approach to control process safety risks: Possible ways forward," *Reliability Engineering & System Safety*, vol. 117, pp. 21–29, 2013.

[25] Y. Peng, M. Dong, M. Zuo, "Current status of machine prognostics in condition-based maintenance: A review," *The International Journal of Advanced Manufacturing Technology*, vol. 50, no. 1–4, pp. 297–313, 2010.

[26] N. Raghavan, D. Frey, "Particle filter approach to lifetime prediction for microelectronic devices and systems with multiple failure mechanisms," *Microelectronics Reliability*, vol. 55, no. 9–10, pp. 1297–1301, 2015.

[27] N. Ramzali, M. R. M. Lavasani, J. Ghodousi, "Safety barriers analysis of offshore drilling system by employing fuzzy event tree analysis," *Safety Science*, vol. 78, pp. 49–59, 2015.

[28] J. Sikorska, M. Hodkiewicz, L. Ma, "Prognostic modelling options for remaining useful life estimation by industry," *Mechanical Systems and Signal Processing*, vol. 25, no. 5, pp. 1803–1836, 2011.

[29] J. Sun *et al.*, "Prognostics uncertainty reduction by fusing on-line monitoring data based on a state-space-based degradation model," *Mechanical Systems and Signal Processing*, vol. 45, no. 2, pp. 396–407, 2014.

[30] T. Sutharssan *et al.*, "A review on prognostics and health monitoring of proton exchange membrane fuel cell," *Renewable and Sustainable Energy Reviews*, DOI:10.1016/j.rser.2016.11.009, 2016.

[31] X. Tan *et al.*, "A novel approach of testability modeling and analysis for PHM systems based on failure evolution mechanism," *Chinese Journal of Aeronautics*, vol. 26, no. 3, pp. 766–776, 2013.

[32] H. Wang, "Prognostics and health management for complex system based on fusion of model-based approach and data-driven approach," *Physics Procedia*, vol. 24, pp. 828–831, 2012 (International Conference on Applied Physics and Industrial Engineering 2012).

[33] Y. Wang, M. Xie, "Approach to integrate fuzzy fault tree with Bayesian network," *Procedia Engineering*, vol. 45, pp. 131–138, 2012.

[34] P. Weber *et al.*, "Overview on Bayesian networks applications for dependability, risk analysis and maintenance areas," *Engineering Applications of Artificial Intelligence*, vol. 25, pp. 671–682, 2012.

[35] J. Yu, "Machine health prognostics using the Bayesian-inference-based probabilistic indication and high-order particle filtering framework," *Journal of Sound and Vibration*, vol. 358, pp. 97–110, 2015.

[36] M. Zaidan *et al.*, "Bayesian Hierarchical Models for aerospace gas turbine engine prognostics," *Expert Systems with Applications*, vol. 42, no. 1, pp. 539–553, 2015a.

[37] M. Zaidan, R. Relan, A. Mills, R. Harrison, "Prognostics of gas turbine engine: An integrated approach," *Expert Systems with Applications*, vol. 42, no. 22, pp. 8472–8483, 2015b.

[38] E. Zio, F. Di Maio, M. Stasi, "A data-driven approach for predicting failure scenarios in nuclear systems," *Annals of Nuclear Energy*, vol. 37, no. 4, pp. 482–491, 2010.

Chapter 9

Fault Diagnosis for a Solar-Assisted Heat Pump System Under Incomplete Data and Expert Knowledge

Fault diagnosis for a solar-assisted heat pump (SAHP) system in the presence of incomplete data and expert knowledge is discussed in this chapter. A method for parameter learning of Bayesian networks (BNs) from incomplete data based on the back-propagation (BP) neural network and maximum likelihood estimation (MLE), which is called BP-MLE method, is presented. The BP neural network is utilized to impute the missing data and the complete data sets are addressed with MLE to obtain the parameters of BN. A method for parameter estimation under incomplete expert knowledge based on BP neural networks and fuzzy set theory is also presented, which is called BP-FS method. Similarly, the missing information is imputed by the trained BP neural network. Fuzzy set theory is employed to quantify the parameters of BN based on the complete qualitative expert knowledge. The presented methods are applied to parameter learning of diagnostic BN for an SAHP system with incomplete simulation data and expert knowledge. The developed BN can perform fault diagnosis with complete or incomplete symptoms.

9.1. Introduction

A solar-assisted heat pump (SAHP) system integrates a heat pump with solar collectors to take advantage of solar energy as an evaporating heat source, which can achieve a high coefficient of performance [1]. Until now, plenty of theoretical and experimental researches have been done on SAHP systems. Mohanraj *et al.* [2] developed an artificial neural network (ANN) to predict the performance of a direct expansion SAHP and the reported results demonstrate that the proposed method is acceptable. Liang *et al.* [3] presented a new SAHP system with flexible operational modes to improve the performance of the heating system and the developed system validates the established mathematical model. Li and Yang [4] investigated the application of the SAHP system for hot water production in Hong Kong. Chow *et al.* [5] described a case study with a new design of SAHP for indoor swimming pool space- and water-heating purpose.

Since failures in pump system will cause the occurrence of abnormal operation and degradation in performance, fault diagnosis of the system is beneficial for saving energy and operating cost. Some fault diagnosis methods have been developed by the researchers. Zhao *et al.* [6] presented a new fault detection and diagnosis method based on support vector regression and the exponentially-weighted moving average control charts for centrifugal chillers of building air-conditioning systems. Chen and Lan [7] proposed a fault detection method based on the principle component analysis model to detect the faults in air-source heat pump water chillers/heaters. Zogg *et al.* [8] developed a model-based fault diagnosis system for commercial heat pumps, which is based on parameter identification and vector clustering techniques. Zhao *et al.* [9] presented a three-layered diagnostic Bayesian network (BN) to make use of more useful information of the chiller concerned and expert knowledge. Cai *et al.* [10] presented a multi-source information fusion-based fault diagnosis method for ground-source heat pump system to increase the diagnostic accuracy. Najafi *et al.* [11] have developed diagnostic algorithms for air handling units using machine-learning techniques.

Recently, BNs for fault diagnosis have been widely developed in a variety of fields including electrical power systems [12], telecommunication networks [13, 14], rotating machinery [15], airplane engine [16], and others. A BN is a directed acyclic graph composed of nodes and arcs among the nodes. In a BN, nodes denote random variables and the directed arcs mean the conditional dependencies among variables [17].

A BN consists of parameters and structure, which can be defined by expert knowledge or obtained by machine learning with data sets. The former can reflect experts' knowledge, but the process is difficult and time-consuming. Besides, it is unsure if the network presented by the experts is really the most proper model. Although learning from data is able to overcome the problems of the former, data are not always available because they cannot be ready at all times when the BN is constructed. In addition, the change of environments is not considered [18]. Generally, there are two kinds of Bayesian learning problems: parameter learning and structure learning. For parameter learning, one is to learn BN parameters when its structure is known, and the other is to learn both the structure and parameters at the same time. This chapter focuses on BN parameter learning with a known BN structure. Parameter learning is divided into two categories. If the data is complete, maximum likelihood estimation (MLE) or Bayesian estimation method can be used to learn the parameters. However, in the real world, to find the complete data for learning is difficult for various reasons. Some of the variables may be difficult or even impossible to observe. The presence of missing data leads to analytical intractability and complex computation compared with the complete data scenario since the incomplete sample sets might reduce the accuracy of parameters. The easiest way to deal with incomplete data is to delete it directly. Instead of discarding incomplete data sets, only relevant information may be deleted [19]. In addition, throwing away data can lead to estimates with larger standard errors due to the reduced sample size. Rather than deleting the incomplete data sets, another approach is to impute the missing values. This method keeps the full sample size, which can be advantageous for bias and precision [20]. Many methods have been proposed to learn BN parameters when the data are incomplete. The most common learning algorithm is expectation maximization (EM) [21]. Recently, many researchers have proposed some new approaches for parameter learning under incomplete data. Pernkopf and Bouchaffra [22] present a genetic-based EM algorithm for learning Gaussian mixture models for multivariate data and the algorithm is less sensitive to the initialization compared to the standard EM. Majdi-Nasab *et al.* [23] proposed new approaches based on genetic algorithms, simulated annealing, and EM for parameter learning of the mixture Gaussian model. Huda *et al.*

[24] presented a hybrid algorithm for the estimation of the hidden Markov model in automatic speech recognition using a constraint-based evolutionary algorithm and EM and the presented algorithm overcomes the problem of EM converging to a local optimum. Liao and Ji [25] proposed a learning algorithm incorporating qualitative constraints of domain knowledge on some of the parameters into the learning process and this algorithm is able to regularize the otherwise ill-posed problem, limit the search space, and avoid local optima.

ANN is a powerful tool in the modeling of nonlinear multivariate systems. It is able to capture the complicated nonlinear relationships between inputs and outputs by proper training. Among ANN methods, back-propagation (BP) neural network is the most widely used training algorithm. BP neural network is a multilayer feed-forward that is trained by the error BP algorithms. A classical BP neural network has three layers: input layer, hidden layer, and output layer. Input layer receives and distributes the input pattern. Hidden layer establishes the nonlinearities of the input and output relationship. Output layer produces the output pattern [26]. BP neural network has provided effective solutions to quality prediction, prediction of the mechanical properties, prediction of various stock indices and oil reservoir prediction, etc. [27–30].

Fuzzy sets were introduced to represent/manipulate data and information processing nonstatistical uncertainties [31]. To mathematically represent uncertainty and vagueness, it is a formalized tool for dealing with the imprecision intrinsic to many problems. Fuzzy set theory has been widely used in different fields of application including risk assessment, rock mass classification, radiation therapy, decision support system, and pattern recognition [32–36].

In the real system, incomplete data is a common phenomenon, which could be caused by a sudden mechanical breakdown, hardware sensor failure or data acquisition system malfunction, etc. [37]. Another increasing common source for the incomplete data problem is the integration of communication networks and the subsequent potential for data losses and packet dropouts [38]. Sensor failures are only one of the possible reasons that lead to incomplete data.

The data sets for learning BN parameters refer to statistical samples of the equipment or systems. There might be unreliable data samples in the complete data sets due to undetected sensor faults. However, preventive maintenance of and repair actions on the equipment can greatly reduce the existence of various faults in practice. Besides, there is a large quantity of samples. Therefore, unreliable samples account for only a tiny proportion of the whole data sets. It is rational that most of the data sets can be considered to be reliable. Great efforts should be paid to collect the sample sets. It is possible that there might be unreliable samples due to various faults. But even if there are none of these faults, human error might also lead to some unreliable samples. Therefore, uncertainty is unavoidable in the data sets. Fortunately, BN is a very powerful tool in uncertainty representation and reasoning. In order to delete the samples with wrong value or outlier, some data analysis methods in statistics can be used for data preprocessing, such as expert judgments, Chebyshev's theorem, distance-based clusters, pattern recognition, etc. In addition, experts might not be able to provide complete qualitative knowledge because they are not familiar with the concerned issues. In this chapter, fault diagnosis of an SAHP system in the presence of incomplete data and expert knowledge is discussed. Based on the BP neural network and MLE, BP-MLE method is presented for the parameter learning of BN from incomplete data. The BP neural network is used to impute the missing data and then the complete data sets are addressed with MLE to obtain the parameters of BN. Based on the BP neural network and fuzzy set theory, BP-FS method is proposed for parameter estimation under incomplete expert knowledge. Similarly, the missing information is imputed by the trained BP neural network. Fuzzy set theory is used to quantify the parameters of BN based on the complete qualitative expert knowledge.

Firstly, the missing information is reconstructed to be complete data sets. Then, all the data sets are used to determine the parameters of the developed diagnostic BN. Finally, based on the developed BN, fault diagnosis can be performed by using observed symptoms. Due to the powerful reasoning capacity of the

BN, it can perform fault diagnosis based on complete or incomplete information. The reminder of this chapter is organized as follows. Section 9.2 provides a description of the proposed methods. The presented methods are applied to parameter estimation of an SAHP system in Sec. 9.3. Section 9.4 performs fault diagnosis based on developed BNs. Section 9.5 summarizes the chapter.

9.2. The Proposed Methods

9.2.1. *BP-MLE method under incomplete data*

Figure 9.1 shows the flow chart of the proposed parameter learning method under incomplete data. The incomplete data is composed of complete and incomplete sample sets. The complete sample sets are utilized to train BP neural networks. When a BP network is trained,

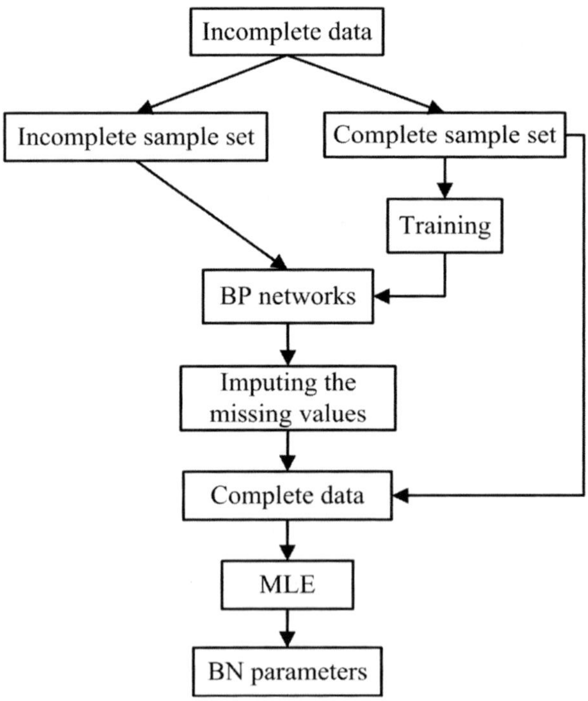

Fig. 9.1. Flow chart of the proposed parameter learning method under incomplete data.

parameters such as the number of hidden layers, hidden neurons, input neurons, and output neurons are needed. The incomplete sample sets are entered into the trained networks and the missing values will be imputed. The incomplete samples with the same missing variable are grouped into one category. Hence, the same trained BP neural network is used to predict the missing values in a category, through which the efficiency can be improved. After imputation, the initially incomplete data becomes complete data. With the MLE method, the parameters of the BN are obtained.

MLE is a frequency estimation approach, which determines parameters by calculating the available frequencies from data [39]. Let r_i represent the cardinality of X_i, and q_i denote the cardinality of the parent set of X_i. The kth probability value of a conditional probability distribution $P(X_i|\text{pa}(X_i) = j)$ can be denoted as $\theta_{ijk} = P(X_i = k|\text{pa}(X_i) = j)$, where $\theta_{ijk} \in \theta$, $1 \leq i \leq n$, $1 \leq j \leq q_i$, and $1 \leq k \leq r_i$. Assuming $D = \{D_1, D_2, \ldots, D_N\}$ is a dataset of complete cases for a BN, then the lth case D_l is a vector of values of each variable. The log-likelihood function of θ given data D is

$$l(\theta|D) = \log P(D|\theta) = \log \prod_l P(D_l|\theta) = \sum_l \log P(D_l|\theta). \tag{9.1}$$

Let N_{ijk} be the number of data records in sample D, where X_i takes its kth value and its parent $\text{pa}(X_i)$ takes its jth value. Then $l(\theta|D)$ can be rewritten as $l(\theta|D) = \sum_{ijk} N_{ijk} \log \theta_{ijk}$. The MLE tries to estimate θ by maximizing $l(\theta|D)$. Therefore, the estimation of each parameter is obtained as follows:

$$\theta^*_{ijk} = \frac{N_{ijk}}{\sum_{k=1}^{r_i} N_{ijk}}. \tag{9.2}$$

9.2.2. *BP-FS method under incomplete expert knowledge*

A key problem is that it is difficult to get experts with domain knowledge to provide explicit (and accurate) probability values. Recent research has shown that experts feel more comfortable providing qualitative judgments and these are more robust than their numerical assessments [40, 41]. To solve this difficulty, linguistic variables are adopted to represent the parameters of BN. Using linguistic expression to estimate subjective events is natural. A linguistic variable is defined as a variable whose values are sentences in a natural or artificial language. For example, the values of parameters can be expressed as very low, low, fair low, medium, fair high, high, and very high. In this study, expert knowledge on the parameters of BN is represented by linguistic variables.

Figure 9.2 shows the flow chart of the proposed method for parameter learning under incomplete expert knowledge. Before imputation, the linguistic values are denoted by numbers. Complete sample sets are used to train BP neural networks, which are employed

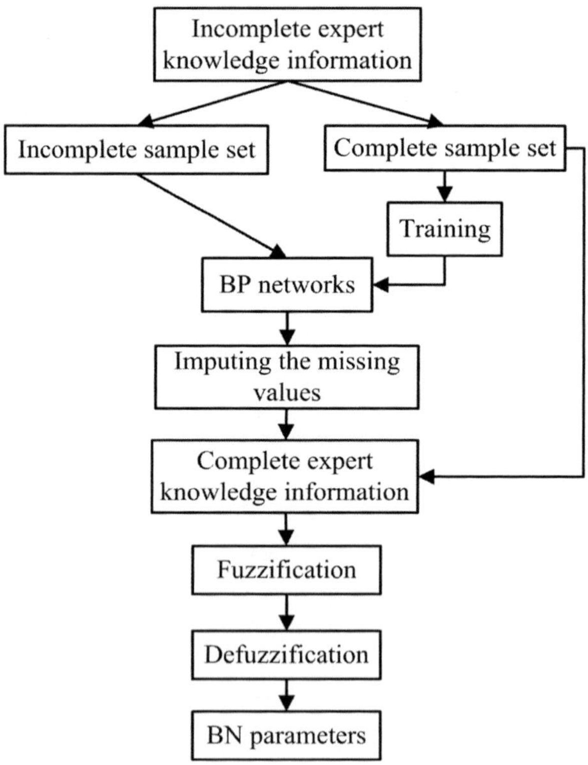

Fig. 9.2. Flow chart of the proposed parameter estimation method under incomplete expert knowledge.

to impute the missing values of the incomplete sample sets. After fuzzification and defuzzification, parameters of the BN are obtained according to fuzzy set theory. The details of the method are described in the following section.

Figures 9.1 and 9.2 show the main procedures of the presented BP-MLE and BP-FS methods, respectively. In both the methods, BP networks are used to impute the missing values of the incomplete sample sets and expert knowledge. For BP-MLE method in Fig. 9.1, the MLE is used to obtain the parameters after the missing data are imputed. However, Fig. 9.2 shows that fuzzy set theory is applied to calculate the BN parameters after the imputation. Therefore, they have different input data and different methods to calculate the BN parameters after the imputation.

9.3. Application of the Proposed Methods in an SAHP System

As known, SAHP system has two different types: direct and indirect expansion [42]. In this chapter, an indirect expansion SAHP system is studied. Its schematic diagram is shown in Fig. 9.3. A SAHP system is mainly composed of a solar collector, an evaporator, a compressor, a condenser, and an expansion valve. An expansion valve is a component that controls the amount of refrigerant flowing

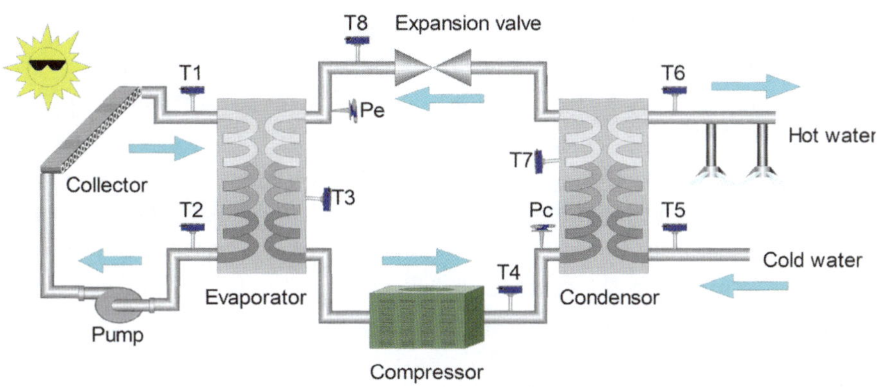

Fig. 9.3. Schematic diagram of SAHP system.

into the evaporator thereby controlling the superheating at the outlet of the evaporator. The two commonly used types are thermal and electronic expansion valves. In this chapter, an electronic expansion valve is used [10]. As shown in the figure, solar energy is collected by the solar collector to heat the water. The evaporator draws energy from solar-heated water and sends the working fluid to the compressor. Compressed by the compressor, the working fluid goes into the condenser. On the condenser, the working fluid releases heat to the space heating water, which is ready for the users. Finally, the refrigerant enters into the expansion valve and gets transported to the evaporator afterwards [43].

9.3.1. *Structure of the BN*

The relevant faults occurring in a heat pump are divided into hard faults and soft faults. Generally, hard faults are simple to detect and identify, but soft faults are more difficult to diagnose. In this chapter, six common soft faults are imposed as follows: (a) refrigerant leakage (RL); (b) refrigerant overcharge (RO); (c) fouling of the condenser (FC); (d) fouling of the evaporator (FE); (e) excessive lift of expansion valve (ELEV); and (f) blocking of liquid pipeline (BLP) [7, 10]. When a fault appears, status of the system will change, which can be used for identifying the fault. As shown in Fig. 9.3, eight temperature sensors are used for monitoring the status of key locations in the SAHP system, which are denoted by T1–T8. Besides, two pressure sensors are used to monitor the evaporating and condensing pressure in the system. The fault symptoms include: (a) evaporator water temperature difference (EWTD); (b) evaporator temperature (ET); (c) compressor discharge temperature (CDT); (d) condenser temperature (CT); (e) condenser water temperature difference (CWTD); (f) expansion valve discharge temperature (EVDT); (g) evaporating pressure (EP); and (h) condensing pressure (CP). In the system, temperature sensors are stuck on the evaporator and condenser to obtain the temperature. Therefore, the ET and CT do not refer to the temperature of the refrigerant inside the evaporator and condenser. The

Table 9.1. Relationship between faults and symptoms.

	Faults					
Symptoms	RL	RO	FC	FE	ELEV	BLP
EWTD	Lower	Normal	Lower	Higher	Lower	Lower
ET	Lower	Higher	Higher	Lower	Higher	Lower
CDT	Higher	Lower	Lower	Lower	Lower	Higher
CT	Lower	Higher	Lower	Lower	–	Lower
CWTD	Lower	Higher	Lower	Lower	Normal	Higher
EVDT	Lower	Higher	Higher	Lower	Higher	Lower
EP	Lower	Higher	Lower	Lower	Higher	Lower
CP	Lower	Higher	Higher	Lower	Higher	Higher

relationship between faults and symptoms obtained from sensor data is given in Table 9.1, which is obtained based on the engineering judgment and reference reviews about the heat pump systems [7, 10].

Based on the relationship between faults and symptoms of an SAHP system, its BN structure for fault diagnosis is developed in Fig. 9.4. The established BN is a two-layered network, composed of six fault nodes and eight symptom nodes. The relationship is denoted by arcs between the faults and symptoms. For example, fault node RL is related to all the eight symptoms. Each fault node has two states: present and absent. Each symptom node has three states: lower, normal, and higher. After the structure is developed, parameters of the nodes need to be determined. In the real world, data or expert knowledge is often incomplete. It is tempting to discard the incomplete data sets, but the relevant information may be deleted and the reduced sample size will lead to estimates with larger standard errors. In order to keep the full sample size and reduce the estimate errors, the incomplete data sets and expert knowledge are used to obtain the BN parameters with the proposed methods in this chapter. Therefore, conditional probabilities of symptom nodes are learned by the proposed BP-MLE method and prior probabilities of the fault nodes are estimated by the presented BP-FS method.

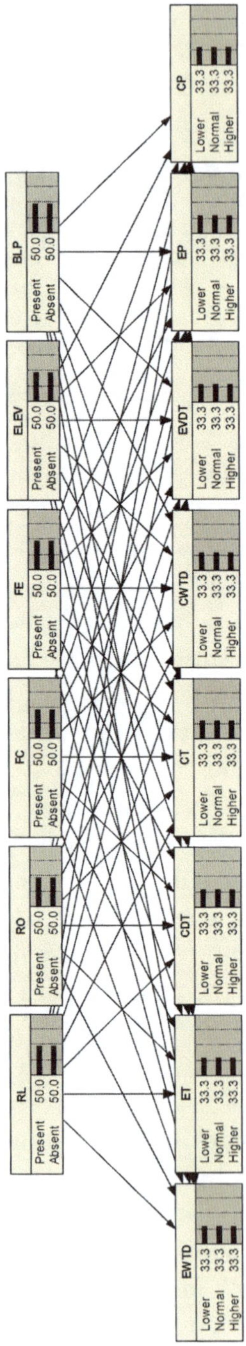

Fig. 9.4. Developed BN structure for fault diagnosis of SAHP system.

9.3.2. *Parameter learning of conditional probabilities with incomplete data*

Assume that there is incomplete data composed of 4000 samples shown in Table 9.2 for determining the parameters of this case. The values in brackets are the missing ones. The imputed values are listed in braces. For the fault nodes, "present" is denoted by 2 and "absent" is denoted by 1. For the symptom nodes, lower, normal, and higher are denoted by 1, 2, and 3, respectively. With a missing rate of 20%, there are 800 incomplete sample sets and 3200 complete sample sets. Only using the complete sample sets will reduce the sample size and decrease the diagnostic accuracy. Consequently, deletion technique should be used only in situations where the amount of missing values is very small [44]. For the incomplete data with more than 5% missing rate, imputation technique is usually used [45, 46]. In this chapter, 4000 data samples with 20% missing rate are used to establish 1536 conditional probabilities for the developed BN. Therefore, it is appropriate to use the incomplete data samples to establish the parameters. For sample S2, the state of "EWTD" is missing. It might be caused by a sudden hardware sensor failure or data acquisition system malfunction. Other reasons are data losses and packet dropouts in communication networks. Hence, a BP neural network with the output of "EWTD" and inputs of the other thirteen nodes is needed to develop for sample S2. With 3200 complete sample sets, the BP neural network can be trained to predict the missing value. Similarly, a BP neural network with the output of "FC" and inputs of the other variables is needed for sample S3. In fact, the

Table 9.2. The incomplete data and imputed values.

	RL	RO	FC	FE	ELEV	BLP	EWTD	ET	CDT	CT	CWTD	EVDT	EP	CP
S1	1	2	1	2	1	2	3	3	3	3	3	3	1	3
S2	1	1	2	1	1	2	[2] {2.0}	2	3	2	3	3	2	2
S3	1	2	[1] {0.98}	1	1	1	1	3	2	1	3	3	3	3
⋮	⋮	⋮	⋮	⋮	⋮	⋮	⋮	⋮	⋮	⋮	⋮	⋮	⋮	⋮
S3999	2	1	1	2	2	1	3	3	3	2	2	3	3	3
S4000	2	2	1	1	2	2	2	3	3	3	3	3	1	3

incomplete samples with the same missing variable are classified as one category, which use the same trained BP neural network for predicting the missing values.

In this model, fourteen BP neural networks are established. A three-layered BP neural network with one hidden layer is developed. The number of neurons in the hidden layer is selected by an empirical formula and is ten. Thirteen input neurons and one output neuron are determined. MATLAB software is used to develop the BP neural network.

After the missing values are imputed, the complete data is obtained. With the MLE method, parameter θ^*_{ijk} can be calculated from the complete data by Eq. (9.2). In order to test the accuracy and efficiency of parameter estimation, the proposed method is compared with the EM method and MLE method. The EM algorithm is a widely used method for finding the maximum likelihood estimate of the parameters from a given data set when the data is incomplete or has missing values. The MLE method is employed to calculate the parameters after deleting the incomplete sample sets. The time and mean errors of the three methods with different samples and missing rates are listed in Table 9.3. Obviously, MLE method has the shortest computing time, but the greatest mean error, which means that deleting the incomplete data samples increases the error. Therefore, discarding the incomplete sample sets is not recommended. As shown in Table 9.3, the time increases as the missing rate or sample size

Table 9.3. Mean error and time of the three methods for parameter learning.

	Missing rate (%)	Mean error		Time (s)	
		2000	4000	2000	4000
BP-MLE	10	0.0327	0.0252	12	20
	20	0.0335	0.0261	13	21
EM	10	0.0337	0.0259	102	200
	20	0.0358	0.0270	108	215
MLE	10	0.0536	0.0377	0.1772	0.3768
	20	0.0583	0.0419	0.1581	0.3227

increases for BP-MLE and EM methods. The time of BP-MLE method, especially, increases less slowly than that of EM as the sample size increases under the same missing rate. EM method needs much more time than BP-MLE method. Table 9.3 also shows that the mean error increases as the missing rate or sample size increases for BP-MLE and EM methods. Besides, the BP-MLE method has a smaller mean error than the EM method under the same missing rate. From the aspect of efficiency, BP-MLE is better than EM. From the aspect of accuracy, the proposed BP-MLE method is not obviously better than the EM method. One of the reasons is that the BP networks cannot learn all the missing patterns of the incomplete data sets. Therefore, BP-MLE is effective for parameter learning with incomplete data.

9.3.3. *Parameter estimation of prior probabilities with BP-FS method*

A prior probability is often the purely subjective assessment of an experienced expert. Assuming there are six experts describing the prior probabilities of fault nodes, the parameters of a BN are represented by linguistic variables and the values of parameters could be expressed as very low, low, fairly low, medium, fairly high, high, and very high. The incomplete expert knowledge about fault nodes in the "present" state are given in Table 9.4. The values in brackets are the missing ones.

It shows that the incomplete expert knowledge information are composed of four complete sample sets and two incomplete sample

Table 9.4. The incomplete expert knowledge.

Experts	RL (Present)	RO (Present)	FC (Present)	FE (Present)	ELEV (Present)	BLP (Present)
1	Very low	Very low	low	Very low	Very low	Very low
2	Low	[Low]	Very low	Very low	Very low	Very low
3	Very low	Very low	Very low	low	Very low	Very low
4	Low	Low	Very low	Very low	Very low	Very low
5	Very low	Very low	[Very low]	Low	Fairly low	Very low
6	Low	Very low	Very low	Very low	Very low	Low

sets. The values very low, low, fair low, medium, fair high, high, and very high are denoted by 1, 2, 3, 4, 5, 6, and 7, respectively. Using the complete sample sets, a BP neural network with the inputs of nodes RL, FC, FE, ELEV, and BLP, and output of node RO is trained. The missing value of the node RO can be predicted with the trained BP neural network. The imputed value is 2.0, namely low. Similarly, a BP neural network is trained for sample 5 and its missing value can be obtained. The imputed value of node FC in sample 5 is 0.98, which means very low. Therefore, the imputed values are the same as the missing ones.

With the complete expert knowledge information, fuzzy set theory is used to determine the parameters of BN. In this chapter, triangular fuzzy number and trapezoidal fuzzy number are used to represent the natural language. As shown in Fig. 9.5, triangular fuzzy number is notated as $A = (a, b, c)$ and its membership function has the following form [47]:

$$A(t) = \begin{cases} 1 - (a-t)/\alpha & \text{if } a - \alpha \le t \le a, \\ 1 - (t-a)/\beta & \text{if } a \le t \le a + \beta, \\ 0 & \text{otherwise.} \end{cases} \tag{9.3}$$

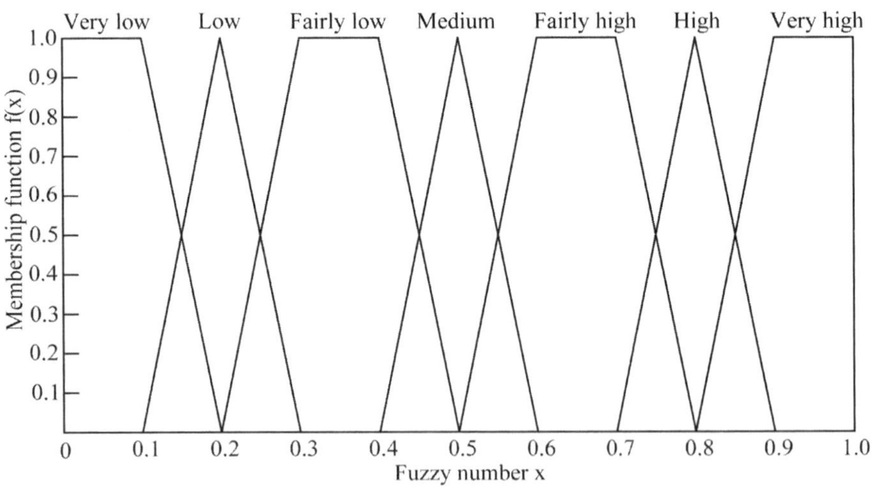

Fig. 9.5. Fuzzy number representing natural language.

The trapezoidal fuzzy number is notated as $A = (a, b, \alpha, \beta)$ and its membership function has the following form:

$$A(t) = \begin{cases} 1 - (a-t)/\alpha & \text{if } a - \alpha \leq t \leq a, \\ 1 & \text{if } a \leq t \leq b, \\ 1 - (t-a)/\beta & \text{if } a \leq t \leq a + \beta, \\ 0 & \text{otherwise.} \end{cases} \tag{9.4}$$

For fuzzy numbers, λ-cut values could be obtained from the Table 9.5. With the description of several experts, it is necessary to aggregate different experts' opinions into one. There are several ways to aggregate fuzzy numbers, such as mean, median, max, min, and mixed operators. In this chapter, the mean operator is employed because it is the most commonly used method [48]. With n experts, the average fuzzy number of event i can be obtained based on Eq. (9.5) as follows:

$$M_i = \frac{A_{i1} \oplus A_{i2} \oplus \cdots \oplus A_{in}}{n}, \quad i = 1, 2, \ldots, m, \tag{9.5}$$

where M_i is the average fuzzy number of event i, and A_{ij} is the linguistic expression being given event i by expert j. According to the extension principle, M_i is also triangular or trapezoidal fuzzy number.

In fuzzy set theory, the process to determine a value to represent the fuzzy set is called defuzzification. In this chapter, integral value

Table 9.5. Fuzzy number and its λ-cut.

Natural language	Fuzzy number	λ-cut
Very low	AVL $= (0, 0, 0.1, 0.2)$	$A^{\gamma}_{VL} = [0, -0.1\lambda + 0.2]$
Low	AL $= (0.1, 0.2, 0.3)$	$A^{\gamma}_{L} = [0.1\lambda + 0.1, -0.1\lambda + 0.3]$
Fair low	AFL $= (0.2, 0.3, 0.4, 0.5)$	$A^{\gamma}_{FL} = [0.1\lambda + 0.2, -0.1\lambda + 0.5]$
Medium	AM $= (0.4, 0.5, 0.6)$	$A^{\gamma}_{M} = [0.1\lambda + 0.4, -0.1\lambda + 0.6]$
Fair high	AFH $= (0.5, 0.6, 0.7, 0.8)$	$A^{\gamma}_{FH} = [0.1\lambda + 0.5, -0.1\lambda + 0.8]$
High	AH $= (0.7, 0.8, 0.9)$	$A^{\gamma}_{H} = [0.1\lambda + 0.7, -0.1\lambda + 0.9]$
Very high	AVH $= (0.8, 0.9, 1.0, 1.0)$	$A^{\gamma}_{VH} = [0.1\lambda + 0.8, 1.0]$

method proposed by Liou and Wang [49] is used. With this method, index of optimism $\varepsilon \in [0,1]$ reflects the degree of optimism of the experts. The total integral value is defined as

$$I^{\varepsilon}(A) = \varepsilon I_R(A) + (1-\varepsilon)I_L(A), \tag{9.6}$$

where $I_L(A)$ and $I_R(A)$ are the integral values of the left membership function and right membership function, respectively. A larger value of ε represents a higher degree of optimism. For $\varepsilon = 0$ and $\varepsilon = 1$, the values of $I^{\varepsilon}(A)$ indicate the upper and lower bounds, respectively. When $\varepsilon = 0.5$, $I_L(A)$ is the representative value of deffuzification of fuzzy number A. The total integral value of fuzzy number A becomes:

$$I^{0.5}(A) = (I_R(A) + I_L(A))/2. \tag{9.7}$$

For a triangular or trapezoidal fuzzy number, $I_L(A)$ and $I_R(A)$ can be calculated according to the following equations:

$$I_R(A) = \frac{1}{2}\left[\sum_{\lambda=0.1}^{1} \lambda_R(A)\Delta\lambda + \sum_{\lambda=0}^{0.9} \lambda_R(A)\Delta\lambda\right], \tag{9.8}$$

$$I_L(A) = \frac{1}{2}\left[\sum_{\lambda=0.1}^{1} \lambda_L(A)\Delta\lambda + \sum_{\lambda=0}^{0.9} \lambda_L(A)\Delta\lambda\right], \tag{9.9}$$

where $\lambda_R(A)$ and $\lambda_L(A)$ are the upper and lower bounds of λ-cut of the fuzzy number, respectively. $\Delta\lambda = 0.1$ and $\lambda = 0, 0.1, 0.2, \ldots, 1.0$. Based on Eqs. (9.7)–(9.9), probabilities of the nodes can be calculated. For example, six experts describe the probabilities of the node RL in the state "present" using a natural language as shown in Table 9.4. According to the Table 9.5, the fuzzy number and its λ-cut of each expert can be obtained. Based on the Eq. (9.5), we obtain the average fuzzy number $\text{MRL} = [0.05\lambda + 0.05, -0.1\lambda + 0.25]$. According to Eqs. (9.7)–(9.9), the prior probabilities are $P(\text{RL} = \text{present}) = 0.1375$ and $\text{P}(\text{RL} = \text{absent}) = 0.8625$. Similarly, the prior probabilities of the other nodes in this BN are calculated in Table 9.6.

Table 9.6. Calculated prior probabilities of the nodes.

Nodes	State	Prior probability
RL	Present	0.1375
	Absent	0.8625
RO	Present	0.1167
	Absent	0.8833
FC	Present	0.0958
	Absent	0.9042
FE	Present	0.1167
	Absent	0.8833
ELEV	Present	0.1208
	Absent	0.8792
BLP	Present	0.0958
	Absent	0.9042

9.4. Result and Discussion

After parameter estimation, the complete BN for fault diagnosis of an SAHP system is determined under incomplete data and expert knowledge. A fault can be diagnosed based on the known status of the symptoms from sensors.

9.4.1. *Fault diagnosis using complete symptoms*

Two cases with complete symptoms are presented to evaluate the BN. Case One is shown in Fig. 9.6(a). The states of symptom nodes are marked by 100% and they are lower, lower, higher, lower, lower, lower, lower, and lower, respectively. The poster probabilities of the faults are calculated based on these evidences. Figure 9.6(a) shows that the most suspected fault is RL with a probability of 96.2%. Probabilities of the other faults are low. For Case Two shown in Fig. 9.6(b), the evidences are lower, higher, lower, lower, lower, higher, lower, and higher for EWTD, TT, CDT, CT, CWTD, EVDT, EP, and CP, respectively. The diagnostic results show that the FC has the greatest probability (98.0%). These two cases demonstrate that the BN can perform fault diagnosis with complete symptoms.

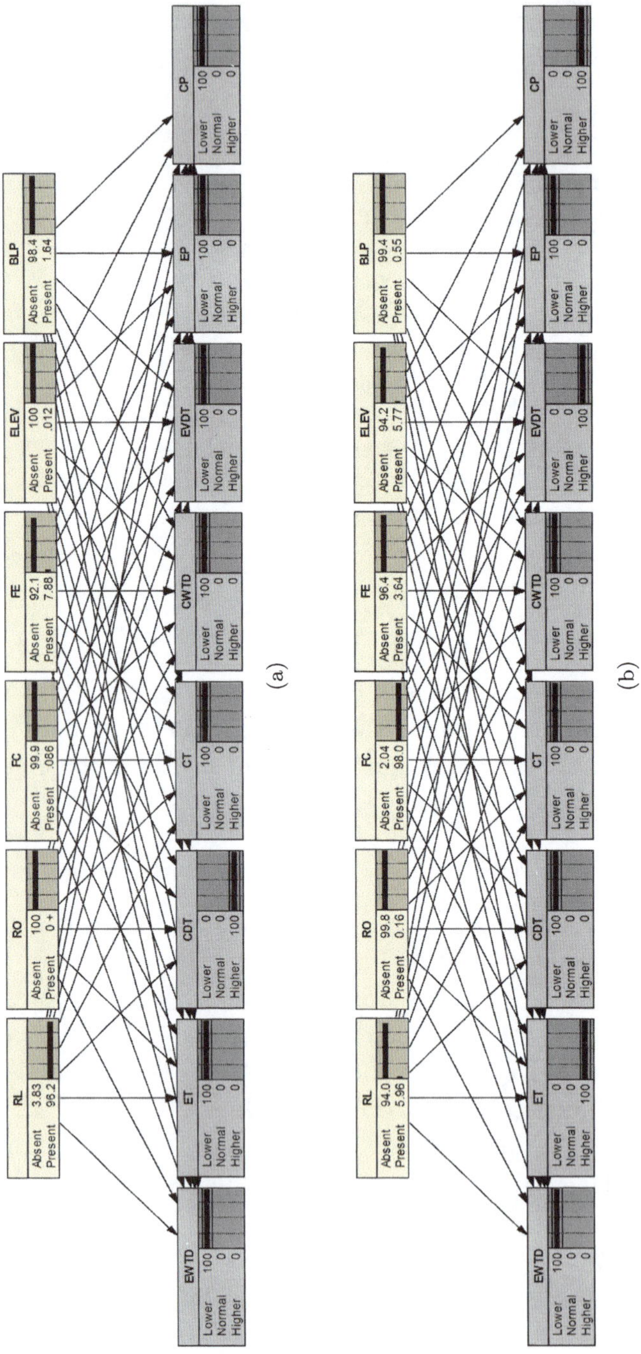

Fig. 9.6. Fault diagnosis using complete symptoms: (a) Case One; and (b) Case Two.

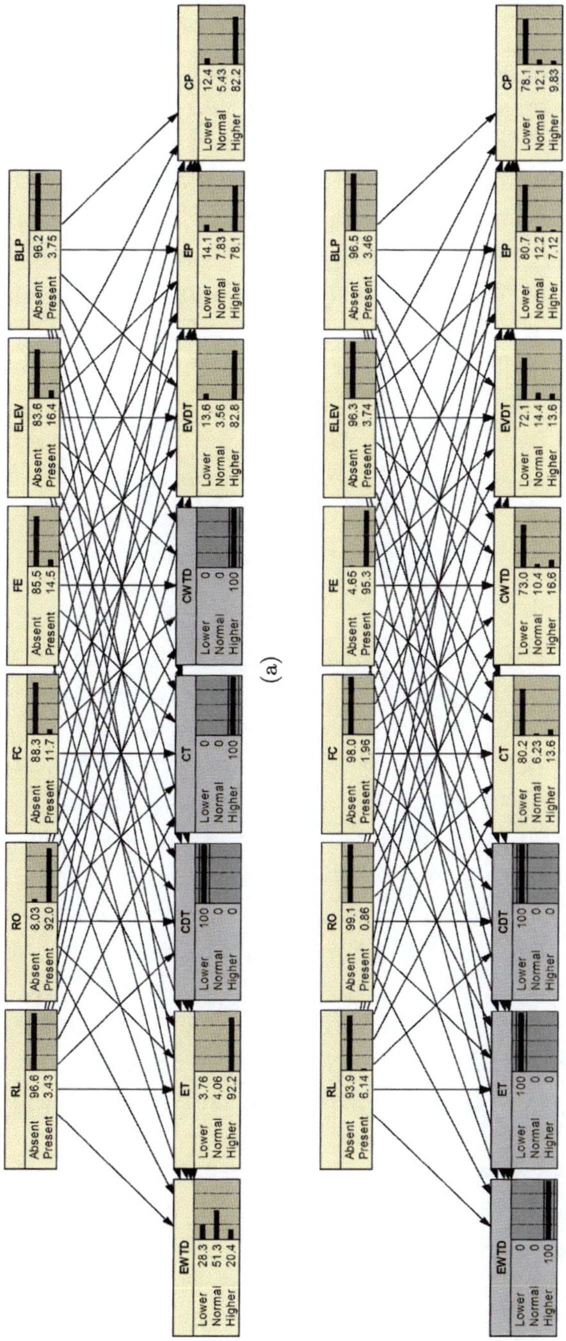

Fig. 9.7. Fault diagnosis using incomplete symptoms: (a) case one; and (b) case two.

9.4.2. *Fault diagnosis using incomplete symptoms*

Two cases are investigated to evaluate the developed BN using incomplete symptoms. Case one assumes that symptoms CDT, CT, and CWTD are obtained. With one symptom (evidence: CDT is lower), RO (27.3%), FC (22.5%), FE (28.6%), and ELEV (28.0%) are suspected faults. With the addition of an evidence (CT is higher), the fault probability of RO increases to 68.9% and the probabilities of the other suspected faults decrease. Based on these three evidences shown in Fig. 9.7(a), RO (92.0%) is the most suspected fault. Case two is shown in Fig. 9.7(b). Based on only three evidences (EWTD is higher, ET is lower, and CDT is lower), FE (95.3%) is also diagnosed. These two cases demonstrate that the developed BN can perform fault diagnosis for an SAHP system based on incomplete symptoms.

9.5. Conclusion

Failures in an SAHP system will cause the occurrence of abnormal operation and performance degradation. Therefore, fault diagnosis for the system is beneficial for saving energy and cost. This chapter presents the BP-MLE method for the parameter learning of BN from incomplete data based on the BP neural network and MLE. The BP neural is used to impute the missing values and then the complete data sets are addressed with MLE to obtain the parameters of the BN. A method for parameter estimation under incomplete expert knowledge based on the BP neural network and fuzzy set theory is also presented, which is called the BP-FS method. Similarly, the missing information is imputed by the trained BP neural network. Complete qualitative expert knowledge is processed by fuzzification. Then, defuzzification is performed to quantify the parameters of the BN.

The presented methods are applied to the parameter learning of diagnostic BN for an SAHP system with incomplete simulation data and expert knowledge. Conditional probabilities of the BN are learned by the proposed BP-MLE method and prior probabilities are estimated with the presented BP-FS method. Compared with EM,

BP-MLE has higher efficiency and accuracy. Finally, several cases with complete or incomplete evidences are studied to evaluate the performance of the BN for fault diagnosis. The results demonstrate that the developed BN can perform fault diagnosis with complete or incomplete symptoms.

References

[1] H. D. Fu, G. Pei, J. Ji, H. Long, T. Zhang, T. T. Chow, "Experimental study of a photovoltaic solar-assisted heat-pump/heat-pipe system," *Applied Thermal Engineering*, vol. 40, pp. 343–350, 2012.

[2] M. Mohanraj, S. Jayaraj, C. Muraleedharan, "Performance prediction of a direct expansion solar assisted heat pump using artificial neural networks," *Applied Energy*, vol. 86, pp. 1442–1449, 2009.

[3] C. H. Liang, X. S. Zhang, X. W. Li, X. Zhu, "Study on the performance of a solar assisted air source heat pump system for building heating," *Energy and Buildings*, vol. 43, pp. 2188–2196, 2011.

[4] H. Li, H. X. Yang, "Study on performance of solar assisted air source heat pump systems for hot water production in Hong Kong," *Applied Energy*, vol. 87, pp. 2818–2825, 2010.

[5] T. T. Chow, Y. Bai, K. F. Fong, Z. Lin, "Analysis of a solar assisted heat pump system for indoor swimming pool water and space heating," *Applied Energy*, vol. 100, pp. 309–317, 2012.

[6] Y. Zhao, S. W. Wang, F. Xiao, "A statistical fault detection and diagnosis method for centrifugal chillers based on exponentially-weighted moving average control charts and support vector regression," *Applied Thermal Engineering*, vol. 51, pp. 560–572, 2013.

[7] Y. M. Chen, L. L Lan, "A fault detection technique for air-source heat pump water chiller/heaters," *Energy and Buildings*, vol. 41, pp. 881–887, 2009.

[8] D. Zogg, E. Shafai, H. P. Geering, "Fault diagnosis for heat pumps with parameter identification and clustering," *Control Engineering Practice*, vol. 14, pp. 1435–1444, 2006.

[9] Y. Zhao, F. Xiao, S. Wang, "An intelligent chiller fault detection and diagnosis methodology using Bayesian belief network," *Energy and Buildings*, vol. 57, pp. 278–288, 2013.

[10] B. P. Cai, Y. H. Liu, Q. Fan, Y. W. Zhang, Z. K. Liu, S. L. Yu, R. J. Ji, "Multi-source information fusion based fault diagnosis of ground-source heat pump using Bayesian network," *Applied Energy*, vol. 114, pp. 1–9, 2014.

[11] M. Najafi, D. M. Auslander, P. L. Bartlett, P. Haves, M. D. Sohn, "Application of machine learning in the fault diagnostics of air handling units," *Applied Energy*, vol. 96, pp. 347–358, 2012.

[12] B. Ricks, O. J. Mengshoel, "Diagnosis for uncertain, dynamic and hybrid domains using Bayesian networks and arithmetic circuits," *International Journal of Approximate Reasoning*, vol. 55, pp. 1207–1234, 2014.

[13] Á. Carrera, C. A. Iglesias, J. García-Algarra, D. Kolarík, "A real-life application of multi-agent systems for fault diagnosis in the provision of an Internet business service," *Journal of Network and Computer Applications*, vol. 37, pp. 146–154, 2014.

[14] R. Barco, P. Lázaro, V. Wille, L. Díez, S. Patel, "Knowledge acquisition for diagnosis model in wireless networks," *Expert Systems with Applications*, vol. 36, 4745–4752, 2009.

[15] B. G. Xu, "Intelligent fault inference for rotating flexible rotors using Bayesian belief network, "*Expert Systems with Applications*, vol. 39, pp. 816–822, 2012.

[16] F. Sahin, M. Ç. Yavuz, Z. Y. Arnavut, O. Uluyol, "Fault diagnosis for airplane engines using Bayesian networks and distributed particle swarm optimization," *Parallel Computing*, vol. 33, pp. 124–143, 2007.

[17] D. Marquez, M. Neil, N. Fenton, "Improved reliability modeling using Bayesian networks and dynamic discretization," *Reliability Engineering & System Safety*, vol. 95, pp. 412–425, 2010.

[18] S. Lim, S. B. Cho, "Online learning of Bayesian network parameters with incomplete data," *Lecture Notes in Computer Science*, vol. 4114, no. LNAI-II, pp. 309–314, 2006.

[19] M. L. Wong, Y. Y. Guo, "Learning Bayesian networks from incomplete database using a novel evolutionary algorithm," *Decision Support Systems*, vol. 45, pp. 368–383, 2008.

[20] T. J. Cleophas, A. H. Zwinderman, "Missing data imputation," In: *Statistical Analysis of Clinical Data on a Pocket Calculator, Part* 2, pp. 529–543, Springer, 2012.

[21] C. Riggelsen, "Learning parameters of Bayesian networks from incomplete data via importance sampling," *International Journal of Approximate Reasoning*, vol. 42, pp. 69–83, 2006.

[22] F. Pernkopf, D. Bouchaffra, "Genetic-based EM algorithm for learning Gaussian mixture models," *IEEE Transactions on Pattern Analysis and Machine Intelligence*, vol. 27, 1344–1348, 2005.

[23] N. Majdi-Nasab, M. Analoui, E. J. Delp, "Decomposing parameters of mixture Gaussian model using genetic and maximum likelihood algorithms on dental images," *Pattern Recognition Letters*, vol. 27, pp. 1522–1536, 2006.

[24] S. Huda, J. Yearwood, R. Togneri, "A constraint-based evolutionary learning approach to the expectation maximization for optimal estimation of the hidden Markov model for speech signal modeling," *IEEE Transactions on Systems, Man, and Cybernetics, Part B (Cybernetics)*, vol. 39, pp. 182–197, 2009.

[25] W. H. Liao, Q. Ji, "Learning Bayesian network parameters under incomplete data with domain knowledge," *Pattern Recognition*, vol. 42, pp. 3046–3056, 2009.

[26] B. H. M. Sadeghi, "BP-neural network predictor model for plastic injection molding process," *Journal of Materials Processing Technology*, vol. 103, pp. 411–416, 2000.

[27] W. C. Chen, P. H. Tai, M. W. Wang, W. J. Deng, C. T. Chen, "A neural network-based approach for dynamic quality prediction in a plastic injection molding process," *Expert Systems with Applications*, vol. 35, pp. 843–849, 2008.

[28] Q. Li, J. Y. Yu, B. C. Mu, X. D. Sun, "BP neural network prediction of the mechanical properties of porous NiTi shape memory alloy prepared by thermal explosion reaction," *Materials Science and Engineering: A*, vol. 419, pp. 214–217, 2006.

[29] Y. D. Zhang, L. N. Wu, "Stock market prediction of S&P 500 via combination of improved BCO approach and BP neural network," *Expert Systems with Applications*, vol. 36, pp. 8849–8854, 2009.

[30] S. W. Yu, K. J. Zhu, F. Q. Diao, "A dynamic all parameters adaptive BP neural networks model and its application on oil reservoir prediction," *Applied Mathematics and Computation*, vol. 195, pp. 66–75, 2008.

[31] A. ZadehL, "Fuzzy sets," *Information and Control*, vol. 8, pp. 338–353, 1965.

[32] A. John, D. Paraskevadakis, A. Bury, Z. L. Yang, R. Riahi, J. Wang, "An integrated fuzzy risk assessment for seaport operations," *Safety Science*, vol. 68, pp. 180–194, 2014.

[33] A. Aydin, "Fuzzy set approaches to classification of rock masses," *Engineering Geology*, vol. 74, pp. 227–245, 2004.

[34] S. B. Park, J. I. Monroe, M. Yao, M. MacHtay, J. W. Sohn, "Composite radiation dose representation using fuzzy set theory," *Information Sciences*, vol. 187, pp. 204–215, 2012.

[35] V. Bugarski, T. Backalic, U. Kuzmanov, "Fuzzy decision support system for ship lock control," *Expert Systems with Applications*, vol. 40, pp. 3953–3960, 2013.

[36] Y. M. Kim, C. K. Kim, G. H. Hong, "Fuzzy set based crack diagnosis system for reinforced concrete structures," *Computers & Structures*, vol. 85, pp. 1828–1844, 2007.

[37] J. Deng, B. Huang, "Identification of nonlinear parameter varying systems with missing output data," *AIChE Journal*, vol. 58, pp. 3454–3467, 2012.

[38] Z. Zhang, F. Dong, "Fault detection and diagnosis for missing data systems with a three time-slice dynamic Bayesian network approach," *Chemometrics and Intelligent Laboratory Systems*, vol. 138, pp. 30–40, 2014.

[39] Y. Zhou, N. Fenton, M. Neil, "Bayesian network approach to multinomial parameter learning using data and expert judgments," *International Journal of Approximate Reasoning*, vol. 55, pp. 1252–1268, 2014.

[40] E. M. Helsper, L. C. Van Der Gaag, F. Groenendaal, "Designing a procedure for the acquisition of probability constraints for Bayesian networks," *Lecture Notes in Artificial Intelligence*, vol. 3257, pp. 280–292, 2004.

[41] A. Feelders, L. C. Van Der Gaag, "Learning Bayesian network parameters under order constraints," *International Journal of Approximate Reasoning*, vol. 42, pp. 37–53, 2006.

[42] H. Li, H. X. Yang, "Potential application of solar thermal systems for hot water production in Hong Kong," *Applied Energy*, vol. 86, pp. 175–180, 2009.

[43] I. Atmaca, S. Kocak, "Theoretical energy and exergy analyses of solar assisted," *Thermal Science*, vol. 18, pp. S417–S427, 2014.
[44] Q. Song, M. Shepperd, "A new imputation method for small software project data sets," *Journal of Systems and Software*, vol. 80, pp. 51–62, 2007.
[45] V. Ravi, M. Krishna, "A new online data imputation method based on general regression auto associative neural network," *Neurocomputing*, vol. 138, pp. 106–113, 2014.
[46] W. L. Junger, A. Ponce de Leon, "Imputation of missing data in time series for air pollutants," *Atmospheric Environment*, vol. 102, pp. 96–104, 2015.
[47] V. F. Yu, L. Q. Dat, "An improved ranking method for fuzzy numbers with integral values," *Applied Soft Computing*, vol. 14, pp. 603–608, 2014.
[48] C. T. Lin, M. J. J. Wang, "Hybrid fault tree analysis using fuzzy sets," *Reliability Engineering & System Safety*, vol. 58, pp. 205–213, 1997.
[49] T. S. Liou, M. J. J. Wang, "Ranking fuzzy numbers with integral value," *Fuzzy Sets and Systems*, vol. 50, pp. 247–255, 1992.

Chapter 10

An Approach for Developing Diagnostic Bayesian Network Based on Operation Procedures

In this chapter, a novel approach of developing the Bayesian network for fault diagnosis based on operation procedures is presented. The proposed Bayesian network consists of operation procedure layer, fault layer, and fault symptom layer. First, operation procedure layer containing procedure nodes and state decision nodes is developed. Second, the fault layer is determined based on the state decision nodes in the operation procedure layer. Then fault symptoms sensitive to the concerned faults are developed. Finally, the entire Bayesian network is established by integrating the three layers. The presented approach is applied to the hydraulic control system of subsea blowout preventer (BOP). Taking an example of closing the BOP, the operation procedures are illustrated. The entire Bayesian network for the fault diagnosis of closing the BOP is established. Several cases possible to appear during the closing process are studied to evaluate the developed model.

10.1. Introduction

Fault diagnosis and prognostics has obtained a lot of attention because of the growing demands for the safety and reliability of engineering systems. Bayesian network is a powerful tool in knowledge representation and reasoning, suitable for the modeling of casual processes with uncertainty. A Bayesian network is an acyclic directed graph, consisting of nodes and arcs between the nodes [13, 20]. In the networks, nodes represent random variables and directed arcs define the probabilistic dependences between the variables [11]. The probabilistic dependences are quantified by a conditional probability table for each node [2]. Each conditional probability table contains the probability of a node, given any possible combination of its parent nodes. Without parent nodes, root nodes only have priori probabilities. Given the values of the observed variables as evidences, the posterior probabilities of the unobserved variables could be obtained by inferences.

Recently, Bayesian networks for fault diagnosis have been widely used in various fields. Barco *et al.* [3] presented an automatic diagnosis system for the radio access network of wireless systems and experimental results have shown the feasibility of the proposed methods. Sahin *et al.* [16] developed a fault diagnosis system for airplane engines using the Bayesian network and distributed particle swarm optimization, which is used for learning the structure of the model from a large data set. Riascos *et al.* [15] presented a fault diagnosis system to diagnose different types of faults during the operation of a proton exchange membrane fuel cell based on the online monitoring of variables easy to measure in the machine, such as voltage, electric current, and temperature. Cruz-Ramírez *et al.* [6] evaluated the effectiveness of seven Bayesian network classifiers as potential tools for the diagnosis of breast cancer using two real-world database and average accuracies of 93.04% for the former and 83.31% for the latter are obtained. Alaeddini and Dogan [1] developed a hybrid intelligent method-based Bayesian networks for fault detection and diagnosis in control charts, which describes the cause–effect relationship among chart patterns, process information, and possible root/assignable causes. Zhao *et al.* [24] proposed a three-layer Bayesian network to simulate the actual diagnostic thinking of chiller experts and the developed model includes fault layer and fault symptom layer, and additional information layer. Sun *et al.* [18] developed a mild cognitive impairment (MCI) expert system to address MCI's prediction and inference question to assist the diagnosis of doctor and the experimental results indicate that the developed model achieved better results than some existing methods in most instances. Verron *et al.* [21] presented a methodology for industrial process diagnosis with Bayesian network and the performances of the method are evaluated on the data of an example.

The structure of Bayesian networks is a graphical and qualitative illustration of relationships among different nodes using directed arcs. There are two ways for establishing the structure of a Bayesian network. The first way is machine learning using data sets. Based on complete or incomplete data, many structure learning algorithms have been proposed [7, 12, 17, 22]. The second way is manually developed by experts. Generally, the experts develop the network based on the cause–effect relationship among the defined variables [5, 23]. Within the scope of the second way for constructing Bayesian network, some researchers have proposed several approaches by using the existing models of the system. Bobbio *et al.* [4] took advantage of the developed fault tree of the system to build the Bayesian network for fault diagnosis. Lo *et al.* [10] proposed a novel approach for constructing the Bayesian network structure based on a bond graph model. For the systems without fault trees or bond graphs, these approaches are inapplicable.

To achieve a specific function, several procedures are needed to perform in order and the following procedure depends on the previous one. Hence, state decision is needed to determine whether the previous procedure is completed successfully or not. So, fault diagnosis models can be established for state decision. This chapter proposes a generic fault diagnosis method for constructing the Bayesian network structure based on the operation procedures. The proposed Bayesian network for diagnosis has three layers: operation procedure layer, fault layer, and fault symptom layer. Based on the operation procedure layer, it is convenient to build one entire Bayesian network for fault diagnosis, which can be integrated by developing the diagnostic subsystems of state decisions. The operation procedure layer makes the diagnostic process more clear and organized. The remainder of this chapter is organized as follows. Section 10.2 describes the proposed approach. In Sec. 10.3, a case study is presented to illustrate the implementation steps. Section 10.4

performs fault diagnosis based on the developed Bayesian networks. Section 10.5 summarizes the chapter.

10.2. The Proposed Fault Diagnosis Methodology

Figure 10.1 shows a generic framework of structuring Bayesian network for fault diagnosis based on operation procedures. The proposed methodology consists of three layers: operation procedure layer, fault layer, and fault symptom layer. The operation procedure layer is developed based on the operation procedures, which perform a function of the system. To achieve a specific function, several procedures are needed to perform in order and the next procedure depends on the previous one. State decision is used to determine whether the previous one is completed successfully or not. Therefore, the developed operation procedure layer in Bayesian network is composed of procedure nodes and state decision nodes. Fault and fault symptom layers are developed based on the state decision nodes in the operation procedure layer. Each state decision node is usually related with several faults. Through the operation procedure layer, all the faults leading to failures of a function can be connected together.

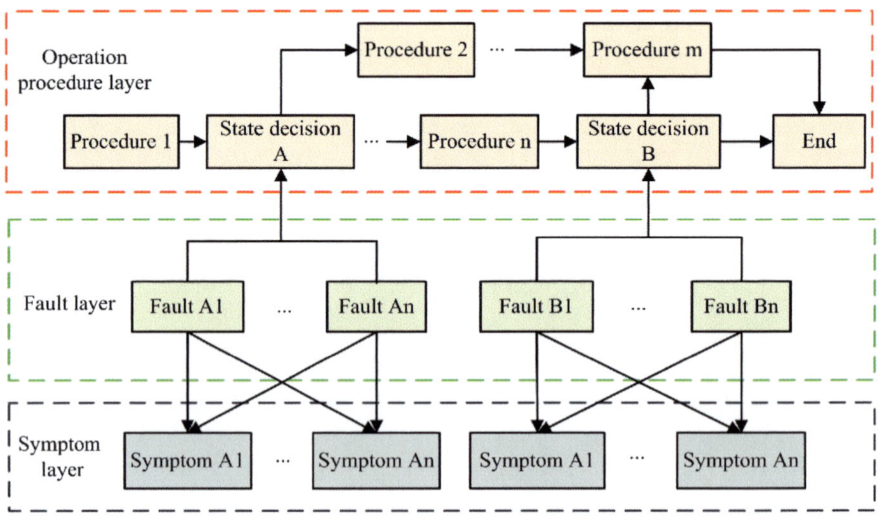

Fig. 10.1. A generic Bayesian network framework for the proposed methodology.

Fault symptom layer includes sensor measurement information, which is sensitive to the faults. Usually, several symptoms are used to identify the different faults.

The proposed methodology is a real-time fault diagnosis model in the process of performing a function. A fault detection system is needed for monitoring the important signals sensitive to the concerned faults. In addition, the diagnostic result is helpful to determine the next procedure to perform. For example, if a fault is diagnosed in the fault diagnosis system of "state decision A", the operation will move to procedure 3. If not, the operation will go on with procedure 2. To develop the structure of Bayesian network based on the proposed methodology, operation procedure layer is constructed firstly. Then the fault layer and fault symptom layer are developed according to the state decision nodes in the operation procedure layer. Finally, the entire Bayesian network is developed by integrating the operation procedure layer, fault layer, and fault symptom layer.

A Bayesian network contains two elements: structure and parameters. After the development of the Bayesian network structure, parameters for each node are needed to be determined. Root nodes have prior probabilities and child nodes have conditional probabilities based on the combination of their parent nodes. Prior probability of a node is the probability of the event occurring without new evidence or information. Conditional probability is the probability that an event occurs for the given new evidence. Generally, theses parameters can be defined by expert knowledge or learned from data sets (including faulty and normal data). The two methods can be used individually or jointly [24].

10.3. Case Study

10.3.1. *Hydraulic control system of subsea blowout preventer*

In this section, the proposed methodology is applied to the hydraulic control system of subsea blowout preventer (BOP), the schematic diagram of which is shown in Fig. 10.2. For redundancy, a subsea

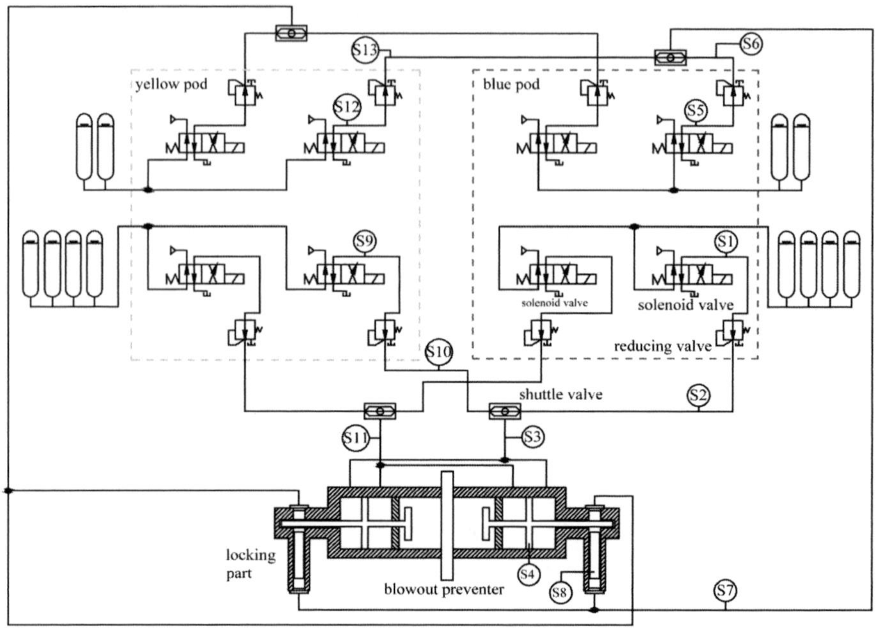

Fig. 10.2. Schematic diagram of Hydraulic control system of subsea BOP.

BOP system has two identical control pods: blue pod and yellow pod. Each pod is able to perform all necessary functions on the BOP, containing solenoid valves and reducing valves. When the control system is in operation, one pod is active and the other one is standby. Once there is something wrong with the active pod, the standby one will replace it to continue working. Each pod has a main control system and a locking system. The main control system is responsible for opening or closing the ram of subsea BOP and the locking system is used for locking the ram when it is in position. Accumulators provide high and low pressure fluid to control BOP and lock the ram. An example of closing the subsea BOP is used to demonstrate the proposed approach.

Fault diagnosis is a reasoning process from symptoms to faults. Usually, it is not a fixed one-to-one correspondence between the faults and symptoms. One fault may lead to several different symptoms while one symptom may be caused by two or more faults. Therefore, it is more reasonable to give the probabilities of faults at given symptoms in the diagnostic results. A deterministic-based model reports the diagnostic results in the Boolean format, i.e., Yes/Faulty. It does not consider the uncertainties when developing the model. However, uncertainties widely exist in sensors, faults, symptoms, fault–symptom relationships, the interconnection between a fault and other faults/symptoms [8, 23]. For example, the collected diagnostic information for the same fault is not always the same every time due to sensor bias or observation error. In the Bayesian network, the uncertainties are reflected by the probabilities [14, 19]. Assuming $P(\text{Symptom 1} = \text{abnormal} \mid \text{Fault 1} = \text{present}) = 95\%$, it means that symptom 1 is very possible to be abnormal if Fault 1 is present. The probability 95% takes into account uncertain factors such as sensor accuracy, induced electrical noise, etc.

10.3.2. *Establish Bayesian networks for fault diagnosis*

10.3.2.1. *Develop Bayesian networks of operation procedures*

Assuming the blue pod is working, the yellow pod is standby at present. The operation procedures to close the BOP are described in Fig. 10.3. When the command to close the BOP is initiated, the solenoid valve in blue pod is activated. Then high pressure fluid will drive the ram closed if no faults appear in the process. When the ram is closed successfully, it will be locked by blue pod. However, if the ram fails to be closed, yellow pod will continue to perform the function. Figure 10.3 shows that the function of closing the BOP will fail if the yellow pod cannot close the ram or failures of locking ram are present.

Figure 10.4 shows the developed Bayesian network of the operation procedures to close the BOP. In the Bayesian network, each node representing an event has two states: Yes and No. "Close_BOP" denotes receiving the command to close the BOP. "SV_Blue" denotes that the activation of the solenoid valve of the main control system in the blue pod. "Ram_Close_Blue" means that the ram is closed by the blue pod. "Lock_BOP_Blue" denotes the activation of the solenoid

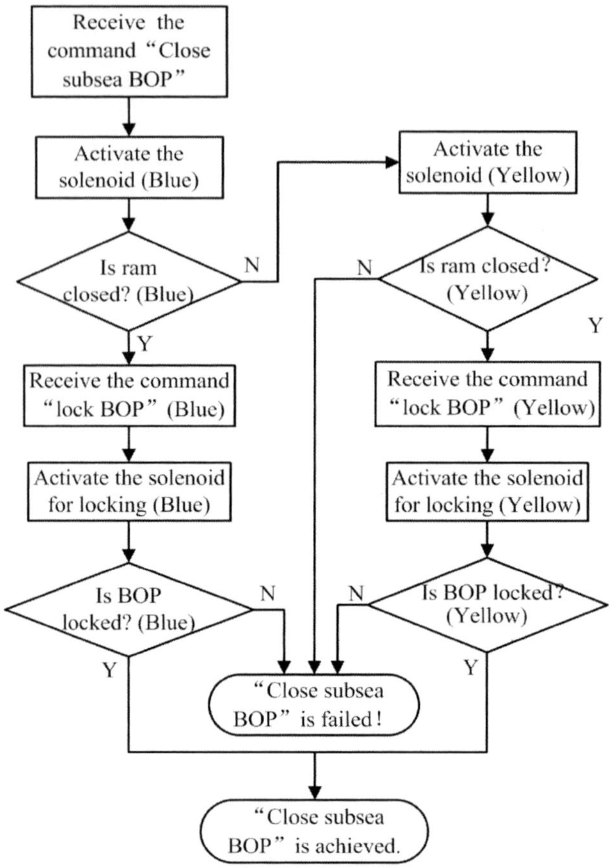

Fig. 10.3. Operation procedures of closing BOP.

valve of locking system in the blue pod. "SV_Lock_Blue" means that the ram is locked by the blue pod. "Close_BOP_OK" denotes that the function of closing the BOP is completed. The other nodes of the yellow pod have the same meanings as those of the blue pod. The Bayesian network is constructed based on the order of procedures. So, the next procedure is the parent node of the previous one. As shown in Fig. 10.4, there are four state decision nodes. Parameters of the developed Bayesian network are established according to the logic relationship of the procedures. To determine whether the function of closing the ram or locking the ram is completed successfully

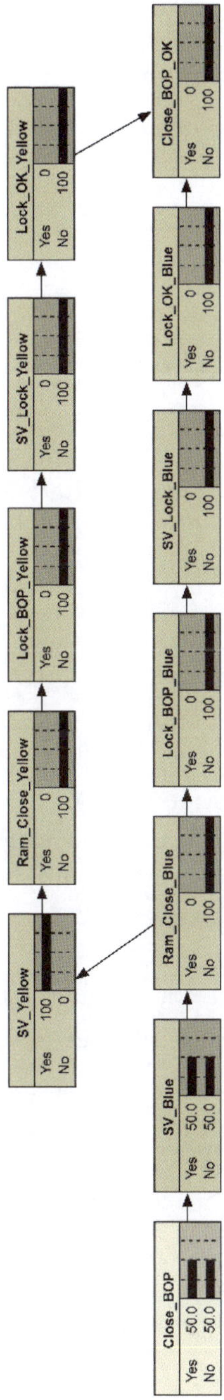

Fig. 10.4. Developed Bayesian network of operation procedures.

or not, fault diagnosis models are established in the following section.

10.3.2.2. *Establish the Bayesian network of state decision nodes*

Firstly, the fault diagnosis model of "Ram_Close_Blue" node is established. The developed Bayesian network is shown in Fig. 10.5. It consists of four fault nodes and four symptom nodes. In this chapter, only blocking faults are taken into account: blocking of solenoid valve in blue pod (BSV_B), blocking of reducing valve in blue pod (BRV_B), blocking of shuttle valve (BSV), and blocking of ram (BR). Each fault node has two states: present and absent. Signals from three pressure sensors (S1, S2, and S3) in hydraulic circuit and one displacement sensor (S4) installed on the ram serve as the fault symptoms for fault diagnosis. When a blocking fault appears, the hydraulic circuit will be divided into two parts. The pressure behind the faulty component will be lower in the hydraulic circuit. But, the pressure in front of the component will be the same as the exit pressure of accumulators. Therefore, each fault symptom node has two states: lower and normal. For example, if BSV_B is present, S1, S2, S3, and S4 will be lower. Based on the different symptoms, the faults are determined. Although more states can be selected for the symptom nodes, the number of parameters will increase greatly. Hence, selecting two states for the symptom nodes can also simplify the demonstration.

Using the machine learning method to obtain the Bayesian parameters is almost impractical to implement because large amount

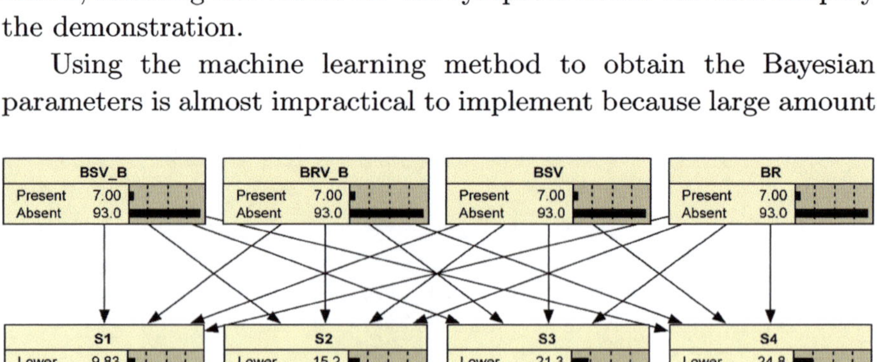

Fig. 10.5. Bayesian network of main control system in blue pod.

of samples are hardly available. The conditional probabilities of a child node depend on all the possible combination of states of its parent nodes. A complete conditional probability table for a child node with x states and n parent nodes contains $(x-1)\Pi_{i=1}^{n} S_i$ probabilities, where S_i is the number of states of parent node i. Each fault symptom node with four parent nodes has 32 parameters, and 128 probabilities in total need to be specified in this model. So, using machine learning to obtain Bayesian parameters needs a large number of data. It is usually difficult to obtain the full-set fault data. Performing so many experiments will be too costly and time-consuming. Although the data can be obtained by collecting daily records of the equipment, it will also be time-consuming. Therefore, the parameters of the developed Bayesian network are specified by the expert knowledge. Definition of the parameters by the experts appears subjective. However, expert knowledge ensures those conditional probabilities are reliable [24]. Because fault nodes are root nodes without parent nodes, prior probabilities are set in Table 10.1. Probabilities of the diagnostic results are influenced by the parameters. Prior probability of the fault is the probability of the event occurring without symptoms. The higher the prior probability of the fault, the more likely the fault is to happen. As the child nodes of fault nodes, fault symptom nodes have conditional probabilities listed in Table 10.2. Conditional probability is the probability that a fault occurs for the given symptoms. Hence, the higher the

Table 10.1. Prior probability of fault nodes.

Node	State	Prior probability (%)
BSV_B	Present	7
	Absent	93
BRV_B	Present	7
	Absent	93
BSV	Present	7
	Absent	93
BR	Present	7
	Absent	93

Table 10.2. Conditional probabilities of fault symptom nodes.

Fault				Lower (%)				Normal (%)			
BSV_B	BRV_B	BSV	BR	S1	S2	S3	S4	S1	S2	S3	S4
present	present	present	present	98	98	98	98	2	2	2	2
present	present	present	absent	97	97	97	97	3	3	3	3
present	present	absent	present	97	96	95	97	3	4	5	3
present	present	absent	absent	96	96	95	96	4	4	5	4
present	absent	present	present	97	92	96	97	3	8	4	3
present	absent	present	absent	96	91	96	95	4	9	4	5
present	absent	absent	present	95	90	85	96	5	10	15	4
present	absent	absent	absent	94	86	90	93	6	14	10	7
absent	present	present	present	8	91	94	97	92	9	6	3
absent	present	present	absent	7	90	94	95	93	10	6	5
absent	present	absent	present	6	90	91	95	94	10	9	5
absent	present	absent	absent	7	92	93	93	93	8	7	7
absent	absent	present	present	6	9	92	95	94	91	8	5
absent	absent	present	absent	5	8	92	92	95	92	8	8
absent	absent	absent	present	4	7	7	91	96	93	93	9
absent	absent	absent	absent	3	3	4	2	97	97	96	98

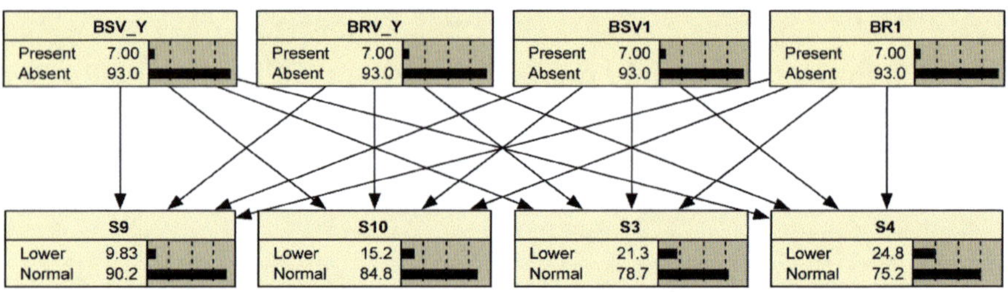

Fig. 10.6. Bayesian network of main control system in yellow pod.

conditional probability, the more likely the fault is to happen given the symptoms. In the same way, the diagnostic Bayesian network of "Ram_Close_Yellow" node is developed in Fig. 10.6. Because the yellow pod is the same as the blue pod, the parameters are also identical with the model in Fig. 10.5.

Hydraulic circuit of the locking system is similar to the main control system. The structure and parameters of diagnostic Bayesian networks for the blue pod and the yellow pod are established in

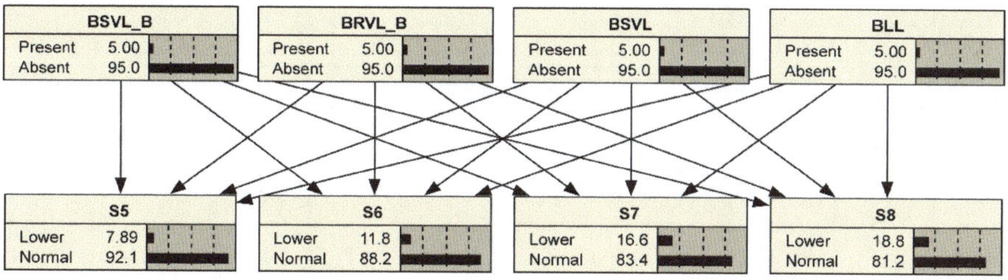

Fig. 10.7. Bayesian network of locking system in blue pod.

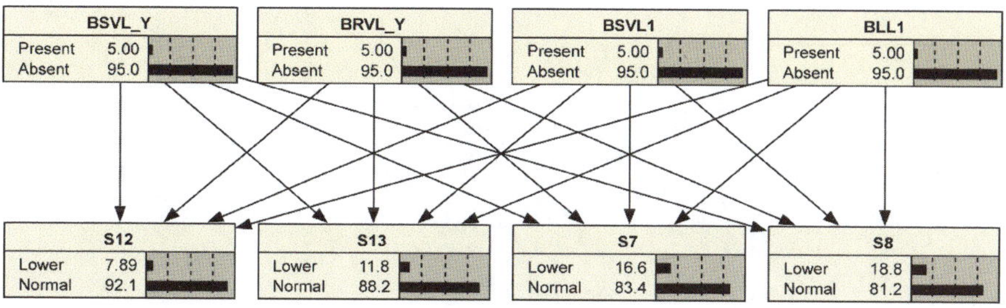

Fig. 10.8. Bayesian network of lock control system in yellow pod.

Figs. 10.7 and 10.8, respectively. As shown in Fig. 10.7, four common faults are considered in the model. For locking system, blocking of solenoid valve in the blue pod, blocking of reducing valve in the blue pod, blocking of shuttle valve, and blocking of latch are denoted by BSVL_B, BRVL_B, BSVL, and BLL, respectively, in the developed Bayesian network. Each fault has two states: present and absent. Three pressure sensors (S5, S6, and S7) are used to monitor pressure signals in the blue pod. Displacement sensor S8 is installed to monitor the displacement of lock latch. Each fault symptom has two states: lower and normal. Similarly, parameters of the developed Bayesian networks are specified.

10.3.2.3. *Develop the entire Bayesian network*

After the development of models for the state decision nodes, the entire Bayesian network structure for the fault diagnosis of closing the BOP is obtained in Fig. 10.9. Based on the operation procedure layer, the fault diagnosis systems of different state decision nodes in the

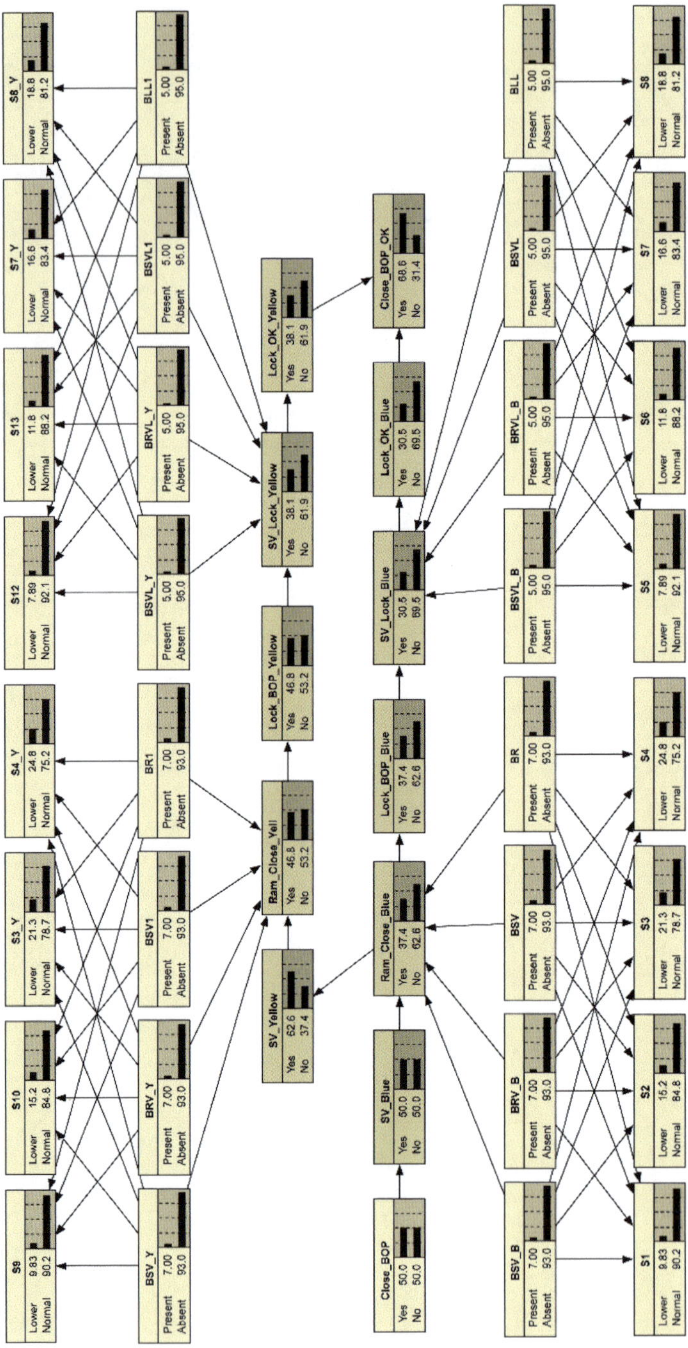

Fig. 10.9. The entire Bayesian network for fault diagnosis of closing BOP.

closing BOP process are integrated into one entire Bayesian network. The diagnostic model is composed of an operation procedure layer, two fault layers, and two fault symptom layers. It is worth mentioning because the four state decision nodes have become the child nodes of fault nodes, their conditional probability table needs to be added.

The presented Bayesian network is a real-time fault diagnosis model in the process of closing the BOP. Once a fault occurs, it will be diagnosed by the fault diagnosis system. Fault diagnosis of the state decision nodes will be performed according to the order of operation procedures. For example, if a fault is diagnosed by the fault diagnosis system of "Ram_Close_Blue", the yellow pod will continue to close the BOP and then the diagnostic model of "Ram_Close_Yellow" will be used to determine whether the ram is closed or not. In addition, the probabilities of all the procedures can be obtained to determine whether the procedure is completed successfully or not.

10.4. Fault Diagnosis and Discussion

In this section, several cases possible to appear during the operation of closing the BOP are studied to evaluate the developed model. The most expected case is that no faults occur in the process of closing the BOP. However, a fault might appear in the main control system or the locking system in the blue pod when it is active. When there is something wrong with the blue pod, the yellow pod will replace it and continue to close the BOP. It is also possible that a fault is present in the main control system or the locking system. These cases can be dealt with the developed Bayesian network.

10.4.1. *No faults in the closing process*

As shown in Fig. 10.10, all the states of S1–S4 in the fault diagnosis system of "Ram_Close_Blue" are normal. The probabilities of the faults inferred by the Bayesian network are very low, which means they are almost impossible to occur. Besides, probability of the ram closed by the blue pod is 99.3%. Therefore, the operation will proceed with the next procedure, namely locking the system by the blue pod. The symptoms in the normal states are entered into the fault diagnosis system of "SV_Lock_Blue". Based on Bayesian

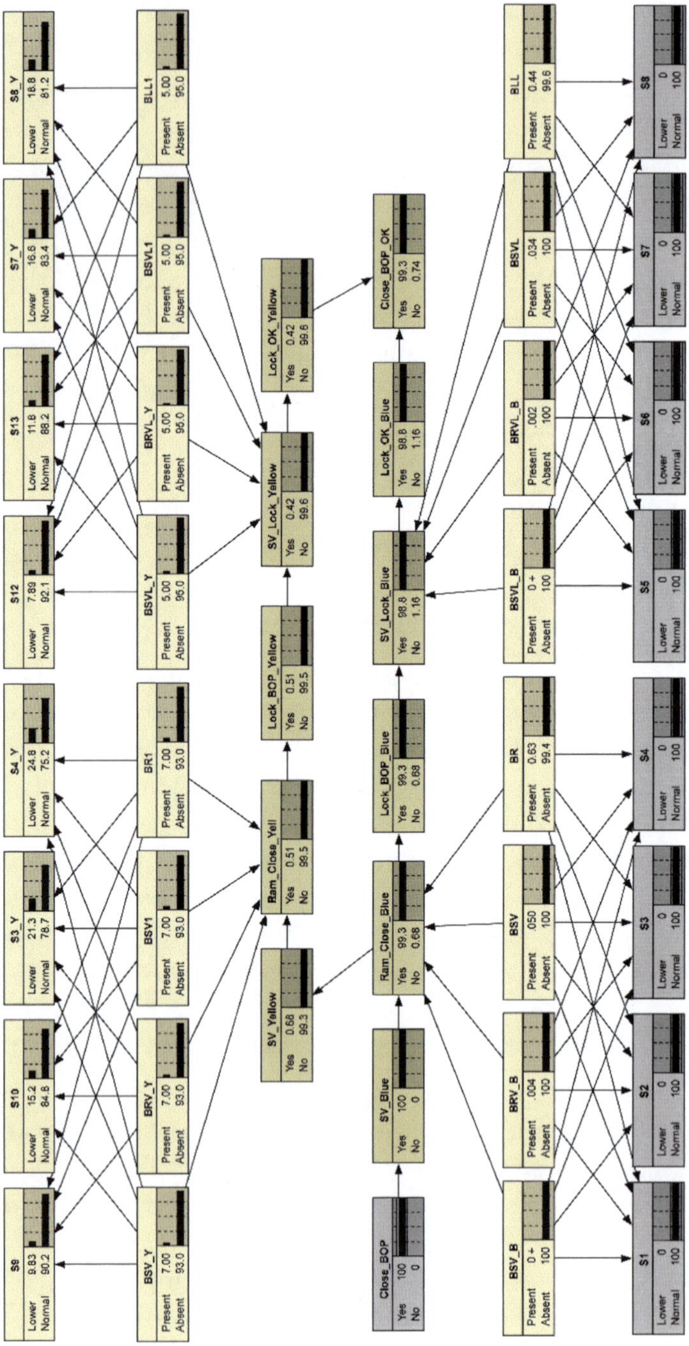

Fig. 10.10. No faults during the closing process.

inference, no faults are diagnosed. The probability of locking the system successfully by the blue pod is 98.8%. Finally, the probability of closing the BOP successfully is 99.3%.

10.4.2. *One fault of main control system in blue pod*

As shown in Fig. 10.11, states of sensors S1–S4 are normal, lower, lower, and lower, respectively. Probabilities of the faults are 5.25%, 87.5%, 13.6%, and 7.36% for BSV_B, BRV_B, BSV, and BR, respectively. The fault of BRV_B can be diagnosed with high confidence. The ram is not closed successfully. Therefore, the closing function will be performed by the yellow pod. Because all the symptoms are normal in the diagnosis systems of "Ram_Close_Yellow" and "SV_Lock_Yellow", no faults are diagnosed because the probabilities are pretty low. The probability of closing the BOP function successfully is 98.8%.

10.4.3. *One fault of main control system in yellow pod*

Based on the states of S1–S4, BSV_B is diagnosed by the fault diagnosis system of "Ram_Close_Blue" in Fig. 10.12. It means that the solenoid valve in the blue pod is blocked. The ram is not closed by blue pod. Then the operation of closing the BOP will be performed by the yellow pod. In the process of closing the ram, fault diagnosis system of "Ram_Close_Yellow" is used to determine whether the ram is closed or not. According to the evidences (S9 is normal, S10 is lower, S3 is lower, and S4 is lower), BRV_Y is diagnosed with the probability of 94.74%. It indicates that the reducing valve in the yellow pod is blocked. Hence, it is believed that the ram is not closed successfully by the yellow pod. The probability of "Close_BOP_OK" is 0.026%, which means that the function of closing the BOP is almost impossible to be achieved.

10.4.4. *One fault of locking system in blue pod*

Figure 10.13 shows that no faults are diagnosed in the process of closing the ram based on the states of S1–S4. After the ram is

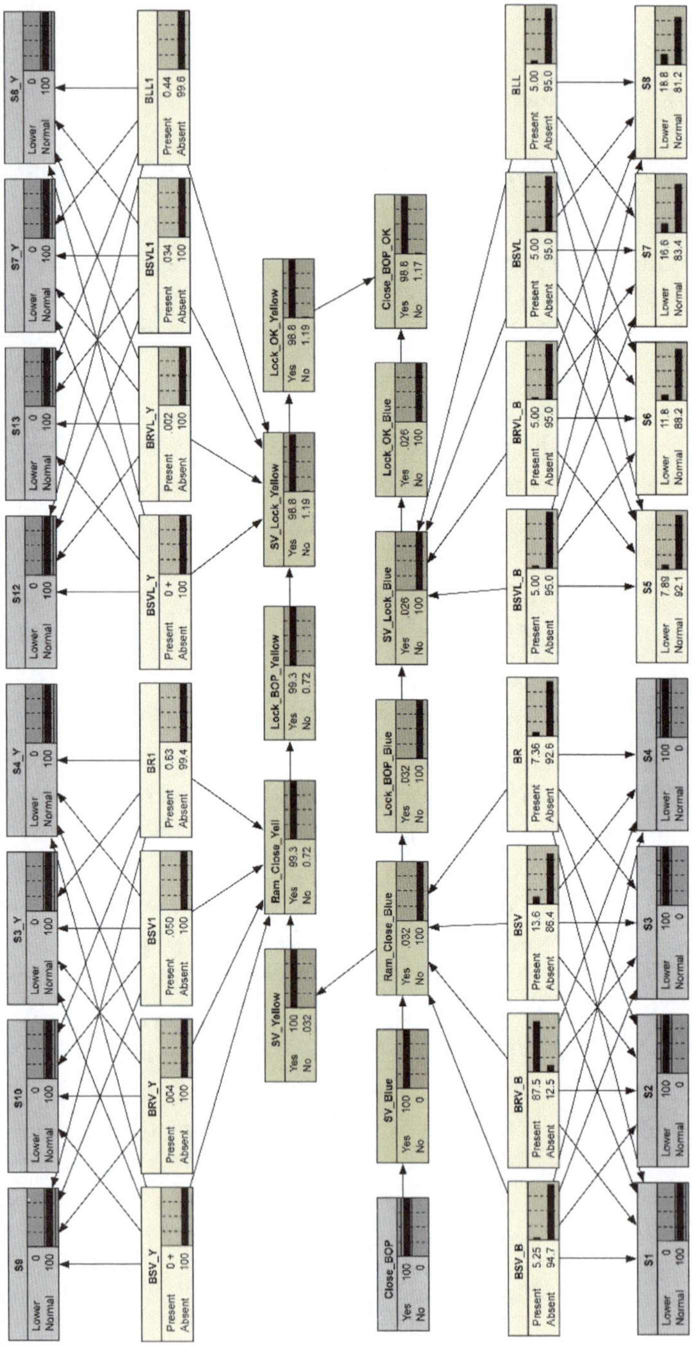

Fig. 10.11. One fault of the main control system in blue pod.

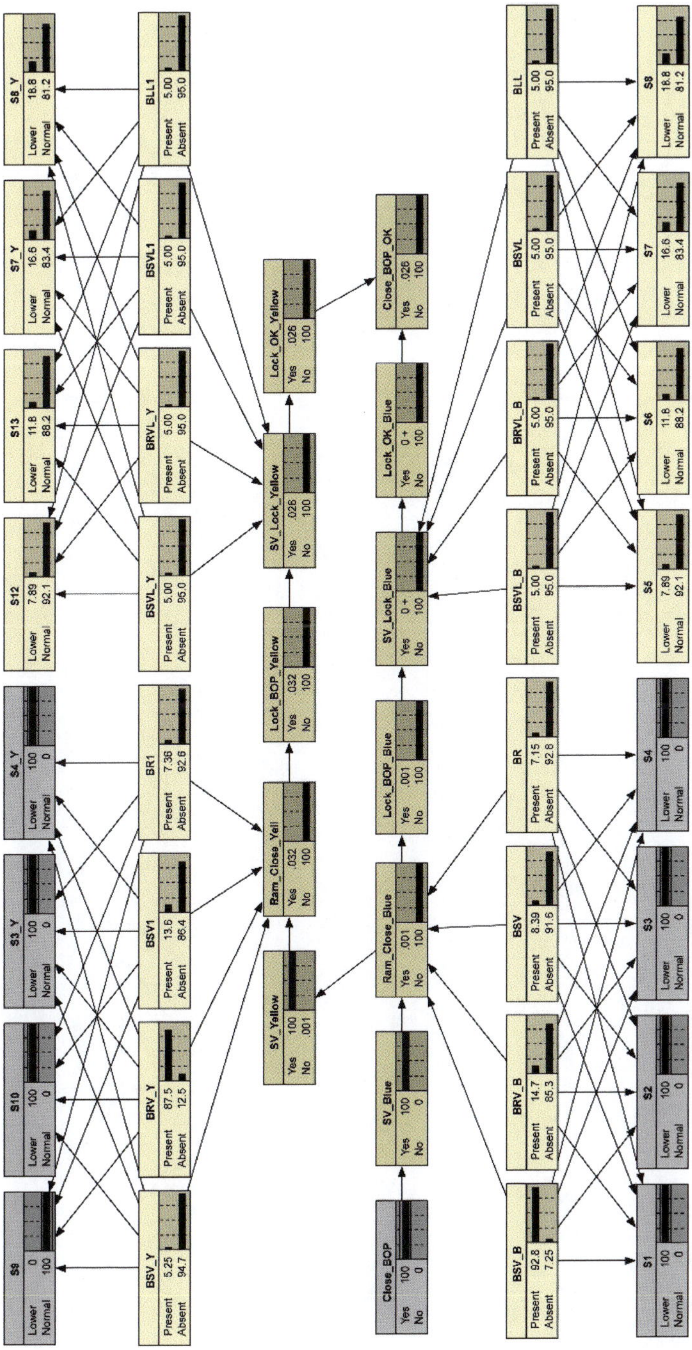

Fig. 10.12. One fault of the main control system in yellow pod.

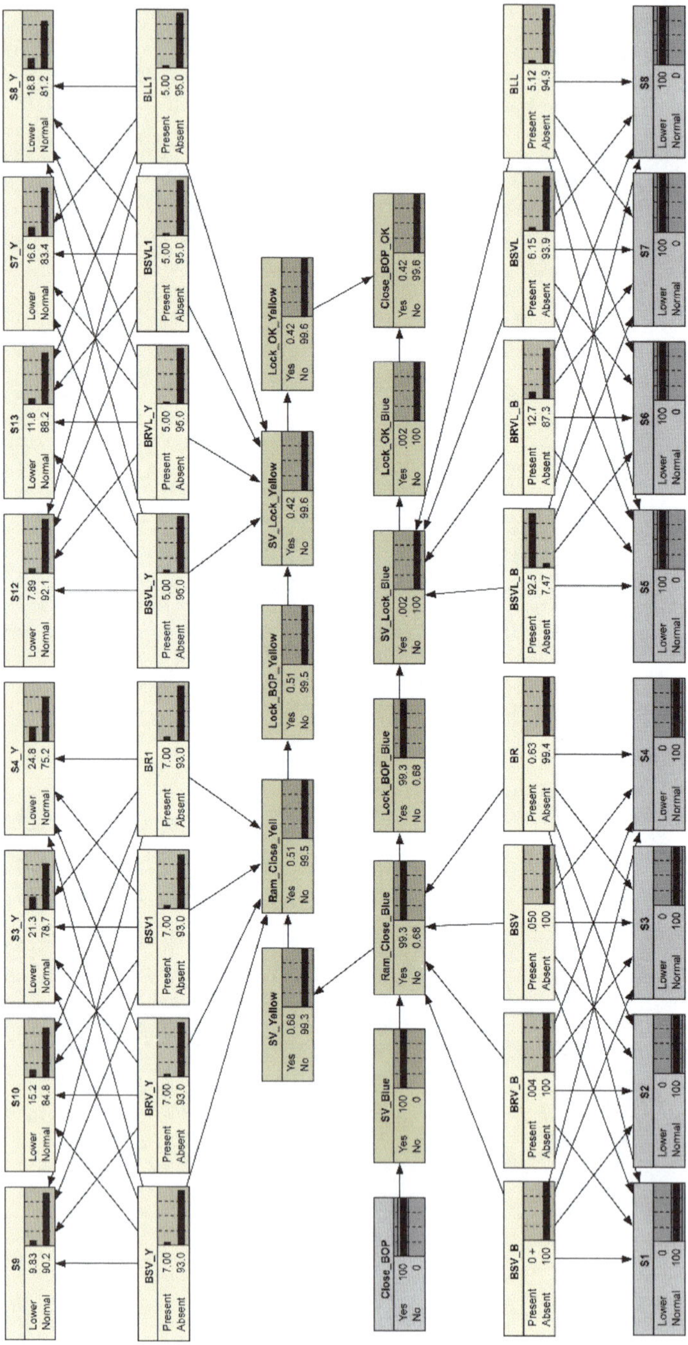

Fig. 10.13. One fault of the locking system in blue pod.

closed, it will be locked. According to the states of S5–S8 in the fault symptom layer, SVVL_B is diagnosed by the fault diagnosis system of "SV_Lock_Blue". It means that the solenoid valve of the locking system is blocked. The probability of "Close_BOP_OK" is 0.42%, which means that the function of closing the BOP is hardly possible to be completed.

10.5. Conclusion

This chapter presents a method to develop the diagnostic Bayesian network based on the operation procedures. The proposed Bayesian network is composed of operation procedure layer, fault layer, and fault symptom layer. There are mainly three steps to establish the Bayesian network for fault diagnosis. First, Bayesian network of the operation procedures is developed. Second, state decision nodes in the operation procedure are determined and their fault diagnosis subsystems consisting of fault layers and symptom layers are presented. Finally, the entire Bayesian network for fault diagnosis is developed by integrating these three layers.

A case study of implementing the method for diagnosing faults of the hydraulic control system in the closing BOP process is presented. Based on the operation procedure layer, four fault diagnosis subsystems of the state decision nodes for closing the BOP are integrated into one entire Bayesian network. In order to evaluate the developed model, several cases possible to appear during the operation of closing BOP are studied. Based on the symptoms from sensor measurement, faults can be diagnosed. In addition, the probabilities of procedures are obtained. The research results demonstrate that the proposed Bayesian network for fault diagnosis is feasible.

There are several areas that require further research. First, some methods or models can be proposed to validate the developed Bayesian networks. Second, in order to evaluate the developed model, a model or an index system can be established. Lastly, the methods to construct the Bayesian network for fault diagnosis based on other models or algorithms can be presented.

References

[1] A. Alaeddini, I. Dogan, "Using Bayesian networks for root cause analysis in statistical process control," *Expert Systems with Applications*, vol. 38, pp. 11230–11243, 2011.

[2] O. Arsene, I. Dumitrache, I. Mihu, "Medicine expert system dynamic Bayesian network and ontology based," *Expert Systems with Applications*, vol. 38, pp. 15253–15261, 2011.

[3] R. Barco, P. Lázaro, V. Wille, L. Díez, S. Patel, "Knowledge acquisition for diagnosis model in wireless networks," *Expert Systems with Applications*, vol. 36, pp. 4745–4752, 2009.

[4] A. Bobbio, L. Portinale, M. Minichino, E. Ciancamerla, "Improving the analysis of dependable systems by mapping fault trees into Bayesian networks," *Reliability Engineering & System Safety*, vol. 71, pp. 249–260, 2001.

[5] A. Bouejla, X. Chaze, F. Guarnieri, A. Napoli, "A Bayesian network to manage risks of maritime piracy against offshore oil fields," *Safety Science*, vol. 68, pp. 222–230, 2014.

[6] N. Cruz-Ramírez, H. G. Acosta-Mesa, H. Carrillo-Calvet, L. Alonso Nava-Fernández, R. E. Barrientos-Martínez, "Diagnosis of breast cancer using Bayesian networks: A case study," *Computers in Biology and Medicine*, vol. 37, pp. 1553–1564, 2007.

[7] M. Gasse, A. Aussem, H. Elghazel, "A hybrid algorithm for Bayesian network structure learning with application to multi-label learning," *Expert Systems with Applications*, vol. 41, pp. 6755–6772, 2014.

[8] B. Huang, "Bayesian methods for control loop monitoring and diagnosis," *Journal of Process Control*, vol. 18, pp. 829–838, 2008.

[9] M. C. Kim, "Reliability block diagram with general gates and its application to system reliability analysis," *Annals of Nuclear Energy*, vol. 38, pp. 2456–2461, 2011.

[10] C. H. Lo, Y. K. Wong, A. B. Rad, "Bond graph based Bayesian network for fault diagnosis," *Applied Soft Computing*, vol. 11, pp. 1208–1212, 2011.

[11] D. Marquez, M. Neil, N. Fenton, "Improved reliability modeling using Bayesian networks and dynamic discretization," *Reliability Engineering & System Safety*, vol. 99, pp. 412–425, 2010.

[12] A. R. Masegosa, S. Moral, "New skeleton-based approaches for Bayesian structure learning of Bayesian networks," *Applied Soft Computing*, vol. 13, pp. 1110–1120, 2013.

[13] A. Pernestål, M. Nyberg, H. Warnquist, "Modeling and inference for troubleshooting with interventions applied to a heavy truck auxiliary braking system," *Engineering Applications of Artificial Intelligence*, vol. 25, pp. 705–719, 2012.

[14] E. Philippot, K. C. Santosh, A. Belaid, Y. Belaid, "Bayesian network for incomplete data analysis in form processing," *International Journal of Machine Learning and Cybernetics*, DOI: 10.1007/s13042-014-0234-4, 2014.

[15] L. A. M. Riascos, M. G. Simoes, P. E. Miyagi, "On-line fault diagnostic system for proton exchange membrane fuel cells," *Journal of Power Sources*, vol. 175, pp. 419–429, 2008.

[16] F. Sahin, M. Ç. Yavuz, Z. Arnavut, O. Uluyol, "Fault diagnosis for airplane engines using Bayesian networks and distributed particle swarm optimization," *Parallel Computing*, vol. 33, pp. 124–143, 2007.

[17] M. Studený, D. Haws, "Learning Bayesian network structure: Towards the essential graph by integer linear programming tools," *International Journal of Approximate Reasoning*, vol. 55, pp. 1043–1071, 2014.

[18] Y. Sun, Y. Y. Tang, S. X. Ding, S. P. Lv, Y. F. Cui, "Diagnose the mild cognitive impairment by constructing Bayesian network with missing data," *Expert Systems with Applications*, vol. 38, pp. 442–449, 2011.

[19] I. Syu, S. D. Lang, "Adapting a diagnostic problem-solving model to information retrieval," *Information Progressing & Management*, vol. 36, pp. 313–330, 2000.

[20] M. Velikova, J. T. Van Scheltinga, P. J. F. Lucas, M. Spaanderman, "Exploiting causal functional relationships in Bayesian network modelling for personalised healthcare," *International Journal of Approximate Reasoning*, vol. 55, pp. 59–73, 2014.

[21] S. Verron, T. Tiplica, A. Kobi, "Fault diagnosis of industrial systems by conditional Gaussian network including a distance rejection criterion," *Engineering Applications of Artificial Intelligence*, vol. 23, pp. 1229–1235, 2010.

[22] E. Villanueva, C. D. Maciel, "Efficient methods for learning Bayesian network super-structures," *Expert Systems with Applications*, vol. 123, pp. 3–12, 2014.

[23] F. Xiao, Y. Zhao, J. Wen, S. W. Wang,"Bayesian network based FDD strategy for variable air volume terminals," *Automation in Construction*, vol. 41, pp. 106–118, 2014.

[24] Y. Zhao, F. Xiao, S. W. Wang, "An intelligent chiller fault detection and diagnosis methodology using Bayesian belief network," *Energy and Buildings*, vol. 57, pp. 278–288, 2013.

Chapter 11

A DBN-Based Risk Assessment Model for Prediction and Diagnosis of Offshore Drilling Incidents

Drilling operations of offshore oil and gas fields are characterized by high technical complexities, high risks, and high costs, since they are always in harsh environments with complicated geological factors. Lost circulation or well "kick" is a typical hazardous event that may occur while drilling wells and it also may develop into a blowout accident without being well handled. It is necessary to identify and analyze the root causes of these events and their consequences in order to prevent serious accidents from happening. In a drilling operation, the risk of blowout may change with time, depending on the operation stage, and such kind of dynamics should be captured in risk assessment. This chapter presents an approach for determining the conditional probabilities of hazardous events and their consequences. The approach includes models that take into account the influence of degradation and (if applicable) new real-time information which represents the changing model parameters (such as state change of mud density). Such an approach is based on the dynamic Bayesian network (DBN) theory and then incorporates additional nodes to address the model uncertainties and parameter uncertainties. In addition, the effect of degradation, which had been ignored in the existing researches, is also taken into account. Given that a hazardous event has occurred, this presented model can be used to predict the risk evolution, as well to reason its root causes during the offshore drilling operation. A bow-tie model is established to link the potential incident scenarios with the pressure regimes and formation load capacity, and then the model is translated into a DBN. DBN inference is adapted to perform prediction and diagnosis for dynamic risk assessment, and then a sensitivity analysis is carried out to find the relative importance of each root cause. A case study focusing on lost circulation during three drilling scenarios is adapted to illustrate the feasibility of the proposed approach.

11.1. Introduction

Drilling into offshore oil and gas fields with high pressure, temperature, H_2S-gases, and weak formations often meets many challenges, such as narrow drilling fluid density window, multiple pressure systems in vertical direction, and high pressure zones. Drilling operations at such fields are more prone to serious problems (here referred to as drilling incidents), such as lost circulation and uncontrolled influx to well ("kick"), and blowouts (to environment) compared to (less demanding) oil and gas fields. These drilling incidents can result in unplanned downtime (which are costly) or may develop into large accidents, with the potential to cause fatalities, environmental damages, and full or partial loss of drilling facility and well [10, 13, 32]. For example, the well-known Macondo blowout that occurred during the last stage of a drilling operation resulted in 11 fatalities and the largest oil spill in the history of offshore oil and gas industry. It is necessary to predict early kick or lost circulation, and then take necessary precautions, so as to avoid such type of disastrous accidents [19, 33].

Kick is the first warning towards a blowout and it is therefore important to detect a kick as early as possible and to implement efficient measures in due time. Mud weight and circulation are the primary barriers to prevent kicks, and lost circulation is an early indication of a kick under development. It is therefore important to direct the attention to avoid and manage this situation. A loss of circulation occurs when the bottom-hole pressure (BHP) in the wellbore is higher than the formation pressure, allowing (or forcing) the drilling fluid to flow into the formation. Several researchers have focused on the effects of lost circulation [30, 38], and proposed measures to reduce such effects [31]. In terms of the causes, lost circulation is usually accompanied by wellbore stability problems, damage of reservoir near well bottom and stuck pipe, and these are the main reasons why kick and even blowout can occur as a consequence. Managed pressure drilling (MPD) technology has been developed and widely used to avoid the flow of drilling fluid into the formation, and the effects of MPD should be included in risk assessments associated with loss of circulation.

Bayesian network (BN) is a flexible approach to analyze the effects of risk influencing factors like well conditions and physical measurements. Abimbola *et al.* [2] have, for example, proposed a BN-based risk model that considers potential scenarios for different pressure regimes. BNs may be derived from other frequently used models, such as bow-ties (BTs), fault trees (FTs), and event trees (ETs), for example, with basis in the BTs, FTs, and ETs developed by Khakzad *et al.* [18]. Bhandari *et al.* [5] have applied the BN method to investigate different risk factors associated with MPD and underbalanced drilling deep water drilling technologies with respect to blowout accidents. Other approaches for modeling risk also exist: Xue *et al.* [37] have proposed a safety barrier-based accident model for blowouts which considers the effects of three-level well control. Ataallahi and Shadizadeh [3] have introduced Delphi and fuzzy approach into a risk analysis model, and used this model to find the main risk influencing factors in different type of wells.

The main weakness of the mentioned risk assessment approaches is their inability to capture dynamic effects of a drilling operation, such as change in well conditions, occurrence of a new event, and the release of new estimation or measurement for the technical state of equipment. FT, ET, and BT models which constitute the main elements of most methods presented are not very effective to evaluate the correlations and dependencies between risk influence factors, and the models cannot be easily updated under changing conditions and handle the uncertainty issues [17]. Models based on BNs have been developed to overcome these modeling deficiencies [7], but cannot explicitly treat temporary relationships between model parameters, i.e., account for the fact that relationships of parameters may change from one drilling phase to the next. These limitations have already been resolved by introducing dynamic Bayesian network (DBN) [8, 14]. DBNs are built on the basis of BNs, but have additional features that allow the incorporation of events, conditions, and inter-relationships that may change over time. Cai *et al.* [9] have explored the use of DBN in performance evaluation of subsea blowout preventer (BOP) considering imperfect repair. DBNs have also been used for the same purpose in other industry sectors, such as for monitoring the risk of tunnel-induced road surface damage [36] and for studying the risk of life extension of fire water pump [27].

However, what seems to be missing in these mentioned DBN models is the possibility to incorporate the effects of uncertainties on both model and parameters. Parameter uncertainty may exist due to parameters of conditional probability that are always assumed to be time-invariant [15] based on prior knowledge being from existing literature, while model uncertainty may relate to uncertainty about the logical relationship between model parameters. Both are relevant in situations where experience is limited and the causal relationship is not well understood. In addition, current DBN-based models assume that failure rates are constant [9], but in practice the mechanical equipment may be subjected to degradation of time as in an ocean environment and most of their failures follow other probability distribution, e.g., Weibull. The main motivation for this chapter is therefore to present a new approach to handle the above-mentioned issues by allowing uncertain logical relationships among root causes, integrating parameter uncertainty of prior knowledge, and introducing failure probability distributions with the DBN theory. This approach can systematically perform incident evolution prediction and root cause reasoning for risk assessment using the predictive, diagnostic, and sensitivity analysis technology.

The rest of this chapter is organized as follows. Section 11.2 presents three drilling scenarios with the MPD technology. In Sec. 11.3, the fundamental theory of BN and DBN will be briefly introduced. In Sec. 11.4, a DBN-based risk assessment model is developed by incorporating some additional nodes to handle the uncertainty issues involving the model uncertainty and relevant parameters' uncertainty and the effect of degradation is also considered for the drilling incidents. The proposed method is applied for incident prediction evolution as well as root cause reasoning for the lost circulation in the case study of Sec. 11.5. Section 11.6 provides the conclusion and research perspectives of this study.

11.2. Manage Pressure Drilling Technology

Manage pressure drilling (MPD) is a powerful drilling hazard mitigation technique for offshore drilling and is defined by International Association of Drilling Contractors (IADC) Underbalanced Operations Committee as "an adaptive drilling process used to precisely control the annual pressure profile throughout the wellbore". The aim of the MPD is to ascertain the downhole pressure environment limits and to manage the annular hydraulic pressure profile accordingly [34]. An MPD system consists of the following main systems: a rotating control device (RCD), an automated dynamic annular pressure control (DAPC) system, a back pressure pump, a DAPC choke manifold, a flowmeter [12]. RCD is also regarded as the first barrier to seal the annulus from drill-string by creating a closed circulation system different from normally open circulation system, and therefore the flow of mud out from the annulus can be controlled by an automated choke. DAPC system is used to maintain the constant BHP by providing the back pressure on the annulus by continuously adjusting the DAPC chocks and back pressure pump. A flowmeter provides the flow-out data and the kick detection is predicted by monitoring flow-in data [35]. MPD drilling techniques include constant bottom-hole pressure (CBHP), pressurized mud-cap drilling, and dual gradient drilling [29]. The case study selected for this chapter focuses on the use of CBHP as a measure to prevent or mitigate drilling hazards, such as differential sticking, lost circulation, and kicks, on a development well in a pressurized, fractured basement with narrow downhole environmental limitation. An MPD system can also optimize the rate of penetration, reduce non-productive time, and the number of casing strings relative to conventional drilling techniques, and deepen casing set points. The typical offshore MPD system is illustrated in Fig. 11.1.

During a drilling operation, it is required to always maintain two functioning well barriers: the primary barrier, which is the active balancing of drilling fluid (i.e., mud) to avoid hydrocarbons escaping from the well, and the secondary barrier, BOP. The BOP mainly consists of BOP control system and BOP stack which are used to seal,

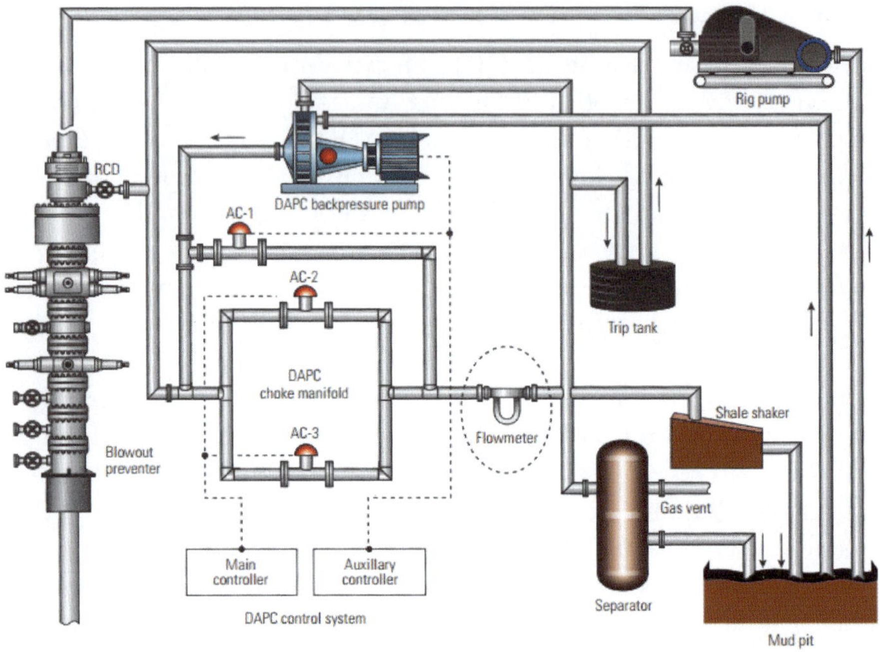

Fig. 11.1. MPD system [12].

control, and monitor O&G wells, preventing the uncontrolled release of crude oil and/or natural gas from the well. The MPD system can be regarded as being part of the primary barrier, as the system applies back pressure control to maintain control with BHP [26]. The hydrostatic pressure of the drilling fluid column must take into account the correct balance between BHP and formation fracture pressure (FFP).

The formula for determining the BHP [29] varies for different types of drilling operations. Three types of drilling operations have been considered in this chapter: not circulating, tripping in, and circulating. When the rig pump is not circulating the drilling fluid, the static BHP is defined as

$$\mathrm{BHP}_{\text{static}} = P_{\text{dfc}} = \rho_d g h + P_b, \tag{11.1}$$

where P_{dfc} is the hydrostatic pressure of the drilling fluid column, ρ_d is the density of drilling fluid and, g stands for gravitational acceleration, h is drilling fluid height, and P_b is the back pressure of wellhead.

When the drillstring is tripping in the wellbore, the dynamic BHP is defined as

$$\text{BHP}_{\text{dynamic}} = P_{\text{dfc}} + P_{\text{sg}} = \rho_d g h + P_{\text{sg}} + P_b, \tag{11.2}$$

where P_{sg} is the surging pressure caused by drillstring tripping in the wellbore.

When the rig pump is on and circulating the drilling fluid, the dynamic BHP is defined as

$$\text{BHP}_{\text{dynamic}} = P_{\text{dfc}} + P_{\text{fc}} = \rho_d g h + P_{\text{fc}} + P_b, \tag{11.3}$$

where P_{fc} is the frictional pressure due to pumping the drilling fluid through the drillstring, and P_{fc} is the back pressure of wellhead.

In this chapter, the main focus is the control of CBHP to avoid drilling fluid loss. If the MPD system fails to perform this function, the result may be serious, such as differential sticking and lost circulation. Lost circulation does not simply mean the loss of a few dollars of drilling mud, but it can be disastrous as a blowout. Drilling crew therefore pays close attention to monitoring of tanks, pits, and flow from the well to quickly assess and control the lost circulation. This chapter studies the causes and effects of lost circulation for the three mentioned, considering the performance of the MPD system and other influencing factors.

11.3. Theoretical Basis for DBNs

This section highlights some selected points about the theories of BNs as well as DBNs.

11.3.1. *Bayesian networks*

A dynamic Bayesian networks (DBNs) is an extension of a Bayesian network (BN). A BN combines graph model and probability theory, consisting of a directed acyclic graph (DAG) and an associated joint probability distribution (JPD) [24]. In a DAG, nodes including parent nodes and child nodes represent random variables, and links determine probabilistic dependences between variables. A conditional probability table (CPT) for discrete variables is defined for the relationship among parent nodes to demonstrate marginal probability. Assuming $\mathrm{Pa}(X_i)$ is the parent node of X_i, the CPT of X_i is denoted by $P(X_i \mid \mathrm{Pa}(X_i))$. Therefore, the JPD, $P(X_1, \ldots, X_N)$, can be rewritten as

$$P(X_1, \ldots, X_N) = \prod P(X_i \mid \mathrm{Pa}(X_i)). \tag{11.4}$$

The quantification of probabilities in BNs includes two steps: assigning prior probabilities to the parent nodes, and defining CPT of child nodes by combining *a priori* knowledge. Such knowledge can be from expert judgment or observations.

11.3.2. *Dynamic Bayesian networks*

A DBN extends a BN by introducing relevant temporal dependencies, so as to model the dynamic behavior of random variables [15]. A DBN consists of a sequence of time slices and temporal links. Each slice represents a static BN to describe variables in the corresponding time step, and temporal links between variables in different time slices represent a temporal probabilistic dependence. A DBN is able to model probability distribution over semi-infinite collection of random variables. The CPT of each variable in the DBN can be calculated independently, facilitating the interpretation of DBN.

In general, there are two assumptions for DBN construction. Firstly, the system is assumed as the first-order Markovian (i.e., $P(X_t \mid X_1, \ldots, X_{t-1}) = P(X_t \mid X_{t-1})$, and secondly the transition probability $P(X_t \mid X_{t-1})$ is the same for all the t. Therefore, a DBN can be defined by a pair of BNs $(B_1, B\rightarrow)$, where B_1 is a BN which defines the prior $P(X_1)$, and $B\rightarrow$ is a two-slice temporal Bayesian net (2TBN) that defines the transition and observation models as a product of the CPTs in the 2TBN [21]:

$$P(X_t \mid X_{t-1}) = \prod_{i=1}^{N} P(X_t^i \mid \mathrm{Pa}(X_t^i)), \tag{11.5}$$

where X_t^i is the ith node in time slice t, $\mathrm{Pa}(X_t^i)$ denotes the parent of X_t^i, which may be in the same time slice t or previous time slice $t-1$, and N indicates the number of random variables in X_t^i.

The nodes in the first time slice of a 2TBN have unconditional initial state distribution, $P(X_1^{1:N})$, while each node in the second time slice has an associated CPT. Then, for a DBN with T slices, the joint distribution can be obtained by "unrolling" the network as

$$P(X_{1:T}^{1:N}) = \prod_{i=1}^{N} P_{B1}(X_1^i \mid \mathrm{Pa}(X_1^i)) \times \prod_{t=2}^{T}\prod_{i=1}^{N} P_{B\rightarrow}(X_t^i \mid \mathrm{Pa}(X_t^i)). \tag{11.6}$$

Several inference algorithms [21, 22] have been proposed for DBN modeling. In this chapter, the forward–backward inference and mutual information are used for Bayesian inference. The main benefit of introducing DBN for risk assessment may be summarized as follows:

All relevant qualitative and quantitative analyses can be carried out in a full probabilistic model, including a broad variety of modeling schemes and a large collection of inference techniques from the BN applied to the dynamical process.

A DBN is more acceptable for predicting values of variables and capable of revealing the system state at any time. At a time slice, new information about model parameters may be incorporated into the model, the value of a variable can be calculated based on probabilistic inference. This information may be in the following formats.

(1) updated probabilities, updating only on the basis of the associated probability distribution and the elapsed time;
(2) updated probabilities, considering new real-time information, such as a change in a state of a model parameter;
(3) a combination of the above two, using Bayesian update.

11.4. Development of a DBN-Based Risk Assessment Model

DBNs and BTs are combined in this chapter as the basis to set up a risk assessment model consisting of factors that may lead to

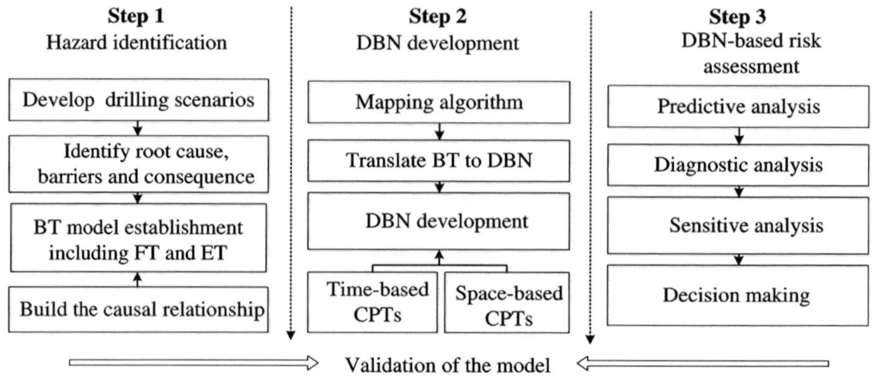

Fig. 11.2. DBN-based risk assessment model for drilling incidents.

drilling incidents, the causal relationship between them, and the effects of measures available to prevent the escalation. The model is used to perform the prediction for the occurrence probability of drilling incidents over time and compare risks among the different drilling processes. The overall workflow needed to derive the model and apply it for risk assessment is shown in Fig. 11.2. As shown in Fig. 11.2, there are three main steps: hazard identification, DBN development, and DBN-based risk assessment.

The particulars of the presenting model are specified as follows:

Step 1: A BT model is integrated, so that the cause–consequence chain or causal relationships can be mostly easily identified and foreseen.

Step 2: Uncertainties on model and parameters related to DBN construction are taken into account. The parameters of CPTs from the previous time to the current time used in the proposed model are assumed both time-variant and time-homogeneous.

Step 3: Estimation not only focuses on forward analysis, but also on dynamics, given any events with occurrence of the drilling process. In addition, the occurrence probability of a hazardous event or the development trend of its underlying consequence as functions of time will be predicted. The root cause reasoning is discussed given the occurrence of a hazardous event in diagnosis analysis.

11.4.1. *Step 1: Hazard identification*

A BT model can provide visual explanation of a complete accident scenario evolution and is widely applied in hazard identification and risk analyses [11]. A simplified BT is shown in Fig. 11.3(a), with three main parts: the left side, the middle, and the right side. On the left side there is an FT, identifying the causes of an unwanted event (which is placed in the middle), and on the right side there is an ET, identifying the possible outcomes given the effects of mitigating measures.

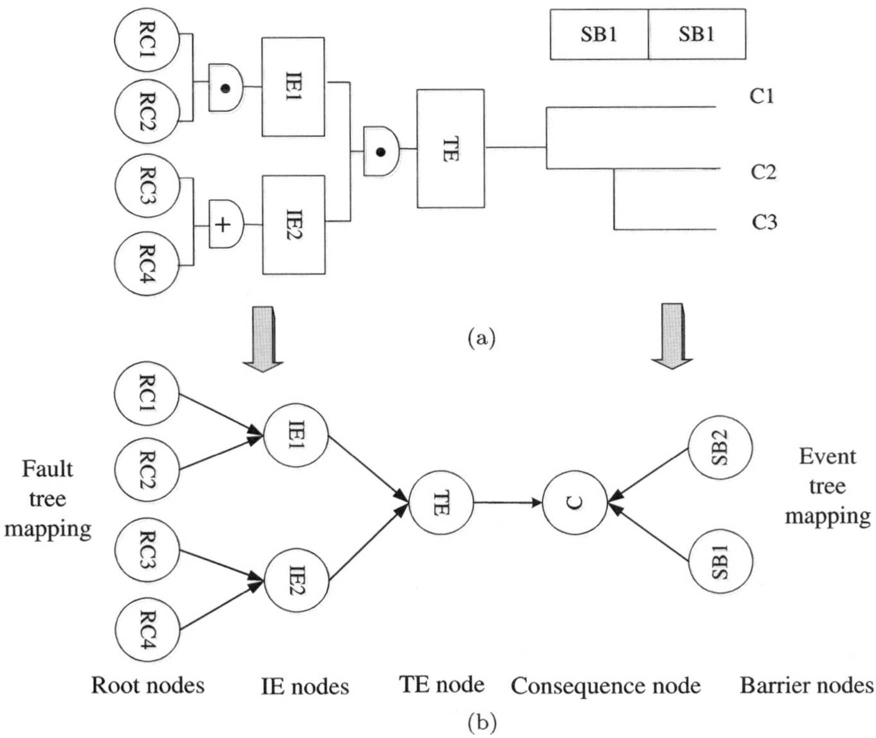

Fig. 11.3. Translating from BT to BN: (a) simplified BT model and (b) simplified BN model.

The following notations will be used in the rest.

- root causes (RC), the basic events of the FT;
- intermediate event (IE), which can be substructures of the FT;
- top event (TE), the unwanted event that is placed in the middle of the BT;
- safety barriers (SB), mitigation measures to reduce the severity of potential consequences (C).

Once hazards have been identified, the bow-tie model can be applied to further build the causal relationships. This process for hazard identification is considered a difficult task for complex offshore wells, especially the drilling with high temperature and pressure information.

The detailed steps for hazard identification are explicitly illustrated as follows:

(1) Develop drilling incidents scenarios based on the drilling operations and pressure regime in Sec. 11.2.
(2) Collect available safety-based information about influence factors including ocean environment factors, geological factors, drilling technology, human factors, etc., incidents (kick, lost circulation, etc.), or underlying consequences (blowout, etc.) associated with the drilling operation in question, using the relative standards, literature, accident reports, and experts input.
(3) Develop BT based on the FT and ET theories for drilling incidents, and the BT should be reviewed by relevant personnel from operations, maintenance, safety and management, etc.
(4) Describe the explicit causal relationships among the root causes, target incidents, and consequences and define the state of each root cause and the corresponding failure data.

But the application of BT in the risk analysis suffers from the limitation of updating probability and cannot take uncertainties into consideration [17]. More importantly, because of being composed of static structures such as FT and ET, BT has not been widely recognized in the context of dynamic analysis. To consider dynamic behavior over time, the BT model needs to be transformed into

DBN for dynamic risk assessment. The dynamic behavior over time consists of three aspects as follows:

- The evolution tendency of the TE can be predicted over time after the actual evidences of root causes are collected in different time slices.
- The occurrence probability of having TE and experiencing corresponding consequences can be predicted given the current status of root causes detected at any time.
- The failure probabilities of any root cause at previous time can be calculated when the status of this cause at current time is detected.

11.4.2. *Step 2: DBN development*

11.4.2.1. *Mapping BT to BN*

The translating algorithm from BT to BN consists of FT mapping and ET mapping [18]. First, the mapping from FT into BN includes a graphical and probability translation based on the previous work [6]. Figure 11.3(b) illustrates the simplified procedure of mapping FT and ET into BN. In this phase, each root cause, intermediate event and top event of FT, are translated into a corresponding root node, intermediate event node and top event node of BN, respectively. The nodes of BN are linked in the same way as the corresponding events in the FT. The failure probabilities of the root causes are assigned to the corresponding parents nodes as prior probabilities. The connections between events such as "AND gate" and "OR gate" are translated into corresponding conditional probability tables (CPTs) in BN.

Bearfield and Marsh [14] presented a mapping algorithm from ET into DBN, which includes safety barriers and consequence translation. Each safety barrier of ET is translated into a corresponding barrier node with two states (success and failure) and the consequences of ET are translated into a corresponding consequence node with multiple states as the number of the event tree consequences. The failure probabilities of safety barriers are assigned to the prior probabilities of corresponding barrier nodes. It is noted that the CPTs of the corresponding consequence node are assigned based on the expert judgment.

11.4.2.2. *Simplified DBN model development*

As the state of the node or the probability of the node, e.g., failure is changing over time, a simplified DBN is established by extending the BN formalism within three time slices from time $t = 0$, $t = t_1$ to time $t = t_2$, as presented in Fig. 11.4(a). The time interval is the same between 0 and t_1 or t_1 and t_2. As indicated in Fig. 11.4(a), the root nodes RC1, RC2, RC3, and RC4, and barrier nodes SB1 and SB2 are extended from zeroth to t_1 or from t_1 to t_2 with inter-slice arcs, respectively. It is noted that the inter-relationship over time is not discussed in this chapter, and so there are no inter-slice arcs assigned for other nodes except for root nodes and barrier nodes.

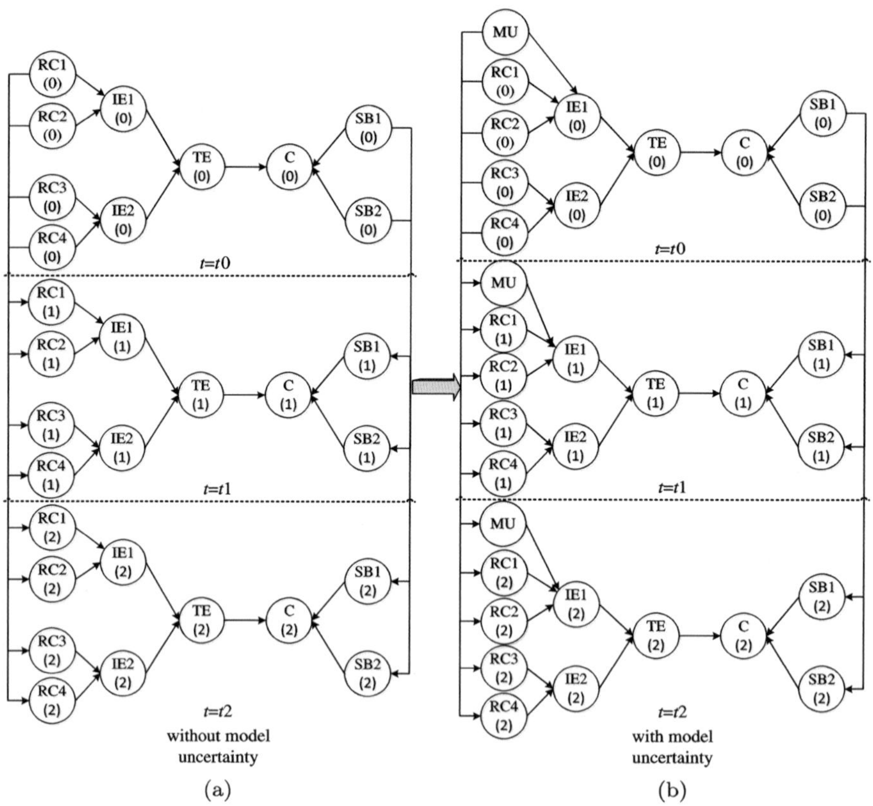

Fig. 11.4. Simplified DBN modeling: (a) without model uncertainty and (b) with model uncertainty for three time slices.

Each root nodes of DBN can have two states, YES and NO. The YES state denotes that an event or a failure occurs, while the NO state means that it does not occur. IE/TE can take True or False state. The True state refers to the occurrence of IE/TE while the False one represents the opposite. Each barrier node in DBNs also involves two states, Success and Failure which refer that the barrier is able to carry out its safety function or not.

This proposed model can handle the uncertainty issues involving the model uncertainty and parameter uncertainty.

(1) Modeling uncertainty is necessary due to the lack of accurate determination of a causal relationship between the nodes and their parents, e.g., the relationship between the nodes RC1 and RC2 cannot completely follow the OR-gate. To handle the model uncertainty, the nodes denoted as MU, as shown in Fig. 11.5(b) are introduced by modifying its CPT and constant with different time slices. MU can take the states OR and AND, which refer to whether IE follows the OR-gate or the AND-gate, respectively. The CPT of IE1 can be assigned as shown in Table 11.1, e.g., $P(\text{IE1} = \text{True} \mid \text{RC1} = \text{YES}, \text{RC2} = \text{YES}, \text{MU} = \text{OR}) = 1$.

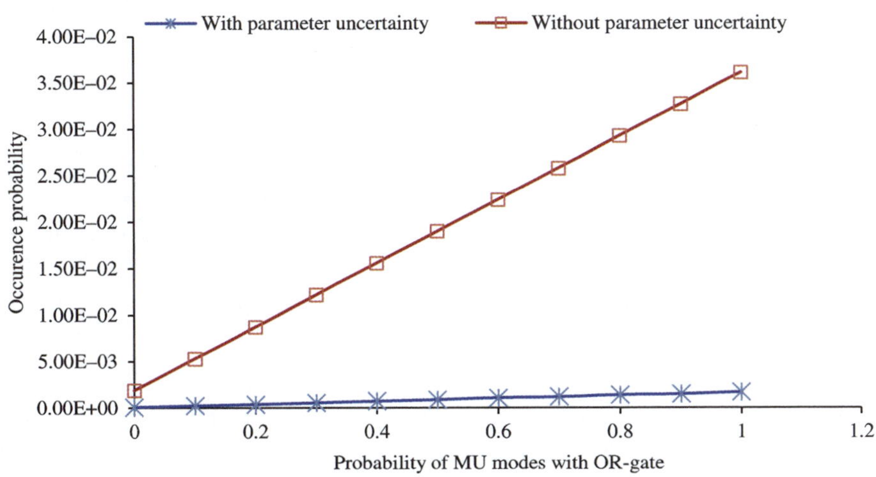

Fig. 11.5. Comparison of occurrence probability for MU nodes with OR-gate probabilities.

Table 11.1. CPT for IE1 node.

RC1		RC2		MU		IE1	
YES	NO	YES	NO	OR	AND	True	False
1	0	1	0	1	0	1	0
0	1	1	0	1	0	1	0
1	0	0	1	1	0	1	0
0	1	0	1	1	0	0	1
1	0	1	0	0	1	1	0
0	1	1	0	0	1	0	1
1	0	0	1	0	1	0	1
0	1	0	1	0	1	0	1

Table 11.2. CPT for IE2 node.

RC3		RC4		IE2	
YES	NO	YES	NO	True	False
1	0	1	0	0.088	0.912
0	1	1	0	0.04	0.96
1	0	0	1	0.05	0.95
0	1	0	1	0	1

(2) Parameter uncertainty can be split into the space-based and the time-based ones.

- The space-based parameter uncertainty occurs when linking the root nodes (R_i) to IE nodes (IE_j), which is based on the uncertainty of the root causes itself, e.g., the nodes RC3 and RC4 which represent the FFP and formation porosity are influenced by the uncertainty effect of geology information for offshore drilling operation. We can handle this uncertainty by using noisy AND-gate or noisy OR-gate algorithm [22]. If we assume that $P[\text{IE2} = \text{True} \,|\, \text{RC3} = \text{YES}, \text{RC4} = \text{NO}) = 0.04$ and $P(\text{IE2} = \text{True} \,|\, \text{RC3} = \text{NO}, \text{RC4} = \text{YES}) = 0.05$, we can get $P(\text{IE2} = \text{True} \,|\, \text{RC3} = \text{YES}, \text{RC4} = \text{YES}) = 0.088$. The CPT can be obtained as shown in Table 11.2.
- The time-based parameter uncertainty occurs when linking the root nodes (R^i_{tj-1}) at the previous time t_{j-1} to the

root nodes (R^i_{tj}) at the current time t_j. The CPTs are assumed time-invariant if prior knowledge is usually obtained in accordance with accident statistic and literature reviews, e.g., the occurrence probability of for a specific change in density is $P(R^i_{tj})$ which is prior probability and assumed to be constant over time. We assume that, when the root cause occurs at time $t-1$, the root cause will not occur at time t. It means that the root cause can be adjusted in a perfect state in the current time interval. The CPTs are therefore obtained as shown in Table 11.3, when $P(R^i_{tj} = \text{YES} \mid R^i_{tj-1} = \text{YES}) = 0$ and $P(R^i_{tj} = \text{NO} \mid R^i_{tj-1} = \text{YES}) = P(R^i_{tj})$.
The CPTs are regarded as time-variant if failures for root causes such as equipment failure and safety barrier failure follow the Weibull distribution. The degradation influence is considered to estimate the parameters of CPTs. If $P(R^i_{tj} = \text{YES} \mid R^i_{tj-1} = \text{YES}) = 1 - \mathrm{e}^{-(\lambda t_{j-1})^\alpha}$ and $P(R^i_{tj} = \text{YES} \mid R^i_{tj-1} = \text{NO}) = 1 - \mathrm{e}^{-(\lambda t_j)^\alpha}$, we have CPTs as listed in Table 11.4, where λ and α denote the scale parameter and shape parameter, respectively.

Table 11.3. CPT for two time slices.

	t_j	
t_{j-1}	YES	NO
YES	0	1
NO	$P(R^i_{tj})$	$1 - P(R^i_{tj})$

Table 11.4. CPT for two time slices with degradation.

	t_j	
t_{j-1}	YES	NO
YES	$1 - \mathrm{e}^{-(\lambda t_{j-1})^\alpha}$	$\mathrm{e}^{-(\lambda t_{j-1})^\alpha}$
NO	$1 - \mathrm{e}^{-(\lambda t_j)^\alpha}$	$\mathrm{e}^{-(\lambda t_j)^\alpha}$

11.4.3. *Step 3: DBN-based risk assessment*

In this step, it is proposed to utilize DBN for the three mentioned decision-support scenarios:

- Predictive analysis: This estimates the risk evolution of drilling operation over time and forecast development in the risk of a drilling operation given the current state of knowledge.
- Diagnostic analysis: This detects and investigates the most likely causes of a drilling incident using backward analysis when the top event occurs.
- Sensitivity analysis: This checks to what extent the result of the predictive or diagnostic analysis is influenced by specific parameters which are regarded as uncertain.

11.4.3.1. *Predictive analysis*

Predictive analysis aims to predict the future risk evolution tendency of drilling operation over time and forecast development in the risk of a drilling operation given the current state of knowledge; using the forward inference technical in DBN. The occurrence probability distribution of a top event at time t is under the combination of root causes $(R_t^1, \ldots, R_t^i)$ and the occurrence probability distribution of the corresponding consequence (C) under the combination of TE and safety barriers $(B_t^1, \ldots, B_t^i)$. The state of each root cause or safety barrier is treated as input by CPTs into DBN model. Probability distribution of TE/C, represented by $P(TE_t = te)/P(C_t = c)$, is calculated in the following two equations:

$$P(TE_t = te) = P(TE_t = te \,|\, IE_t^1 = ie_1, \ldots, IE_t^i = ie_j) \times P(IE_t^1 = ie_1, \ldots, IE_t^i = ie_j, R_t^1 = r_1, \ldots, R_t^i = r_j), \tag{11.7}$$

where te stands for the state of a top event TE_t; ie_j stands for the state of intermediate event IE_t and r_j stands for the state of root nodes R_t^i; $P(TE_t = te \,|\, IE_t^1 = ie_1, \ldots, IE_t^i = ie_j)$ refers to the conditional probability distribution of TE_t; and $P(IE_t^1 = ie_1, \ldots, IE_t^i = ie_j, R_t^1 = r_1, \ldots, R_t^i = r_j)$ refers to the joint probability distribution of *IE* nodes and root nodes.

$$P(C_t = c) = P(C_t = c \,|\, TE_t = te, B_t^1 = b_1, \ldots, B_t^i = b_j) \times P(TE_t = te) \times P(B_t^1 = b_1, \ldots, B_t^i = b_j), \quad (11.8)$$

where c stands for the state of consequence C_t; b_i stands for the state of root nodes B_i; $P(C_t = c \,|\, TE_t = te, B_t^1 = b_1, \ldots, B_t^i = b_j)$ refers to the conditional probability distribution of C_t; and $P(B_t^1 = b_1, \ldots, B_t^i = b_j)$ refers to the joint probability distribution of barrier nodes.

The risk of a drilling operation, given the occurrence of root causes ($R_t^m = r_m$), represented by $P(TE_t = te \,|\, R_t^1 = r_1, \ldots, R_t^m = r_m, \ldots, R_t^i = r_j)$, can also be calculated as follows:

$$P(TE_t = te \,|\, R_t^m = r_m) = P(TE_t = te \,|\, IE_t^1 = ie_1, \ldots, IE_t^i = ie_j) \times P(IE_t^1 = ie_1, \ldots, IE_t^i = ie_j, R_t^1 = r_1, \ldots, R_t^m = r_m, \ldots, R_t^i = r_j), \quad (11.9)$$

where $P(TE_t = te \,|\, IE_t^1 = ie_1, \ldots, IE_t^i = ie_j)$ refers to the conditional probability distribution of *TE*; and $P(IE_t^1 = ie_1, \ldots, IE_t^i = ie_j, R_t^1 = r_1, \ldots, R_t^m = r_m, \ldots, R_t^i = r_j)$ refers to the joint probability distribution of IE nodes and root causes given the occurrence of root causes ($R_t^m = r_m$).

Generally, $P(TE_t = te), P(C_t = c)$ or $P(TE_t = te \,|\, R_t^m = r_m)$ can serve as an indicator to evaluate the risk, providing the basis for decision makers to take proper measures.

11.4.3.2. *Diagnostic analysis*

Diagnostic analysis aims to obtain the posterior probability distribution of each root cause when a TE occurs at certain time, which is performed through the backward analysis of the DBN. The underlying causes with the largest occurrence probability or the occurrence probability above the acceptable safety level can then be detected by means of posterior probability distribution, reminding engineers to pay more attention to these causes. Posterior probability distribution of root nodes R_t^i, represented by $P(R_t^i = r_i \,|\, TE_t = te)$, can be calculated as follows:

$$P(R_t^i = r_i \,|\, TE_t = te) = \frac{P(TE_t = te \,|\, R_t^i = r_i) \times P(R_t^i = r_i)}{P(TE_t = te)}. \tag{11.10}$$

Normally, R_t^i is more likely to become the key root cause at time t leading to the occurrence of a TE when $P(R_t^i = r_i \,|\, TE_t = te)$ is close to 1.

11.4.3.3. *Sensitivity analysis*

Sensitivity analysis checks to what extent the results of the predictive or diagnostic analysis are sensitive to specific parameters regarded as uncertain. The important degree of root cause to the top event can be analyzed by applying Shannon's mutual information (entropy reduction), which is one of the most commonly used measurements for ranking information sources [20]. The mutual information is the total uncertainty-reducing potential of R, given the original uncertainty in R_i prior to consulting R_j. Intuitively, mutual information can measure how much knowing one of these variables reduces our uncertainty about the other. The mutual information of R_i and R_j is given by

$$I(R_i, R_j) = -\sum_i \sum_j P(R_i, R_j) \log \frac{P(R_i, R_j)}{P(R_i)P(R_j)}, \tag{11.11}$$

where $P(R_i, R_j)$ is the joint probability distribution function of root cause R_i and R_j, and $P(R_i)$ and $P(R_j)$ is the probability distribution of root causes R_i and R_j, respectively.

11.4.4. *Validation of the model*

Validation for a newly developed model is a significant process of checking whether it will provide a reasonable amount of confidence to meet its specification and produce the required results in a sound, defensible, and well-grounded way. It seems to become an impractical exercise to gather all the monitored data to perform a fully comprehensive validation for a newly developed model because it ideally requires to cover the complete range of possibilities. The validation in this chapter is carried out from the following aspects:

- Validation of model development process means to verify the model to be constructed in a reasonably, defensibly, and realistically way.
- Validation of model usability means to check sensitivities of results by modeling the change of input data by three-axiom-based validation method by [16].
- Validation of model results means to evaluate the results generated from a developed model involving model parameter inputs, and make the result more reasonable by a comparison with that of another approach such as fault tree and static Bayesian network (SBN) using existing data.

(1) The proposed model is developed based on a bow-tie model illustrated in Fig. 11.3, which is also used for translating to meet the requirement for dynamic risk assessment. Examination of development process, illustrated in Fig. 11.4, consists of checking both effect of uncertainty and degradation. As an example, it is not clear whether the relationship between RC1 and RC2 follows rule of OR-gate or AND-gate and what kind of effect it will bring. The effect of uncertainty on model and parameters is therefore to be checked by comparing results from the model without MU nodes and results from the model with MU nodes by taking probability range of OR-gate [0, 1]. As shown in Fig. 11.5, given that every root cause takes the initial probability 0.1 in the YES state, the comparison of the two cases are conducted: Ignoring the effect of parameter uncertainty, and considering one. The results reveal that the occurrence probability of the top event will be changed from 1.9×10^{-3} to 3.61×10^{-2}, in the former case, and it varies from 9×10^{-5} to 1.7×10^{-3} in the latter one. The validation for the degradation effect, given MU nodes with probabilities (0.3, 0.7), is carried out by comparing the occurrence probabilities without the degradation effect and with the one, as listed in Tables 11.3

Table 11.5. Comparison of occurrence probability for degradation.

Time slice	DBN without degradation effect	DBN with degradation effect
0	0	0
1	5.7E–04	5.7E–04
2	5.6E–04	9.7E–04

and 11.4, respectively. The results (as shown in Table 11.5) also indicate that the occurrence probability of the top event will change slightly initially and increase later. Such changes can be explained reasonably due to the consideration of the causal relationship, parameter uncertainty and the degradation effect.

(2) Validation of model usability, as illustrated in Fig. 11.4, is to check whether sensitivities of results by modeling parameters inputs are expected. At the initial time, the prior probabilities for all root causes are set to 0.1, and probability of MU nodes is taken by (0.3, 0.7). When the probabilities of root causes including RC1, RC2, RC3, and RC4 are set to 1 in sequence and the probability of MU nodes is kept constant, the occurrence probability of the top event will gradually increase from 5.7×10^{-4} to 3.3×10^{-3}, 9×10^{-3}, 4.48×10^{-2}, and 8.8×10^{-2}, respectively. The exercise of increasing the failure probability of each root cause one after another will meet the axiom specification and produce the required results, thus giving a partial verification to the newly developed model.

(3) The results have been validated by the special case with "lost circulation" in the not-circulating scenario using the existing partial data as shown in Fig. 11.6(a). The results from the fault tree (FT), BN with average probability failure on demand (PFDavg) [28] and DBN with probability failure on demand (PFD(t)), and PFDavg [28] are compared for a period with four time slices, which is shown in Table 11.6. The basic events with these occurrence probabilities are listed in Tables 11.7

(a)

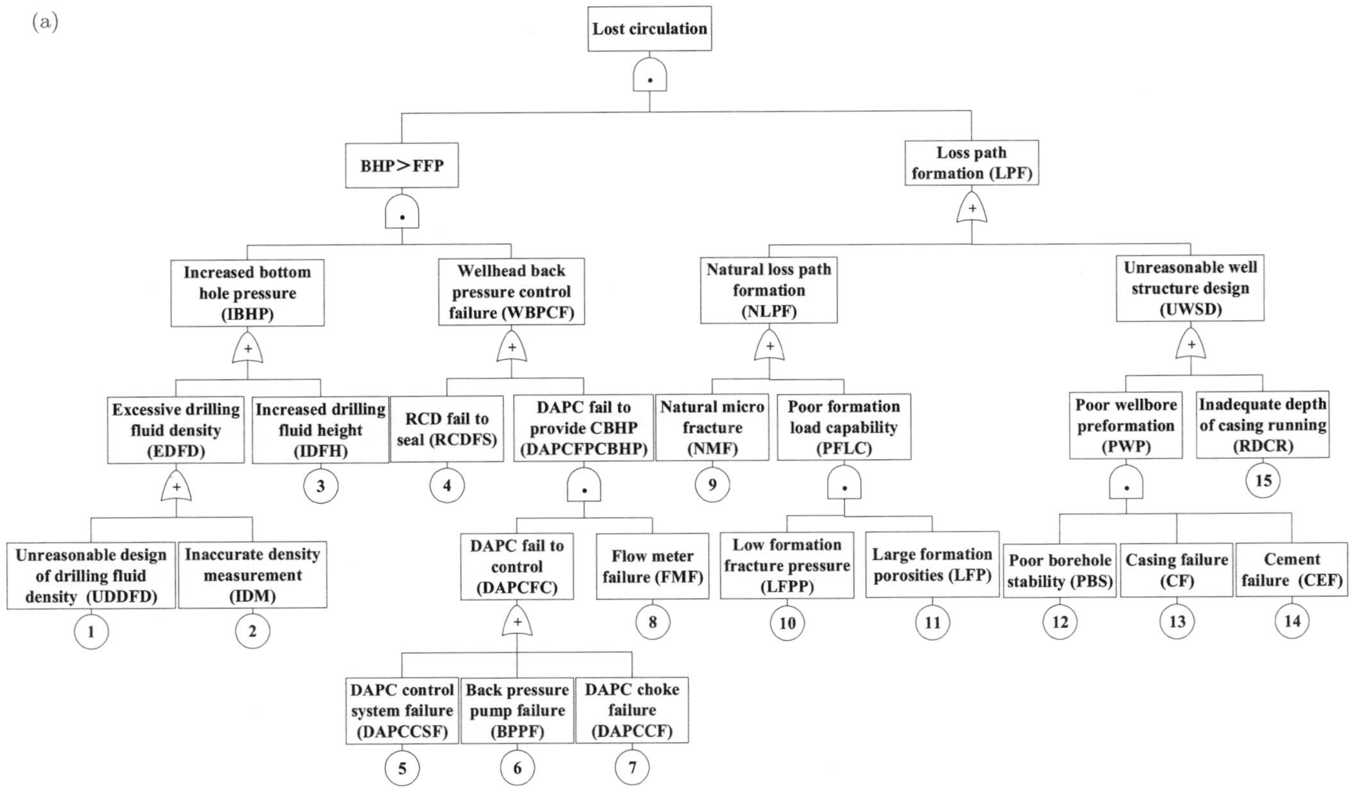

Fig. 11.6. BT model for (a) fault tree of not circulating, (b) fault tree of tripping in, and (c) fault tree of circulating (d) event tree.

(b)

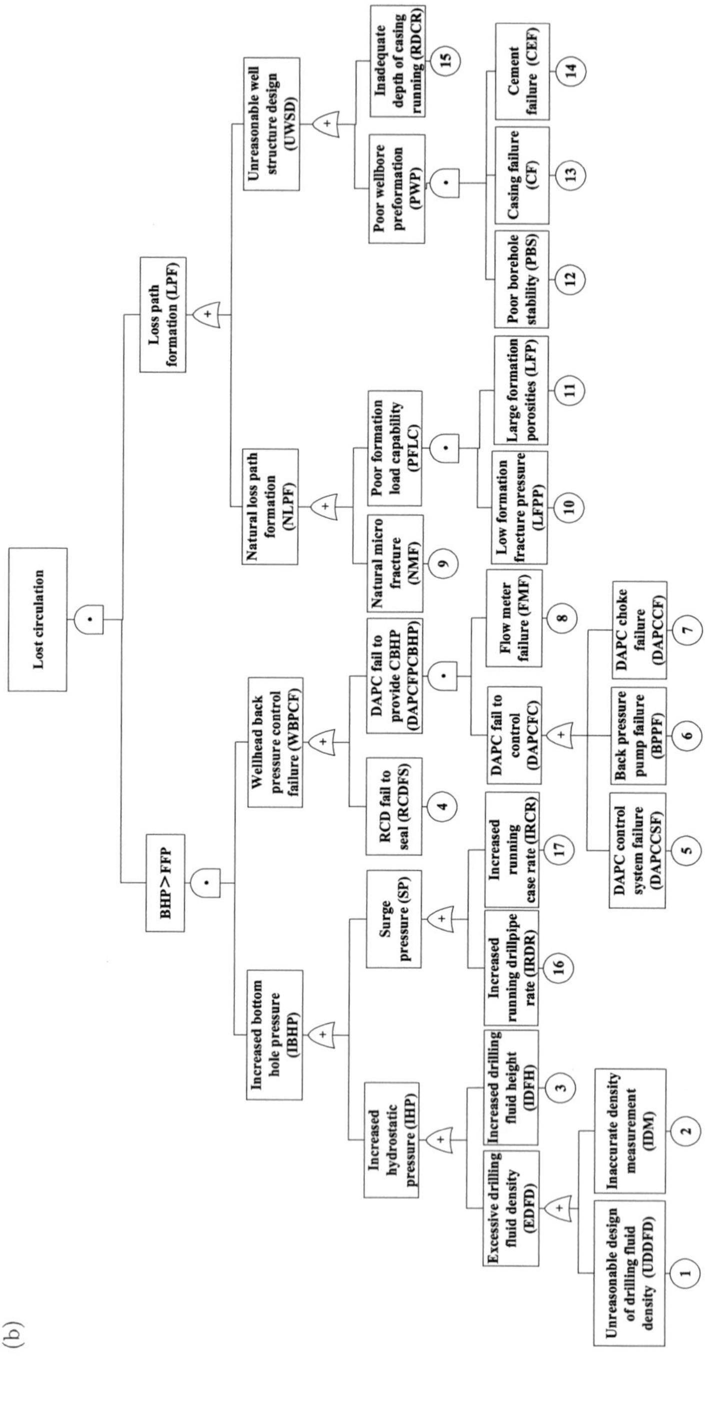

Fig. 11.6. *(Continued)*

(c)

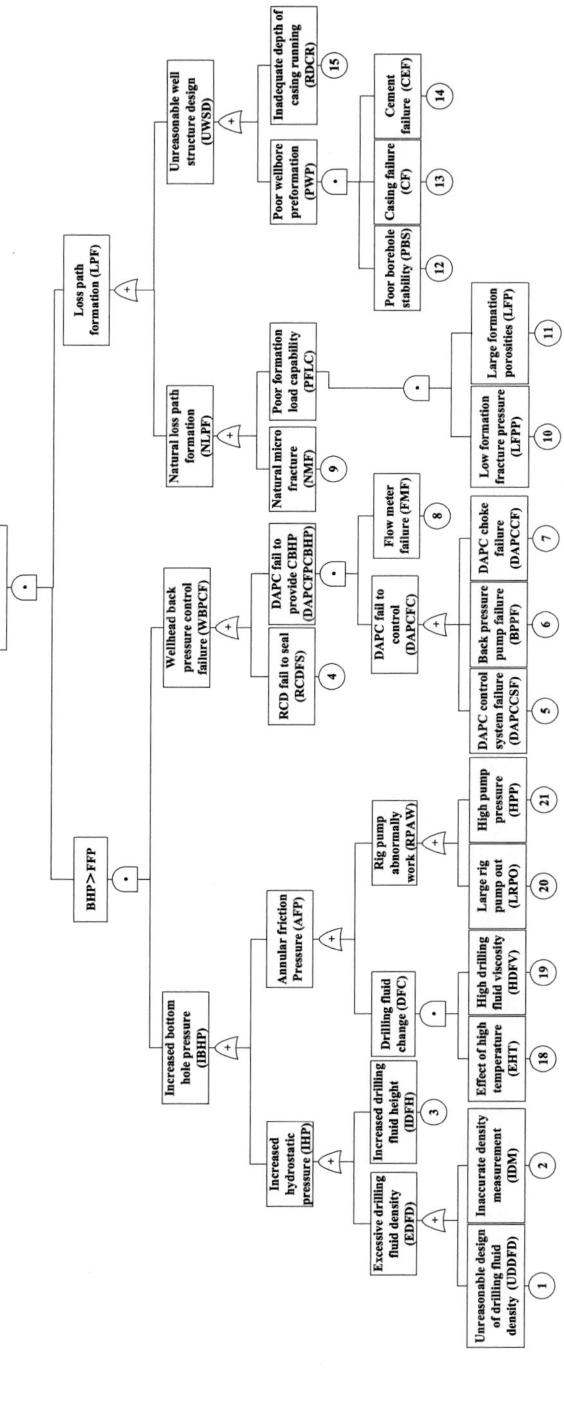

Fig. 11.6. *(Continued)*

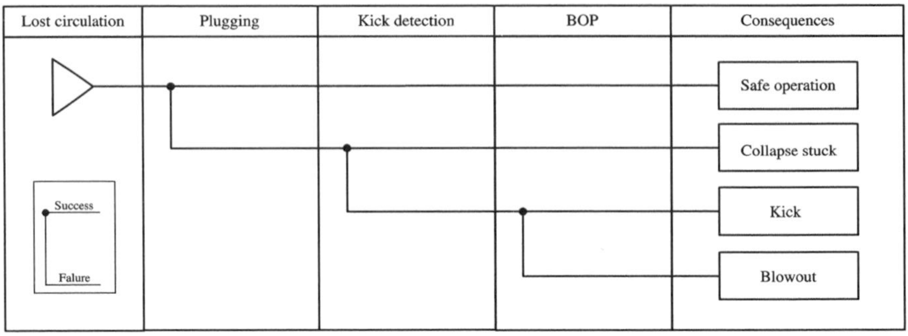

(d)

Fig. 11.6. (*Continued*)

Table 11.6. Comparison of occurrence probability for different methods.

Time interval $[t_{i-1}, t_i]$	FT with PFDavg	BN with PFDavg	DBN with PFD (t_i)	DBN with PFDavg
[0, 360]	8.00E–05	5.00E–05	2.00E–05	9.40E–06
[360, 720]	8.00E–05	5.00E–05	7.00E–05	4.00E–05
[720, 1080]	8.00E–05	5.00E–05	1.40E–04	1.00E–04

and 11.8. The result indicates that the magnitudes of occurrence probabilities almost keep the same. The difference of results between FT and BN is caused by the uncertainty issues, while that between BN and DBN can be explained by the effect of degradation. This part can make the final results from the newly developed model more reasonable.

11.5. Case Study

A case study for an offshore well related to lost circulation is carried out in this section. An offshore drilling well in BD oil and gas field in Madura is considered as the equipment under protection. The highest wind events (thunderstorms) will result in maximum wave heights that are relatively small according to the statistics. The sea condition is therefore relatively safe in this selected area. The interface of the target oil and gas reservoir pressure is approximately 8090 psi,

Table 11.7. Prior and posterior probability for root causes.

			Prior probability		Posterior probability $t = 1440$th		
No.	Basic event	Description	$t = 0$	$t = 1440$th	Not circulating	Tripping in	Circulating
1	UDDFD	Unreasonable design of drilling fluid density	5.00E–02	4.76E–02	9.15E–01	8.29E–01	6.2E–01
2	IDM	Inaccurate density measurement	0	2.90E–03	5.31E–02	4.81E–02	3.59E–02
3	IDFH	Increased drilling fluid height	2E–03	2.00E–03	3.66E–02	3.32E–02	2.48E–02
4	RCDFS	RCD fail to seal	0	1.40E–02	3.89E–01	3.89E–01	3.89E–01
5	DAPCCSF	DAPC control system failure	0	4.8E–03	1.33E–01	1.33E–01	1.33E–01
6	BPPF	Back pressure pump failure	0	1.14E–02	3.16E–01	3.16E–01	3.16E–01
7	DAPCCF	DAPC choke failure	0	4.9E–03	1.36E–02	1.36E–02	1.36E–02
8	FMF	Flow meter failure	0	1.4E–03	3.89E–02	3.89E–02	3.89E–02
9	NMF	Natural microfracture	4.88E–02	4.76E–02	4.79E–01	4.79E–01	4.79E–01
10	LFPP	Low formation fracture pressure	4.88E–02	4.76E–02	8.41E–02	8.41E–02	8.41E–02

(Continued)

Table 11.7. (*Continued*)

No.	Basic event	Description	Prior probability $t = 0$	Prior probability $t = 1440$th	Posterior probability $t = 1440$th: Not circulating	Tripping in	Circulating
11	LFP	Large formation porosities	4.88E–02	4.76E–02	8.8E–02	8.8E–02	8.8E–02
12	PBS	Poor borehole stability	0	1.80E–03	1.80E–03	1.80E–03	1.80E–03
13	CF	Casing failure	0	1.56E–04	1.56E–04	1.56E–04	1.56E–04
14	CEF	Cement failure	0	1.15E–03	1.15E–03	1.15E–03	1.15E–03
15	RDCR	Inadequate depth of casing running	3.00E–03	3.00E–03	4.88E–01	4.88E–01	4.88E–01
16	IRDR	Increased running drillpipe rate	3.00E–03	3.00E–03	—	4.97E–02	—
17	IRCR	Increased running case rate	3.00E–03	3.00E–03	—	4.97E–02	—
18	EHT	Effect of high temperature	2.50E–03	2.50E–03	—	—	3.1E–02
19	HDFV	High drilling fluid viscosity	2.50E–03	2.50E–03	—	—	3.1E–02
20	LRPO	Large rig pump out	0	1.14E–03	—	—	1.4E–01
21	HPP	High pump pressure	0	1.14E–03	—	—	1.4E–01

Table 11.8. Parameters of Weibull distribution.

Basic event	Description	Shape parameter (α)	Scale parameter (λ)
IDM	Inaccurate density measurement	1.0	2.00E–06
RCDFS	RCD fail to seal	2.2	1.00E–04
DAPCCSF	DAPC control system failure	1	3.33E–06
BPPF	Back pressure pump failure	1.7	5.00E–05
DAPCCF	DAPC choke failure	1.6	2.50E–05
FMF	Flow meter failure	1.0	1.00E–06
PBS	Poor borehole stability	1.9	2.50E–05
CF	Casing failure	2.5	2.08E–05
CEF	Cement failure	2.1	2.78E–05
LRPO	Large rig pump out	1.7	5.00E–05
HPP	High pump pressure	1.7	5.00E–05

which is equivalent to the pressure coefficient 1.68, and the formation temperature is about 151.7°C, belonging to the high-temperature and high-pressure systems. Lost circulation or kick is more likely caused by these special geological conditions such as the very light gray and low-density limestone reservoir with narrow drilling fluid density window. Therefore, the MPD technology is adapted in this application.

11.5.1. *Risk identification for lost circulation*

A BT model is firstly developed for risk identification of lost circulation in the three drilling scenarios. Figure 11.6 shows the fault tree and event tree of lost circulation in the BT model. Considering well lost circulation as an undesired event among such drilling incidents, the potential causes and consequences have to be determined. As indicated in Figs. 11.6(a)–(c), three fault trees are established for modeling different drilling operations involving not circulating, tripping in, and circulating processes. The root causes of lost circulation are collected and investigated. According to Sec. 11.2, the overbalanced drilling condition is likely to result in the loss of mud. As drilling encounters limestone and fissure formation, the likelihood of lost circulation will be increased. So, two main reasons could be identified including the larger BHP than

FFP and the leakage path. The increasing BHP and the MPD system failing to maintain a constant BHP will make the larger BHP than FFP possible. The others leading to lost circulation may include excessive drilling fluid density in the not-circulating process, the surging effect caused by tripping activities, and high pump pressure in the circulating process. The formation condition and well structure design can also be taken into consideration in terms of the contribution to the lost circulation. Therefore, totally, 21 potential root causes in the fault tree were found based on the work of Skogdalen and Vinnem [33] and Abimbola *et al.* [2].

Safe operation, collapse stuck, kick, and blowout as potential consequences are emphasized for a weak formation as depicted in Fig. 11.6(d). To forestall the occurrence of these consequences, three safety barriers are installed: plugging barrier, kick detection system, and BOP system. Plugging barriers should be used when massive volume of drilling mud into the formation is losing. The successful plugging plays a critical role in reducing the downtime loss and preventing the wellbore collapse and pipe sticking by utilizing plugging materials, tools, and a series shut or kill operations. Kick detection system has the function to detect the occurrence of kick if the plugging fails to control the loss of mud. The BOP system can prevent the formation fluid getting into the external environment and it will be highlighted when the kick cannot be detected and controlled.

11.5.2. *DBN modeling for the case*

In this study, the DBNs for drilling lost circulation are established using Netica [23] software, which is regarded as a general platform to realize risk assessment in a rapid manner. According to the mapping algorithm described in Sec. 4.2.1, BTs of "lost circulation" combining the root causes and consequences for three drilling operations are translated into corresponding DBNs with three time slices as presented in Fig. 11.7, which is extended from time at t = 0th, t = 720th to t = 1440th hour for modeling.

It is noted that model uncertainty issues can be handled by adding the MU node with two states "OR" and "AND" in the proposed DBN-based model. The states of root/IE nodes, TE nodes,

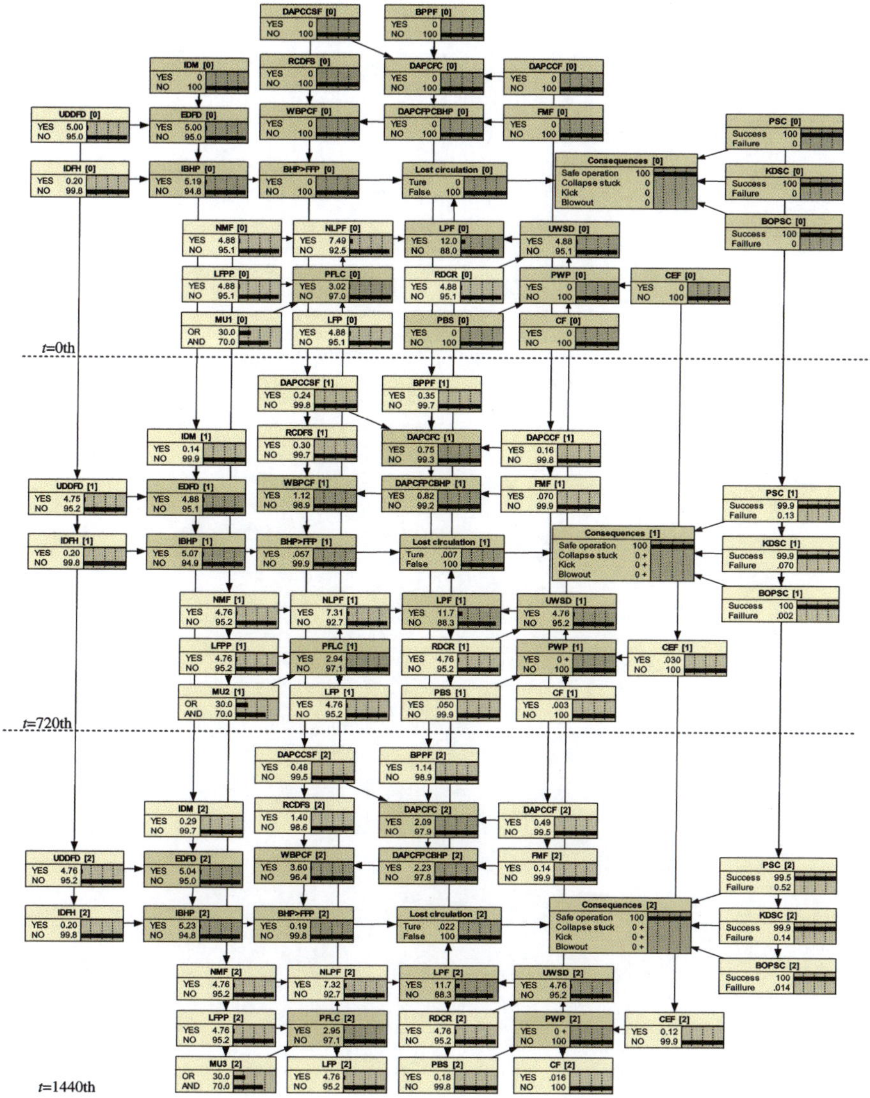

(a)

Fig. 11.7. DBN modeling with three time slices for different drilling scenarios (a) not circulating, (b) tripping in, and (c) circulating.

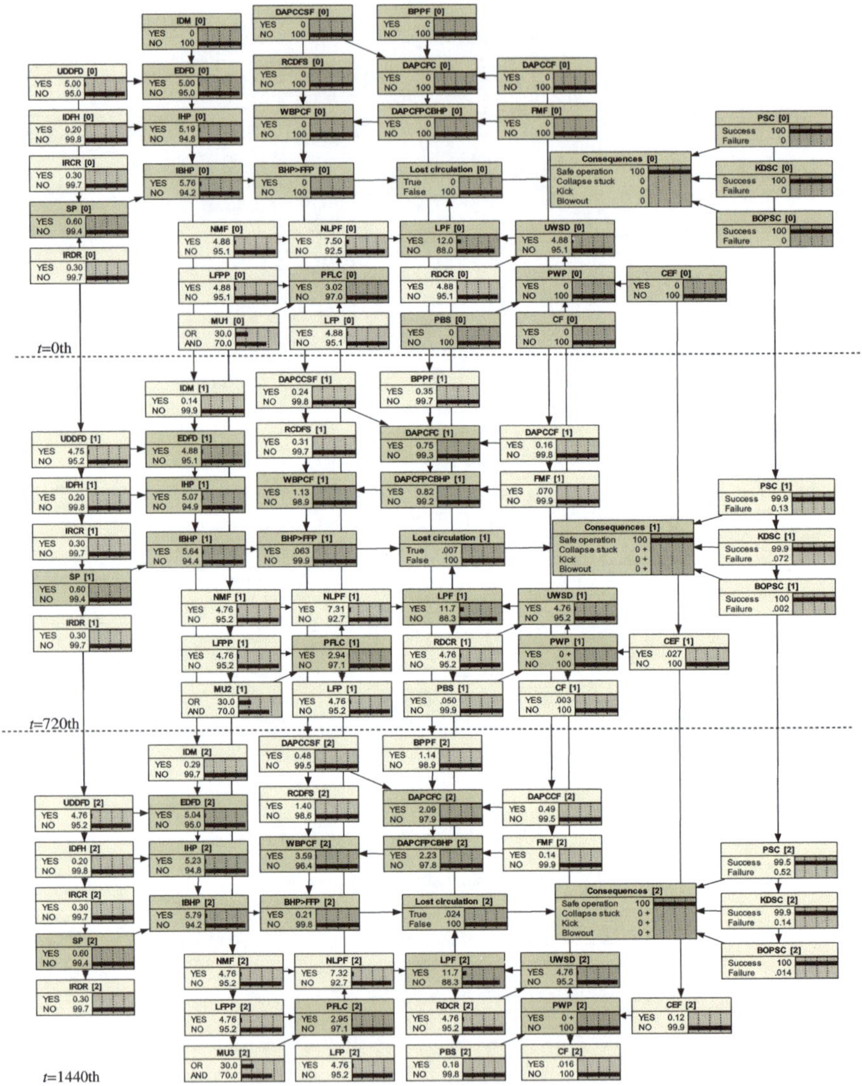

(b)

Fig. 11.7. (*Continued*)

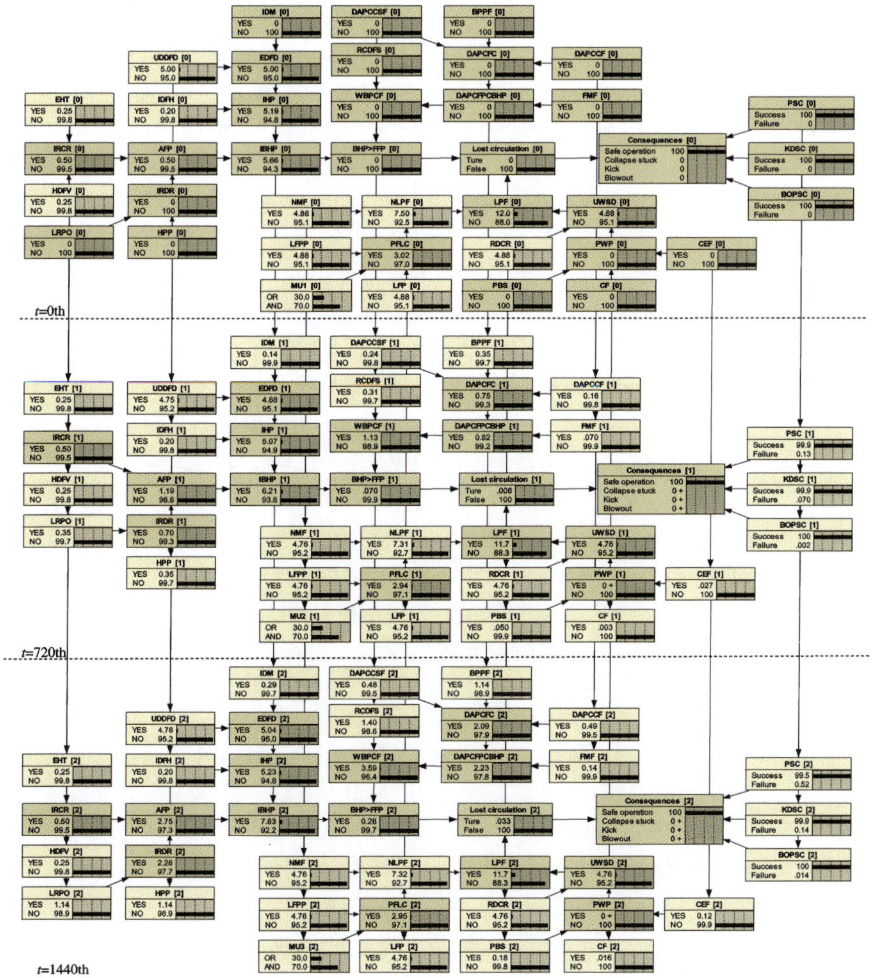

(c)

Fig. 11.7. (*Continued*)

and barriers nodes are assigned “YES/NO”, “True/False”, and “Success/Failure”, respectively, as indicated in Fig. 11.7. Similarly, consequence states are achieved from “safe operation to blowout” according to the availability and reliability performance of safety barriers. The CPTs of nodes considering parameter uncertainty should be assigned to model the DBNs. In the initial time at $t = 0$, the value of the prior probability needs to be assigned to each state of root nodes. If the prior knowledge of root causes such as UDDFD and IDFH is obtained by taking advantage of the available literature [1, 13, 25] and also the expert inputs, if necessary, the prior probabilities of these root causes are assigned as listed in column 4 of Table 11.7. If the probabilities of failure on demand for the equipment and safety barriers such as RCDFS and BOPS are assumed to follow the Weibull distribution, the initial states of these root causes are considered in their perfect functioning state, and the value of failure probability is assigned to 0. The values of scale parameter λ and shape parameter α in Weibull distribution are provided in Table 11.8, and such values are determined by expert's inputs.

The parameters of CPTs should also be assigned to model DBNs. The approach of CPT calculation considering the parameter uncertainty is followed by the discussion in Sec. 4.2.2. There are two examples to illustrate the space-based parameters of CTPs and two examples to explain the time-based parameters of CPTs, respectively. Taking the "DAPC fail to control" as an example, the occurrence of this event is caused by the DAPC system failure, back pressure pump failure, and DAPC choke failure. With the use of Boolean logic relationships, the CPTs can be calculated as listed in Table 11.9. Taking the "NLPF" as an example, the CPT is calculated from the nodes "NMF" and "PFLC" to the node "NLPF" based on experts knowledge and noisy-OR filling-up algorithm in this study,

Table 11.9. CPT for DAPCFC.

DAPCCF		BPPF		DAPCCSF		P_{DAPCFC}	
YES	NO	YES	NO	YES	NO	YES	NO
1	0	1	0	1	0	1	0
1	0	1	0	1	0	1	0
1	0	0	1	1	0	1	0
0	1	0	1	1	0	1	0
1	0	1	0	0	1	1	0
0	1	1	0	0	1	1	0
1	0	0	1	0	1	1	0
0	1	0	1	0	1	0	1

Table 11.10. CPT for NLPF.

NMF		PFLC		NLPF	
YES	NO	YES	NO	YES	NO
1	0	1	0	0.999	0.001
0	1	1	0	0.98	0.02
1	0	0	1	0.95	0.05
0	1	0	1	0	1

Table 11.11. CPT of UDDFD for two time slices.

	t_{720}	
t_0	YES	NO
YES	0	1
NO	0.05	0.95

Table 11.12. CPT of BPPF for two time slices.

	t_{1440}	
t_{720}	YES	NO
YES	0.007	0.993
NO	0.014	0.986

as presented in Table 11.10. The presence of NLPF is caused by NMF and PFLC in the YES state at respective probabilities of 0.02 and 0.05, but not 1 due to the effect of uncertainty. The time-based CPTs, namely the CPTs for two time slices of root causes, follow the rules as depicted in Tables 11.3 and 11.4. Taking the "UDDFD" as an example, the prior probability of UDDFD is 0.05, and the CPT is assigned as listed in Table 11.11. Taking the "BPPF" as an example, the failure probability of BPPF is 0.07 and 0.014 at t = 720th and t = 1440th hour, respectively, and the CPT is assigned as listed in Table 11.12.

11.5.3. *Results and discussion*

11.5.3.1. *Risk evolution prediction*

Figure 11.7 shows DBN modeling results for the three drilling scenarios contributing to lost circulation within three time slices. The predictive results indicate that the occurrence probability of the lost circulation at time t = 720th hour and at time t = 1440th hour for not-circulating, tripping in, and circulating scenarios is 7.0E–05, 7E–05, and 8E–05, and 2.2E–04, 2.4E–04, and 3.3E–04, respectively. Figure 11.8(a) shows the risk comparison for three drilling scenarios and the tendency of the risk evolution within nine time slices. We assume that the time slice interval is the same as 360 hours. It is clear that the occurrence probability of the lost circulation is highest and is growing fastest in the scenario of the circulating process, which means that the lost circulation is much more likely to occur in the circulating process when the rig pump is on. Compared to the static operation, dynamic operations are more vulnerable due to the effect of the surging pressure and the annual friction pressure. In addition, the reliability of the wellhead back pressure control is decreasing over time and it has a great effect on the occurrence probability of the lost circulation. Dynamic operation and the reliability of wellhead back pressure control therefore need to be paid more attention when drilling.

When drilling encounters the formation with the narrow mud density window, the small change of mud density will have a great impact on the occurrence of the lost circulation. Figure 11.8(b) shows the occurrence probability of the lost circulation at different times (at third time slice, fourth time slice, and seventh time slice) given the mud density in the abnormal state in the circulating scenarios. It is clear that the occurrence probability of lost circulation increases fast given the unreasonable change of mud density at third time slice, fourth time slice, and seventh time slice and will decrease at their next time slice. The occurrence probability of the lost circulation increases from 1.8×10^{-4} to 2.6×10^{-3} when the mud density changes at third time slice. According to the assumption in Sec. 4.2.2, when $P(UDDFD_t = \text{YES} \mid UDDFD_{t-1} = \text{YES}) = 0$, the occurrence

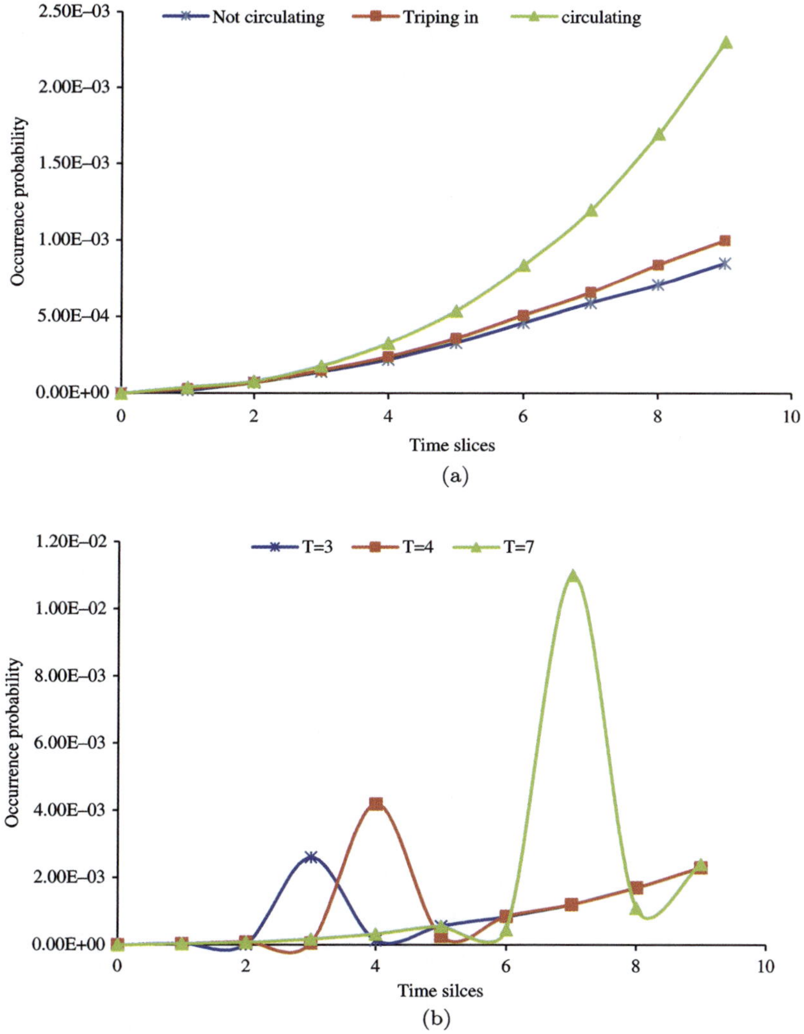

Fig. 11.8. Risk comparison for (a) three drilling scenarios and (b) different time slices given mud density change.

probability of the lost circulation decreases from 3.3×10^{-4} to 1.4×10^{-4} at fourth time slice, given the mud density changed at third time slice. The ratio is largest at seventh time slice compared with that of third time slice and fourth time slice. As a matter of fact, the drilling well goes exactly through more narrow density window as the depth increases over time, and the seventh time slice is mostly considered as a dangerous period during the drilling progress. The likelihood of lost circulation can be estimated with the unreasonable change of mud density.

The blowout may happen at the same time when drilling is encountering the weak formation with gas layers. The occurrence probability is mostly close to 0 in Fig. 11.7 due to the lower probability of the lost circulation and failure of safety barriers. Based on the failure of safety barriers following the Weibull rules, the the reliability of barriers is decreasing along with time. Taking the circulating scenario as an example, their consequence probabilities are calculated when lost circulation occurs as shown in Fig. 11.9. Hence, the consequence "pipe stuck" has the higher likelihood than other consequences. Plugging barrier should be therefore given more attention to meet high-level reliability. It is also worth noting that there is a small change for kick and blowout in occurrence probability because of the higher reliability of kick detection barrier and BOP barrier in the whole drilling process.

11.5.3.2. *Root cause reasoning*

A diagnostic analysis is conducted by assuming the occurrence of lost circulation for three scenarios by setting the state of lost

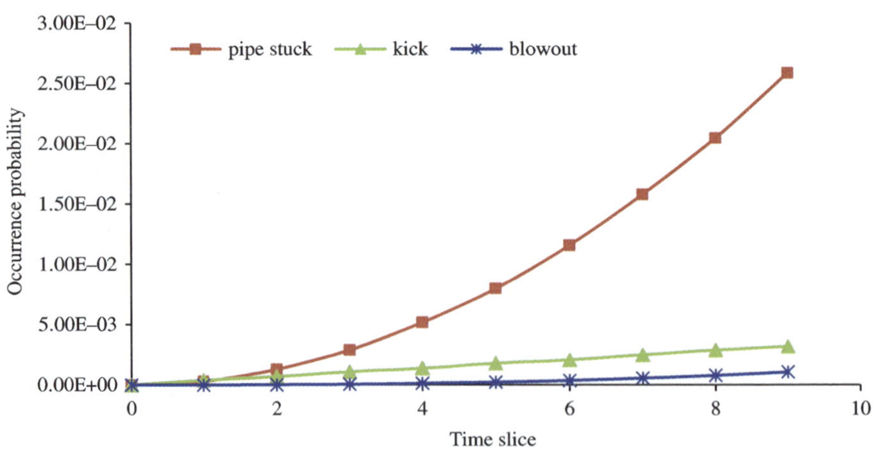

Fig. 11.9. Risk comparison of its consequences for different time slices given lost circulation occurrence.

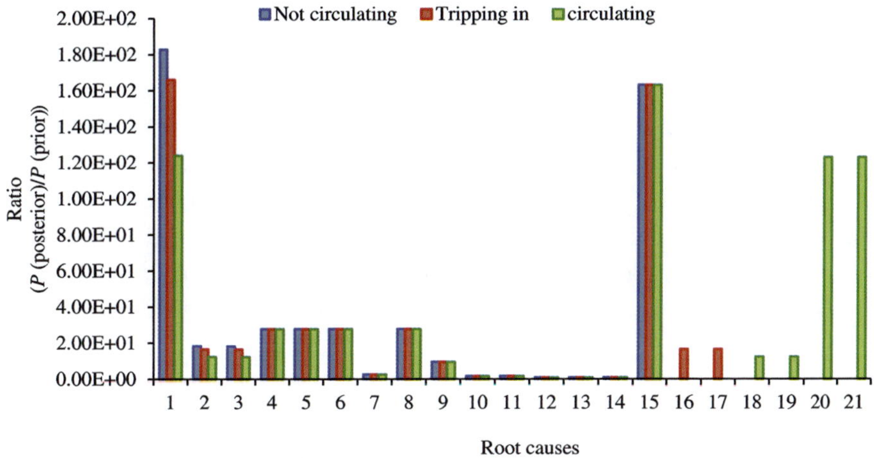

Fig. 11.10. Ratio of posterior probability (P(posterior)) and prior probability (P(prior)) in (a) not circulating (b) tripping in, and (c) circulating.

circulation node to True. The prior and posterior probabilities of these root causes for not circulating, tripping in and circulating scenarios are listed in columns 4 and 5 of Table 11.7, which indicate the available updated failure probability given the occurrence of lost circulation from backward propagation. The comparison of prior probabilities and posterior probabilities by using the ratio as presented in Fig. 11.10 shows that the posterior probabilities are more than 10 times as much as their prior probabilities. In the above diagnostic analysis using DBN probability inference algorithm, the critical roles of drilling fluid density should be highlighted because the ratio of UDDFD (1) is the largest. It is worth noting that the root causes such as UDDFD (1) and RDCD (15) would have been totally dominating as other factors in causing lost circulation in the three scenarios. The other main contributing factors identified are LRPO (20) and HPP (22) in the circulating process. Therefore, the practical diagnosis and checking should then focus on the availability of these root causes until the high risk is controlled in real time.

The occurrence probability of lost circulation at current time can be calculated by the proposed model when the loss of circulation occurred at an earlier time. Taking the circulating scenario as an example, the occurrence probability of lost circulation (LC) at time

t = 1440th given the LC occurred at time t = 720th is calculated, namely $P(\mathrm{LC}_{t=1440\mathrm{th}} = \mathrm{True} \mid \mathrm{LC}_{t=720\mathrm{th}} = \mathrm{True}) = 0.00009$, which become lower than that (0.00033) of LC at time t = 720th. There is a change for root cause (RC) state between time t = 720th and t = 1440th, the $P(\mathrm{RC}_{t=720\mathrm{th}} = \mathrm{YES} \mid \mathrm{LC}_{t=720\mathrm{th}} = \mathrm{True})$ becomes larger and $P(\mathrm{RC}_{t=1440\mathrm{th}} = \mathrm{YES} \mid \mathrm{RC}_{t=720\mathrm{th}} = \mathrm{YES})$ become smaller based on Eqs. (11.10) and (11.6), as shown in Figs. 11.11(a) and 11.11(b). As a result, the posterior probabilities can provide new evidential information for diagnosis analysis, and the values of root causes can be updated in a dynamic manner.

11.5.3.3. *Sensitivity analysis*

Important factors measured by the importance degree sequence of root causes for lost circulation are also calculated by using mutual information, which can measure the information that two variables share and how much uncertainty about one variable is reduced by knowing the other. The individual contribution of each root cause towards lost circulation at time slice $T = 4$ is calculated by comparing three drilling scenarios as shown in Fig. 11.12(a). It is seen that, for three types of operations, UDDFD (1) contributes much to the lost circulation, which is regarded as the most fatal weakness. In Fig. 11.12(a), UDDFD (1) in the not-circulating scenario has a higher contribution for lost circulation compared with other scenarios, whereas RCDFS (4), BPPF (6), NMF (9), and RDCR (15) in the circulating scenario also have higher contributions than those of other scenarios.

The individual contribution of each root cause to lost circulation in circulating scenarios is calculated under different time slices T = 1, T = 4 and T = 9, as shown in Fig. 11.12(b). It is found that UDDFD (1) and RCDFS (4) at time slice T = 1, UDDFD (1) at time slice T = 4, and RCDFS (4) at time slice T = 9, make the highest contribution to the lost circulation, which indicates that these root causes are sensitive to the lost circulation and should be given more attention. In Fig. 11.12(b), the value of mutual information at time slice $t = 9$ has a higher contribution for lost circulation compared with other time slices, such as UDDFD (1), BPPF (6), NMF (9),

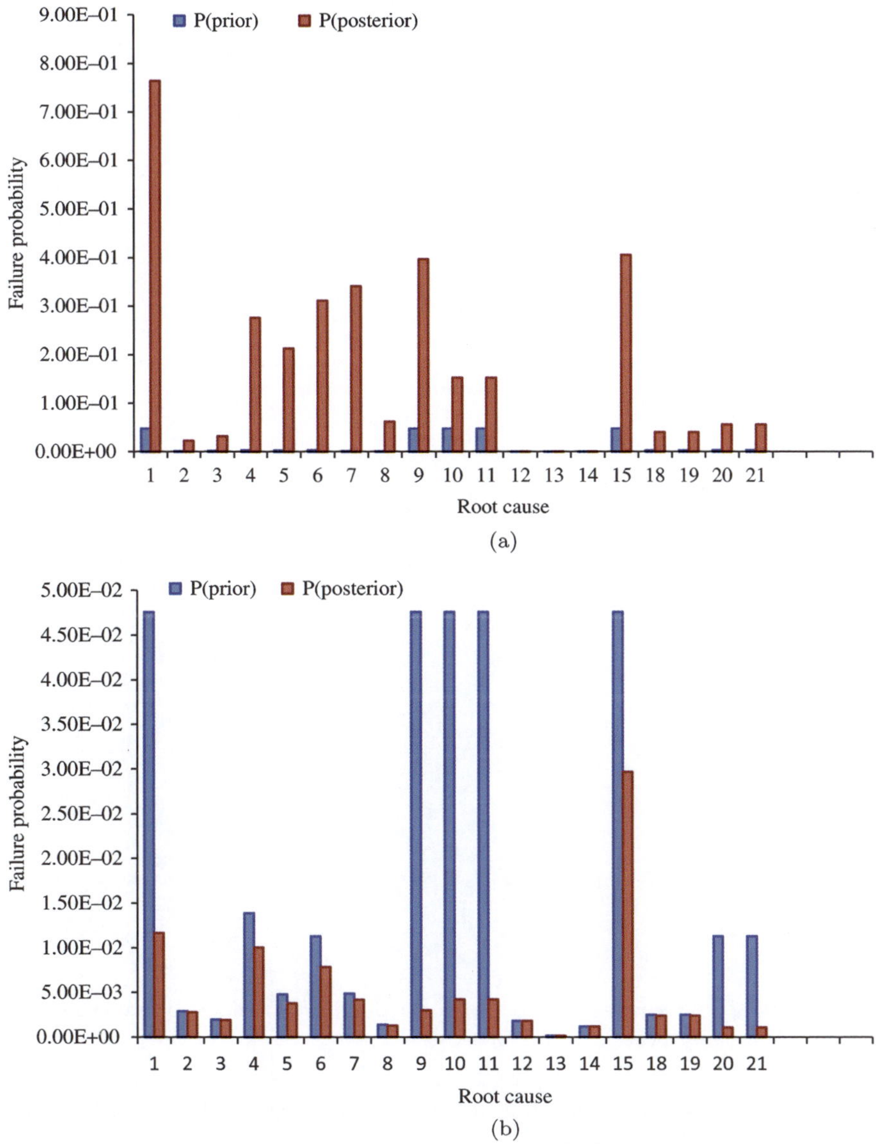

Fig. 11.11. Prior probability (P (prior)) and posterior probability (P (posterior)) of root causes at time: (a) $t = 720$th and (b) $t = 1440$th.

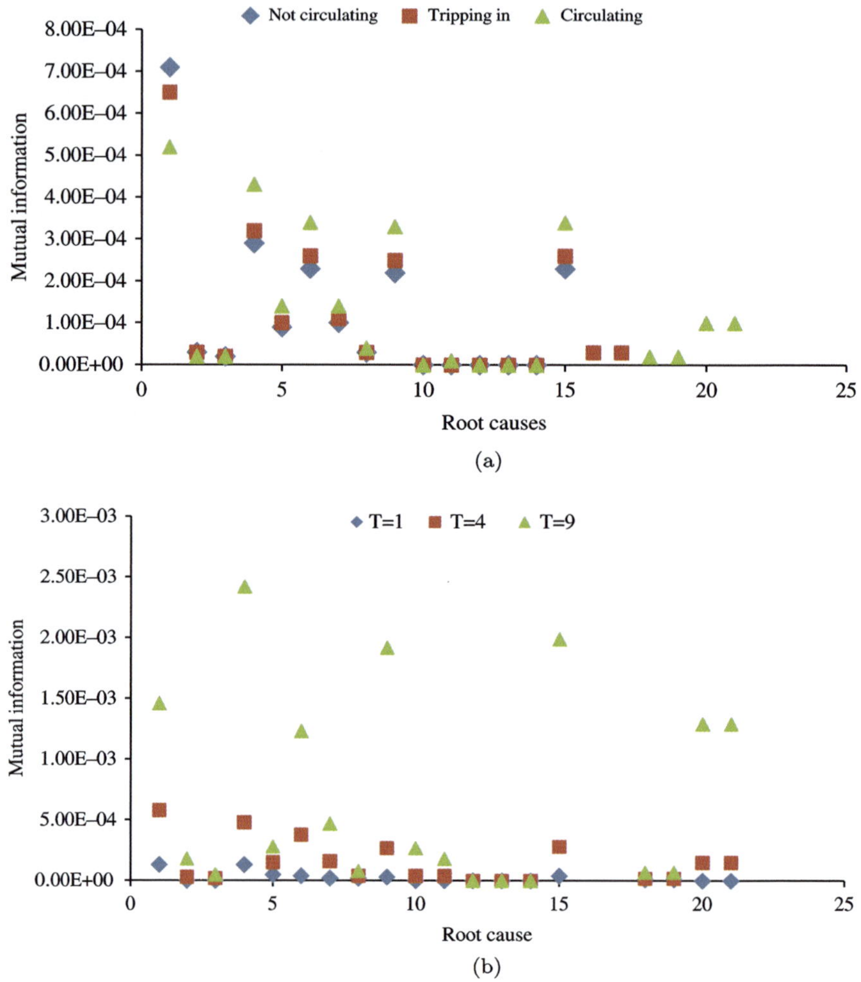

Fig. 11.12. Sensitivity analysis of root causes for (a) different scenarios and (b) different time slices.

and RDCR (15). Therefore, the different root causes should be highlighted at different times of drilling.

11.6. Conclusions and Research Perspectives

This chapter focuses on the safety of drilling operations given the special geological conditions, where the MPD technology is adopted to avoid the drilling incidents. According to the close relationship between hazard factors and the dynamic variance of BHP during drilling, a risk assessment model based on DBN for predict analysis, diagnostic analysis, and sensitivity analysis is proposed.

The application of the proposed model has been presented with a case study on the offshore lost circulation during drilling. In order to provide graphical symbols for the logical causal relationship between factors and the effects of lost circulation, a BT model is established to map different drilling operation scenarios. All potential root causes contributing to the lost circulation and the corresponding possible outcomes identified given the occurrence of this incident are analyzed carefully. Then DBNs are established from BTs. Finally, by the inference mechanism of DBN, the risk evolution tendency of drilling operations are predicted comparing the not circulating, tripping in, and circulating scenarios over time and given the current state of root causes. The root cause reasoning is performed given the occurrence of lost circulation in diagnostic analysis. The most important root causes have been identified with sensitivity analysis based on mutual information for different drilling scenarios and different times. The occurrence probability is highest in the scenario of circulating, which indicates that lost circulation is much more likely to occur in this process. Reasonable drilling fluid density and availability of rotating control device have made the highest contribution to the lost circulation for this scenario, and they may for this reason be regarded as the most important weaknesses that require attention.

The direction of our subsequent work is to extend our model to improve the robustness of probability distribution of root causes from prior knowledge by logging data and apply the method to other oil and gas operations such as production and overwork.

References

[1] M. Abimbola, F. Khan, N. Khakzad, "Dynamic safety risk analysis of offshore drilling," *Journal of Loss Prevention in the Process Industries*, vol. 30, pp. 74–85, 2014.

[2] M. Abimbola, F. Khan, N. Khakzad, S. Butt, "Safety and risk analysis of managed pressure drilling operation using Bayesian network," *Safety Science*, vol. 76, pp. 133–144, 2015.

[3] E. Ataallahi, S. R. Shadizadeh, "Fuzzy consequence modeling of blowouts in Iranian drilling operations: HSE consideration," *Safety Science*, vol. 77, pp. 152–159, 2015.

[4] G. Bearfield, W. Marsh, "Generalising event trees using Bayesian networks with a case study of train derailment," In: *Computer Safety, Reliability, and Security*, Springer, Berlin, pp. 52–66, 2005.

[5] J. Bhandari, R. Abbassi, V. Garaniya, F. Khan, "Risk analysis of deepwater drilling operations using Bayesian network," *Journal of Loss Prevention in the Process Industries*, vol. 38, pp. 11–23, 2015.

[6] A. Bobbio, L. Portinale, M. Minichino, E. Ciancamerla, "Improving the analysis of dependable systems by mapping fault trees into Bayesian networks," *Reliability Engineering & System Safety*, vol. 71, pp. 249–260, 2001.

[7] B. Cai, Y. Liu, Z. Liu, X. Tian, X. Dong, S. Yu, "Using Bayesian networks in reliability evaluation for subsea blowout preventer control system," *Reliability Engineering & System Safety*, vol. 108, pp. 32–41, 2012.

[8] B. Cai, Y. Liu, Y. Ma, Z. Liu, Y. Zhou, J. Sun, "Real-time reliability evaluation methodology based on dynamic Bayesian networks: A case study of a subsea pipe ram BOP system," *ISA Transactions*, 2015.

[9] B. Cai, Y. Liu, Y. Zhang, Q. Fan, S. Yu, "Dynamic Bayesian networks based performance evaluation of subsea blowout preventers in presence of imperfect repair," *Expert Systems with Applications*, vol. 40, pp. 7544–7554, 2013.

[10] M. T. Crichton, K. Lauche, R. Flin, "Incident command skills in the management of an oil industry drilling incident: A case study," *Journal of Contingencies and Crisis Management*, vol. 13, pp. 116–128, 2005.

[11] V. De Dianous, C. Fiévez, "ARAMIS project: A more explicit demonstration of risk control through the use of bow–tie diagrams and the evaluation of safety barrier performance," *Journal of Hazardous Materials*, vol. 130, pp. 220–233, 2006.

[12] D. Elliott, J. Montilva, P. Francis, D. Reitsma, J. Shelton, V. Roes, "Managed pressure drilling erases the lines," *Oilfield Review*, vol. 23, pp. 14–23, 2011.

[13] P. Holland, *Offshore Blowouts: Causes and Control: Causes and Control*, Gulf Professional Publishing, 1997.

[14] J. Hu, L. Zhang, Z. Cai, Y. Wang, A. Wang, "Fault propagation behavior study and root cause reasoning with dynamic Bayesian network based framework," *Process Safety and Environmental Protection*, vol. 97, pp. 25–36, 2015.

[15] J. Hu, L. Zhang, L. Ma, W. Liang, "An integrated safety prognosis model for complex system based on dynamic Bayesian network and ant colony algorithm," *Expert Systems with Applications*, vol. 38, pp. 1431–1446, 2011.

[16] B. Jones, I. Jenkinson, Z. Yang, J. Wang, "The use of Bayesian network modelling for maintenance planning in a manufacturing industry," *Reliability Engineering & System Safety*, vol. 95, pp. 267–277, 2010.
[17] N. Khakzad, F. Khan, P. Amyotte, "Safety analysis in process facilities: Comparison of fault tree and Bayesian network approaches," *Reliability Engineering & System Safety*, vol. 96, pp. 925–932, 2011.
[18] N. Khakzad, F. Khan, P. Amyotte, "Dynamic safety analysis of process systems by mapping bow-tie into Bayesian network," *Process Safety and Environmental Protection*, vol. 91, pp. 46–53, 2013.
[19] F. I. Khan, S. Abbasi, "Major accidents in process industries and an analysis of causes and consequences," *Journal of Loss Prevention in the Process Industries*, vol. 12, pp. 361–378, 1999.
[20] U. B. Kjærulff, A. L. Madsen, *Probabilistic Networks for Practitioners — A Guide to Construction and Analysis of Bayesian Networks and Influence Diagrams*, Department of Computer Science, Aalborg University, HUGIN Expert A/S, 2006.
[21] K. P. Murphy, *Dynamic Bayesian Networks: Representation, Inference and Learning*, University of California, Berkeley, CA, 2002.
[22] R. E. Neapolitan, *Learning Bayesian Networks*, Pearson, 2004.
[23] Netica, Norsys Software Corp, 2015. https://www.norsys.com/legal.html.
[24] T. D. Nielsen, F. V. Jensen, *Bayesian Networks and Decision Graphs*, Springer Science & Business Media, 2009.
[25] OREDA, *OREDA Reliability Data*, 5th edn., OREDA Participants, Norway, 2009.
[26] B. Patel, B. Grayson, H. Gans, "Optimized unconventional shale development with MPD techniques," In: *IADC/SPE Managed Pressure Drilling and Underbalanced Operations Conference and Exhibition*, Society of Petroleum Engineers, 2013.
[27] P. A. P. Ramírez, I. B. Utne, "Use of dynamic Bayesian networks for life extension assessment of ageing systems," *Reliability Engineering & System Safety*, vol. 133, pp. 119–136, 2015.
[28] M. Rausand, *Reliability of Safety-Critical Systems: Theory and Applications*, John Wiley & Sons, 2014.
[29] B. Rehm, J. Schubert, A. Haghshenas, A. S. Paknejad, J. Hughes, *Managed Pressure Drilling*, Elsevier, 2013.
[30] C. Shen, "Transient dynamics study on casing deformation resulted from lost circulation in low-pressure formation in the Yuanba Gasfield, Sichuan Basin," *Natural Gas Industry B*, vol. 2, pp. 347–353, 2015.
[31] L. Sheremetov, I. Batyrshin, D. Filatov, J. Martinez, H. Rodriguez, "Fuzzy expert system for solving lost circulation problem," *Applied Soft Computing*, vol. 8, pp. 14–29, 2008.
[32] J. E. Skogdalen, I. B. Utne, J. E. Vinnem, "Developing safety indicators for preventing offshore oil and gas deepwater drilling blowouts," *Safety Science*, vol. 49, pp. 1187–1199, 2011.

[33] J. E. Skogdalen, J. E. Vinnem, "Quantitative risk analysis of oil and gas drilling, using Deepwater Horizon as case study," *Reliability Engineering & System Safety*, vol. 100, pp. 58–66, 2012.

[34] Ø. N. Stamnes, L. Zhou, G.-O.Kaasa, O. M. Aamo, "Adaptive observer design for the bottomhole pressure of a managed pressure drilling system," In: *47th IEEE Conference on Decision and Control, 2008 (CDC 2008)*, IEEE, pp. 2961–2966, 2008.

[35] A. K. Vajargah, E. van Oort, "Early kick detection and well control decision-making for managed pressure drilling automation," *Journal of Natural Gas Science and Engineering*, vol. 27, pp. 354–366, 2015.

[36] X. Wu, H. Liu, L. Zhang, M. J. Skibniewski, Q. Deng, J. Teng, "A dynamic Bayesian network based approach to safety decision support in tunnel construction," *Reliability Engineering & System Safety*, vol. 134, pp. 157–168, 2015.

[37] L. Xue, J. Fan, M. Rausand, L. Zhang, "A safety barrier-based accident model for offshore drilling blowouts," *Journal of Loss Prevention in the Process Industries*, vol. 26, pp. 164–171, 2013.

[38] L. Yan, H. Wu, Y. Yan, "Application of fine managed pressure drilling technique in complex wells with both blowout and lost circulation risks," *Natural Gas Industry B*, vol. 2, pp. 192–197, 2015.

Chapter 12

A Fault Diagnosis Methodology for Gear Pump Based on EEMD and Bayesian Network

This chapter proposes a fault diagnosis methodology for a gear pump based on the ensemble empirical mode decomposition (EEMD) method and the Bayesian network. Essentially, the presented scheme is a multisource information fusion-based methodology. Compared with the conventional fault diagnosis with only EEMD, the proposed method is able to take advantage of all useful information besides sensor signals. The presented diagnostic Bayesian network consists of a fault layer, a fault feature layer, and a multisource information layer. Vibration signals from sensor measurement are decomposed by the EEMD method and the energy of intrinsic mode functions (IMFs) are calculated as fault features. These features are added into the fault feature layer in the Bayesian network. The other sources of useful information are added to the information layer. The generalized three-layer Bayesian network can be developed by fully incorporating faults and fault symptoms as well as other useful information such as naked eyes' inspection and maintenance records. Therefore, the diagnostic accuracy and capacity can be improved. The proposed methodology is applied to the fault diagnosis of a gear pump and the structure and parameters of the Bayesian network are established. Compared with artificial neural network and support vector machine classification algorithms, the proposed model has the best diagnostic performance when only sensor data is used. A case study has demonstrated that some information from human observation or system repair records is very helpful for fault diagnosis. It is effective and efficient in diagnosing faults based on uncertain, incomplete information.

12.1. Introduction

As the key element of the hydraulic system, a pump is responsible for the mechanical to hydraulic energy conversion process. Gear pumps are widely applied in the modern industry owing to their advantages such as small and compact design, high efficiency, and low manufacturing cost. The working status of a gear pump greatly affects the performance of the whole hydraulic system, thus it is necessary to develop its fault diagnosis technique. The condition of the mechanical equipment is closely associated with vibration signals, which come about during the rotating process for rotating machinery [1]. Therefore, fault diagnosis of gear pumps can be performed with the characteristic information extracted by signal processing techniques such as short-time Fourier transform [2], wavelet transform [3], blind source separation [4], sparse decomposition method [5], and empirical mode decomposition (EMD) [6].

EMD is a time-frequency signal processing technique. Compared with other signal processing methods, EMD is self-adaptive and especially suits the non-stationary and nonlinear signals. For the sparse decomposition method, the algorithm is much more demanding and complex compared to EMD [7]. Based on the local characteristic time scales of a signal, the original vibration signal can be decomposed into several intrinsic mode functions (IMFs). Due to the adaptive analysis and high robustness nature, EMD has been widely applied in the fault diagnosis of rotating machinery [8, 9]. However, EMD suffers from the mode mixing problem, which means a single IMF either consisting of widely disparate scales, or a signal of a similar scale residing in different IMF components [10]. In order to alleviate the problem of mode mixing in EMD, Wu and Huang [11] proposed the ensemble empirical mode decomposition (EEMD) method. Essentially, white noise of finite amplitude is added to the original signal during the EEMD decomposition process. The ensemble means of the corresponding IMFs generated from each trial are defined as the true IMFs of the EEMD [12]. Lei *et al.* [13] employed EEMD in diagnosing rub-impact faults in a power generator and a heavy oil catalytic cracking machine set. Compared with the EMD method, it is demonstrated that EEMD has superiority in fault diagnosis of rotating machinery. Caesarendra *et al.* [14] applied the EEMD method in two real cases of slow speed slewing bearing with natural bearing fault damage and the results show that EEMD is better than

FFT in identifying fault frequencies. Mahgoun *et al.* [15] presented the application of EEMD in purpose to detect localized faults of damage at an early stage.

Although previous research jobs on rotating machinery have produced significant outcomes, only information from sensor measurement is used for fault diagnosis. However, other sources of information besides sensor measurement could be useful for fault diagnosis, which are obviously not fully utilized. For example, the observation information from humans or the maintenance records would make the diagnosis results more reliable. Obviously, the more evidences are used, the more accurate the diagnostic results will be. Recently, multisource information fusion fault diagnosis systems based on Bayesian networks have been developed in some fields. To take advantage of all useful information, a three-layered Bayesian network has been presented for fault diagnosis of a chiller [16] and variable air volume terminals [17]. Xu [18] developed an intelligent expert system of rotating flexible rotors based on the Bayesian network by fully incorporating human experts' knowledge, machine faults, and fault symptoms as well as machine running conditions. Cai *et al.* [19] proposed a multisource information fusion-based fault diagnosis methodology for a ground-source heat pump by making use of sensor information and observation information. Oukhellou *et al.* [20] presented a hybrid diagnosis system based on the combination of local sensor data information and global structural knowledge information for the detection of a broken rail.

Bayesian network is an acyclic directed graph consisting of a set of variables with directed edges between the variables. It is a powerful tool for modeling complex problems in probabilistic knowledge representation and reasoning [21]. It has been widely used for fault diagnosis in a variety of fields. Sun *et al.* [22] developed a mild cognitive impairment (MCI) expert system based on the Bayesian network to address MCI's prediction and inference question and the experimental results indicate that the proposed model achieved better results than some existing methods in most instances. Barco *et al.* [23] proposed a discrete Bayesian network for diagnosis of radio access networks of cellular systems and the research shows that the developed model outperforms the traditional Bayesian network when there is an inaccuracy in the model parameters. Sahin *et al.* [24] developed a Bayesian network for fault diagnosis of airplane engines and the results show that the proposed model can detect the anomalies or faults in the sensor readings. Kariv *et al.* [25] developed a computerized decision support system for the diagnosis of infections among solid organ transplant recipients based on the Bayesian network.

In this chapter, a fault diagnosis methodology for gear pumps based on the EEMD method and Bayesian network is proposed to make full use of multisource information. One of the weak points of a Bayesian network is that there is no specific semantic to guide the model development [26]. This chapter presents a three-layered diagnostic Bayesian network for model development, which is composed of a fault layer, a fault feature layer, and a multisource information layer. With the proposed framework, a Bayesian network for fault diagnosis can be easily developed. Vibration signals from sensor measurement are decomposed by the EEMD method and the features of IMFs are extracted. The obtained fault features and other multisource information are entered into the fault feature layer and multisource information layer, respectively. The remainder of this chapter is organized as follows. In Sec. 12.2, the EEMD method and Bayesian network are introduced. Section 12.3 proposes the fault diagnosis methodology and applies it to a gear pump. In Sec. 12.4, fault diagnosis based on the developed model is performed. Section 12.5 summarizes the chapter.

12.2. EEMD and Bayesian Network

12.2.1. *EEMD algorithm and feature extraction method*

EEMD is proposed to overcome the mode mixing problem, which is defined as a single IMF including oscillations of dramatically disparate scales or a component of a similar scale residing in different IMFs. Essentially, white noise of finite amplitude is added to the original signal during the EEMD decomposition process. In fact, to make the EEMD effective, the amplitude of the added noise should not be too small. In most cases, white noise of an amplitude that is about 0.2 standard deviation of that of the data is suggested. However, when the data is dominated by high-frequency signals, the noise amplitude may be smaller, and when the data is dominated by low-frequency signals, the noise amplitude may be increased. Generally speaking, the range of standard deviation is 0.1–0.4 [11]. The ensemble means of the corresponding IMFs generated from each trial are defined as the true IMFs of the EEMD. The EEMD performance of overcoming the mode mixing problem has been demonstrated [13, 27].

The EEMD algorithm can be given as follows [15].

Determine the number of ensemble M and initialize the amplitude of the added white noise, and $m = 1$.

Perform the mth trial on the investigated signal added white noise. Add a white noise series with the given amplitude to the original signal:

$$x_m(t) = x(t) + n_m(t), \tag{12.1}$$

where $n_m(t)$ represents the mth added white noise series and $x_m(t)$ denotes the noise-added signal of the mth trial.

With the EMD method [28], the noise-added signal $x_m(t)$ is decomposed into N IMFs $c_{n,m}(t)(i = 1, 2, \ldots, I)$, where $c_{n,m}(t)$ represents the nth IMF of the mth trial, and N is the number of IMFs.

If $n < M$ then let $m = m + 1$. Repeat steps (2) and (3) again and again with different white noise series each time until $n = M$.

Calculate the ensemble mean $c_n(t)$ of the M trials for each IMF

$$a_i(t) = \frac{1}{M} \sum_{m=1}^{M} c_{n,m}, \quad n = 1, 2, \ldots, N. \tag{12.2}$$

Report the mean $a_i(t)(i = 1, 2, \ldots, N)$ of each of the N IMFs as the final IMFs.

According to the steps above, a vibration signal measured from a gear pump is decomposed and the decomposition result is given

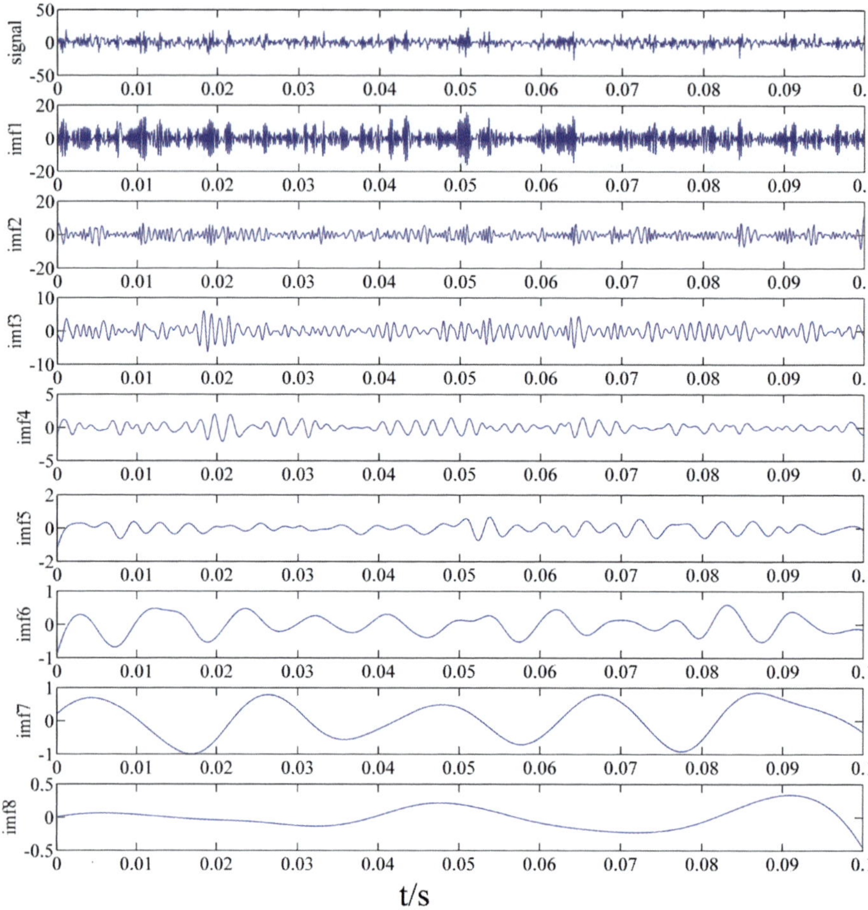

Fig. 12.1. The decomposed components with EEMD and original signal.

in Fig. 12.1. It shows eight IMFs in different frequency bands decomposed by the EEMD algorithm. It can be seen from the figure that the original signal is very complicated and the decomposed IMFs are hard to use for fault diagnosis. Hence, features of the signals need to be extracted.

Feature extraction is an important step for fault diagnosis. IMFs decomposed by EEMD contain valid information for fault diagnosis. The analysis results from EEMD energy of different vibration signals indicate that the energy of a vibration signal will change in different frequency bands when a fault occurs. It means that for the same faults, the decomposed IMFs are similar in the corresponding frequency band. Therefore, the energy of the decomposed IMFs could be used as features for fault diagnosis. E_i is the energy of the ith IMF.

$$E_i = \int_{-\infty}^{+\infty} |a_i(t)|^2 dt, \quad i = 1, 2, 3, \ldots, n. \tag{12.3}$$

Then the feature vector of the investigated signal $T = [E_1, E_2, E_3, \ldots, E_n]$ is obtained.

12.2.2. *Bayesian network*

A Bayesian network is a directed acyclic graph that is composed of a set of variables $\{X_1, X_2, \ldots, X_N\}$ and a set of directed edges between the variables [29, 30]. A variable has several possible states, such as true and false. Bayesian networks are very successful in probabilistic knowledge representation and reasoning. In Bayesian networks, the joint probability distribution function of all nodes can be calculated by

$$P(X_1, X_2, \ldots, X_N) = \prod_{i=1}^{N} P(X_i|\mathrm{Pa}_i), \tag{12.4}$$

where Pa_i is the set of random variables whose corresponding nodes are parent nodes of X_i.

A Bayesian network contains two elements, namely structure and parameters. An example shown in Fig. 12.2 is used to illustrate the basic idea of Bayesian networks. In Fig. 12.2, the nodes (X_1, X_2, X_3, X_4) represent random variables, and arcs represent dependence relationships among them. Each arc starts from a parent node and ends at a child node. $\mathrm{Pa}(X)$ represents the parent nodes of node X, therefore, $\mathrm{Pa}(X_2) = \{X_1\}$, $\mathrm{Pa}(X_3) = \{X_1\}$, $\mathrm{Pa}(X_4) = \{X_2, X_3\}$. X_1 is the root node because it has no input arcs. Each node has two states: state 0 and state 1. Root nodes have prior probabilities. Each child node has conditional probabilities based on the combination of states of its parent nodes.

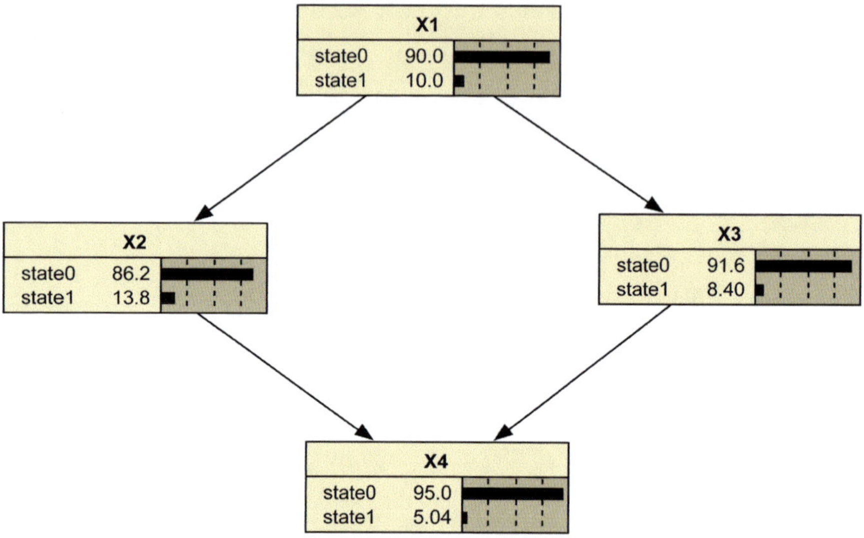

Fig. 12.2. A simple Bayesian network.

12.3. The Proposed Fault Diagnosis Methodology and Its Application

12.3.1. *The proposed methodology*

Fault diagnosis methods in the previous research are mainly based on fault features extracted by some signal processing techniques. In this chapter, a fault diagnosis methodology for a gear pump based on the EEMD method and Bayesian network is proposed. Flow chart of the methodology is shown in Fig. 12.3. To establish the Bayesian network, the investigated faults, fault features from sensor data, and multisource information are integrated into the diagnostic model. According to the evidences obtained from the fault features and multisource information, a fault of the gear pump could be diagnosed.

The proposed methodology consists of fault detection, signal processing, and fault diagnosis. For fault detection, a sensor is responsible for monitoring the vibration signal of the gear pump, and then the data is decomposed by the EEMD method to obtain its IMFs. Fault feature extraction is accomplished by calculating the energy of IMFs according to Eq. (12.3). Actually, the diagnostic model includes three layers: fault layer, fault feature layer, and

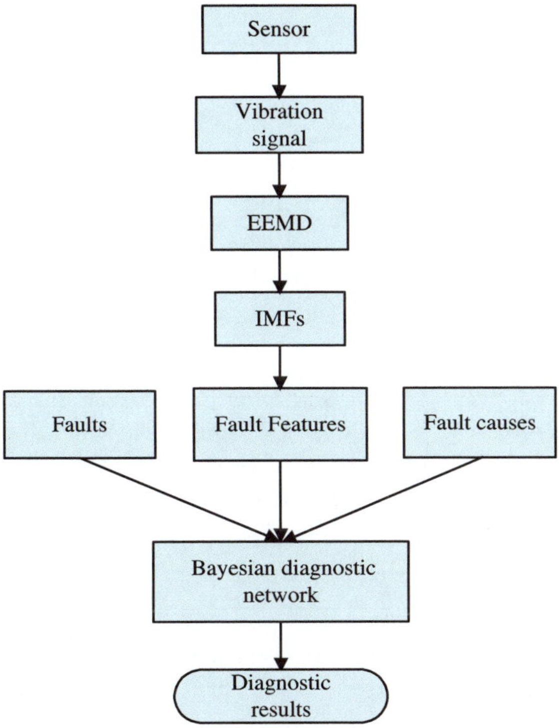

Fig. 12.3. Flow chart of the proposed fault diagnosis methodology.

multisource information layer. Features decomposed by the EEMD method will be entered into the fault feature layer. Fault layer includes the common faults to identify. Obviously, the more faults to diagnose, the more complicated the diagnosis system will be. For multisource information, all the factors such as human observation information, system maintenance information, or abnormal operation records are directly related to the probability of occurrence of the faults. For example, the tooth face wear in the fault layer is less likely to appear if the gear has been replaced by a new one during the recent maintenance.

12.3.2. *Experiment and feature extraction*

The proposed methodology is applied to a gear pump of the type WCB-50. In this chapter, four common fault reasons including tooth face wear (TFW), cavitation (CA), oil pollution (OP), and wear of internal surface of shaft sleeve (WISSS) are investigated. To obtain the data sets, these four fault reasons are simulated respectively. TFW is simulated by grinding one of the meshing surfaces of the driving gear. CA is simulated by loosening the oil pump inlet. By adding pollutants into the working oil, OP is simulated. WISSS is simulated by grinding the internal surface of the shaft sleeve. A piezoelectric acceleration sensor is used for collecting the vibration signal and it is connected to a dynamic test and analysis system of the type DH5923. The sampling frequency is set as 10 kHz. Each fault has 100 training and 50 testing instances. There are 400 training and 200 testing samples in total. Vibration signals of the gear pump with different faults and in normal condition are plotted in Fig. 12.4.

According to the EEMD algorithm and feature extraction process described in Sec. 12.2.1, the vibration signals from different conditions are decomposed. Since the last few IMFs contain very little energy, which are useless for fault diagnosis, only the first eight IMFs are selected for each signal. Therefore, eight features based on the energy of IMFs are calculated, which are used to identify the faults.

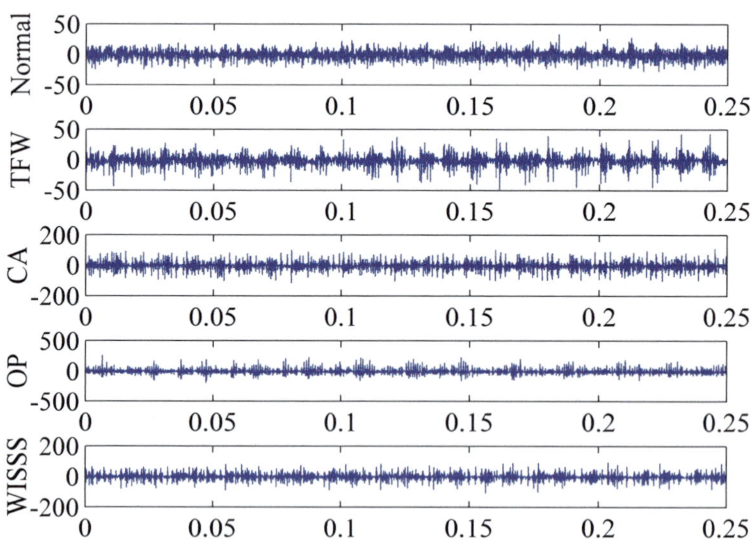

Fig. 12.4. Vibration signals of the gear pump in different conditions.

Table 12.1. Training samples of discretized features of four faults.

Samples	Fea1	Fea2	Fea3	Fea4	Fea5	Fea6	Fea7	Fea8	Conditions
1	4	3	5	5	2	2	4	1	TFW
2	4	3	6	6	1	3	5	1	TFW
3	4	3	6	6	2	3	4	2	TFW
⋮	⋮	⋮	⋮	⋮	⋮	⋮	⋮	⋮	⋮
398	6	1	1	2	1	1	2	1	WISSS
399	6	1	2	2	1	1	1	1	WISSS
400	6	1	1	2	1	1	2	1	WISSS

Before a feature is entered into the Bayesian network, it is discretized according to the range of values of the data samples. Although increasing the number of intervals can improve the accuracy, it will increase the burden for building the Bayesian network. To balance the accuracy and difficulty of developing the Bayesian network, six intervals are determined. After discretization, the extracted feature can be denoted by one of the six numbers $(1, 2, \ldots, 6)$. Table 12.1 shows the training samples of discretized features of four faults. In the table, feature i is denoted by Fea$i$ $(i = 1, 2, \ldots, 8)$. The testing samples can be obtained in the same way.

12.3.3. *Bayesian network structure*

The structure of the Bayesian network is a graphic illustration about the qualitative relationships of nodes in different layers. The Bayesian diagnostic network based on the proposed methodology is shown in Fig. 12.5. The developed Bayesian network has three layers: fault feature layer, fault layer, and multisource information layer. The directed arc denotes that each parent node will cause changes of the child nodes.

The fault layer includes four nodes, indicating the investigated faults. After the nodes are determined, the states of each node should be defined. In the fault layer, each node has two states, namely present and absent, indicating the presence and absence of the fault, respectively. The fault feature layer consists of eight child nodes, indicating eight features extracted from sensor signals using the

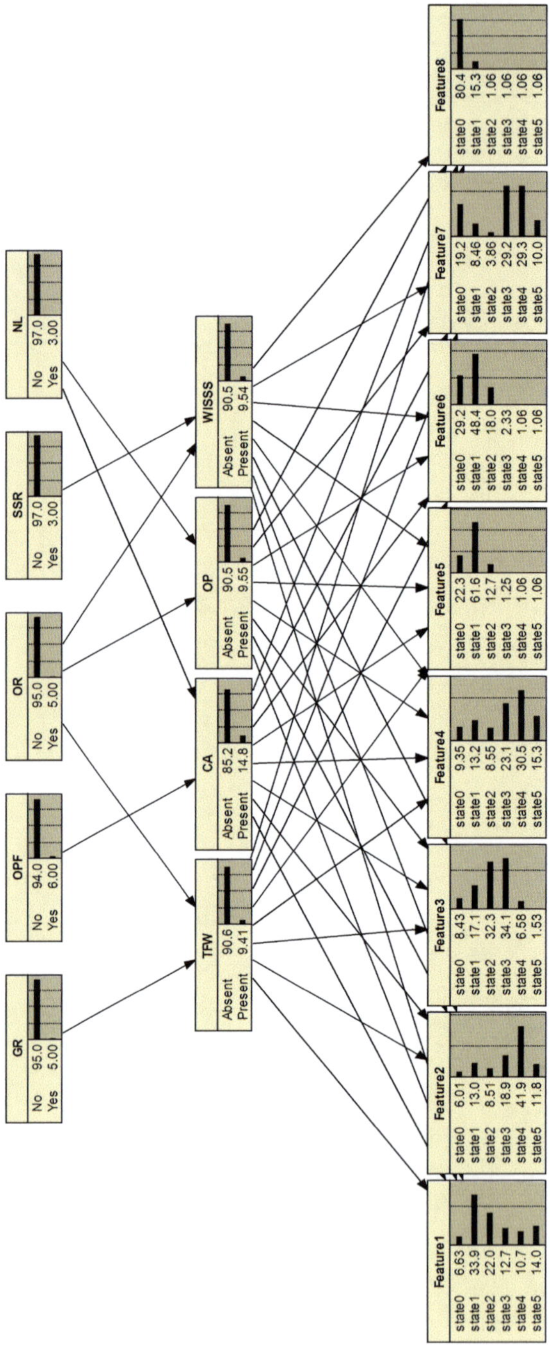

Fig. 12.5. Developed Bayesian network for gear pump.

EEMD method. Each feature node has six states (state1–state6), representing its interval that the energy value of IMF belongs to. Multisource information useful to diagnose the gear pump could be added into the information layer. In this chapter, human observation and repair service information of the gear pump are adopted. Five casual factors are selected, namely gear replacement (GR), oil pipe folding (OPF), oil replacement (OR), shaft sleeve replacement (SSR), and noise level (NL). The events that the nodes represent are listed in Table 12.2. Each node has two states: yes and no.

12.3.4. *Bayesian network parameters*

When the structure of a Bayesian network is established, the prior probabilities and conditional probabilities are required to be specified. A prior probability is the probability that an event occurs without new evidence or information. Usually, prior probabilities are determined by the experts or statistical analysis of historical data. Since historical data is hardly available, prior probabilities are often obtained according to the expert knowledge [16, 17]. It is obvious that the higher the prior probability of an event, the more likely it is to occur. In this chapter, prior probabilities of the nodes in the information layer are determined in Table 12.2.

Table 12.2. Nodes and their states in the multisource information layer.

Node	Event	State	Prior probability (%)
GR	Is the wear replaced by a new one in the last maintenance?	Yes	5
		No	95
OPF	Is the oil pipe of the gear pump folded?	Yes	6
		No	94
OR	Is the oil replaced by new one in the last maintenance?	Yes	5
		No	95
SSR	Is the shaft sleeve replaced by a new one in the last maintenance?	Yes	3
		No	97
NL	Is the noise level high?	Yes	3
		No	97

A conditional probability is the probability that an event occurs for the given new evidence. The conditional probabilities among the nodes in the multisource information layer and the nodes in the fault layer are set according to the knowledge and experience of authors. They are shown in Tables 12.3–12.6.

The conditional probabilities of a child node depend on all possible combinations of the states of its parents. For instance, Feature 1 in the fault feature layer has four parent nodes. Therefore, it has 96 ($6 * 24$) conditional probabilities and the eight feature nodes need 768 parameters in total. To reduce the number of parameters needed, to specify conditional probabilities, noisy-MAX node is

Table 12.3. Conditional probability table of node TFW.

	GR	YES		NO	
Fault	OR	YES	NO	YES	NO
TFW	Present	0.01	0.03	0.05	0.1
	Absent	0.99	0.97	0.95	0.9

Table 12.4. Conditional probability table of node CA.

	OPF	YES		NO	
Fault	NL	YES	NO	YES	NO
CA	Present	0.95	0.9	0.1	0.08
	Absent	0.05	0.1	0.9	0.92

Table 12.5. Conditional probability table of node OP.

	OR	YES		NO	
Fault	NL	YES	NO	YES	NO
OP	Present	0.07	0.05	0.12	0.1
	Absent	0.93	0.95	0.88	0.9

Table 12.6. Conditional probability table of node WISSS.

Fault	SSR / OR	YES: YES	YES: NO	NO: YES	NO: NO
WISSS	Present	0.01	0.03	0.05	0.1
	Absent	0.99	0.97	0.95	0.9

Table 12.7. The conditional probability among nodes in the fault layer and feature layer.

Fault	State	Fea1	Fea2	Fea3	Fea4	Fea5	Fea6	Fea7	Fea8
TFW	1	0.01	0.01	0.01	0.01	0.17	0.3	0.01	0.76
	2	0.01	0.01	0.01	0.01	0.73	0.61	0.01	0.2
	3	0.13	0.61	0.08	0.07	0.07	0.06	0.11	0.01
	4	0.75	0.35	0.51	0.61	0.01	0.01	0.59	0.01
	5	0.09	0.01	0.33	0.25	0.01	0.01	0.26	0.01
	6	0.01	0.01	0.06	0.05	0.01	0.01	0.02	0.01
CA	1	0.01	0.05	0.26	0.76	0.76	0.95	0.95	0.95
	2	0.01	0.86	0.7	0.2	0.2	0.01	0.01	0.01
	3	0.01	0.06	0.01	0.01	0.01	0.01	0.01	0.01
	4	0.01	0.01	0.01	0.01	0.01	0.01	0.01	0.01
	5	0.65	0.01	0.01	0.01	0.01	0.01	0.01	0.01
	6	0.31	0.01	0.01	0.01	0.01	0.01	0.01	0.01
OP	1	0.01	0.01	0.02	0.03	0.04	0.89	0.95	0.81
	2	0.03	0.01	0.69	0.68	0.59	0.07	0.01	0.15
	3	0.3	0.21	0.26	0.26	0.32	0.01	0.01	0.01
	4	0.57	0.58	0.01	0.01	0.03	0.01	0.01	0.01
	5	0.08	0.18	0.01	0.01	0.01	0.01	0.01	0.01
	6	0.01	0.01	0.01	0.01	0.01	0.01	0.01	0.01
WISSS	1	0.01	0.7	0.68	0.01	0.87	0.93	0.08	0.76
	2	0.01	0.26	0.28	0.6	0.09	0.03	0.88	0.2
	3	0.01	0.01	0.01	0.36	0.01	0.01	0.01	0.01
	4	0.01	0.01	0.01	0.01	0.01	0.01	0.01	0.01
	5	0.01	0.01	0.01	0.01	0.01	0.01	0.01	0.01
	6	0.95	0.01	0.01	0.01	0.01	0.01	0.01	0.01

applied [31]. The nodes in the fault feature layer are set as noisy-MAX nodes. Hence, conditional probabilities calculated statistically using the training samples are used as parameters for those noisy-MAX nodes. They are listed in Table 12.7.

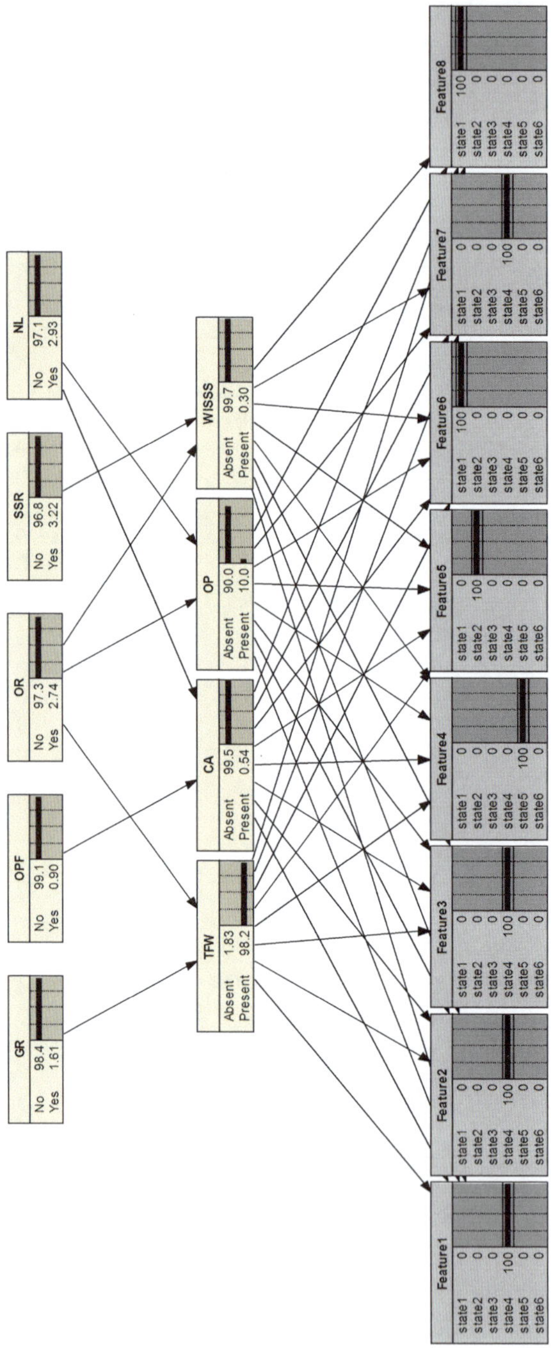

Fig. 12.6. Diagnostic results of a testing sample using only fault features.

12.4. Fault Diagnosis and Discussion

12.4.1. *Fault diagnosis only using fault features*

Take a feature set of TFW faults as a testing sample, $T = \{4, 4, 4, 5, 2, 1, 4, 1\}$, and perform diagnosis only using these features. The diagnostic results are shown in Fig. 12.6. It indicates that the most suspected fault is TFW (98.2%). The diagnostic result is accurate. It demonstrates that the developed Bayesian network only using the sensor data has good performance on identifying the faults.

To test the effectiveness of the model only using evidences from the fault feature layers like other researchers usually did, 200 testing instances are used. Each type of fault has 50 samples. In order to reflect the model superiority, it is necessary to build other models to compare with the proposed model. Recently, some intelligent classification algorithms, such as artificial neural network (ANN) and support vector machine (SVM), have been successfully applied to the intelligent fault diagnosis of mechanical equipment [32–34]. Features of the investigated signals are dealt with ANN or SVM to recognize the health conditions of the objects. In this chapter, ANN and SVM are applied to train and test the same samples as the Bayesian network did. The test results are shown in Table 12.8. It demonstrates that the proposed method based on the Bayesian network achieves the best diagnostic performance. Besides, the average diagnostic accuracies of ANN, SVM, and Bayesian network are 94%, 95%, and 98.5%, respectively. The developed Bayesian network improves the average diagnosis accuracy by 4.5% and 3.5%, respectively, compared with ANN and SVM. The comparison result indicates that the proposed method outperforms the other two common methods in diagnosing different categories of gear pump faults.

Table 12.8. Testing accuracy ANN, SVM and Bayesian network.

Fault	Samples	ANN (%)	SVM (%)	Bayesian network (%)
TFW	50	90	100	100
CA	50	96	100	98
OP	50	98	98	100
WISSS	50	92	82	96

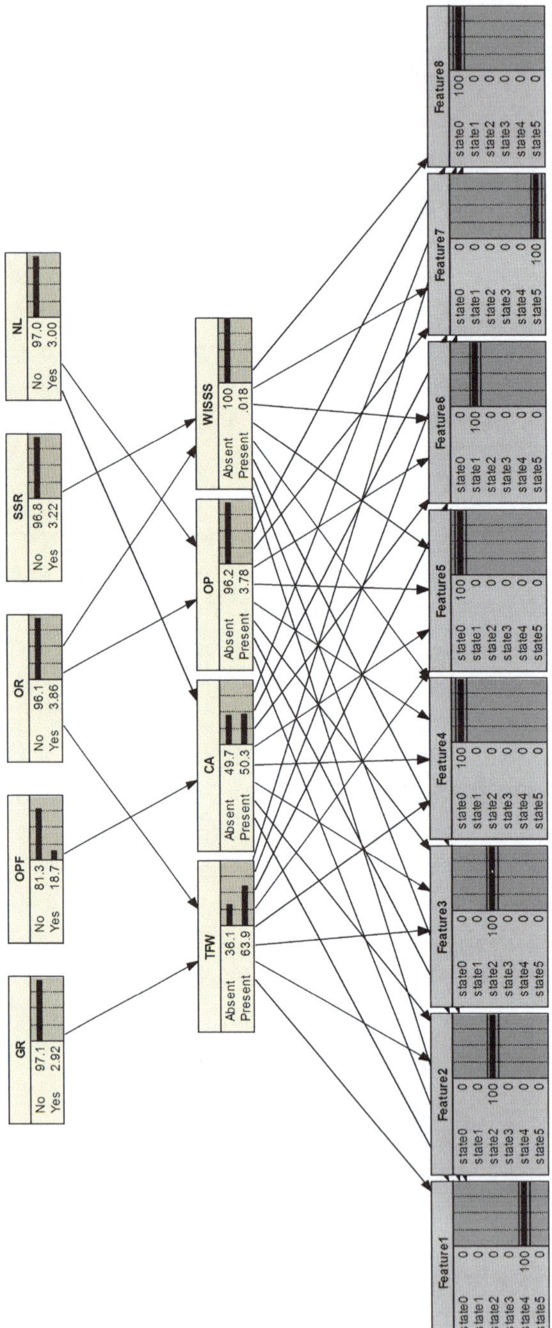

Fig. 12.7. Step 1 of the fault diagnosis for the case.

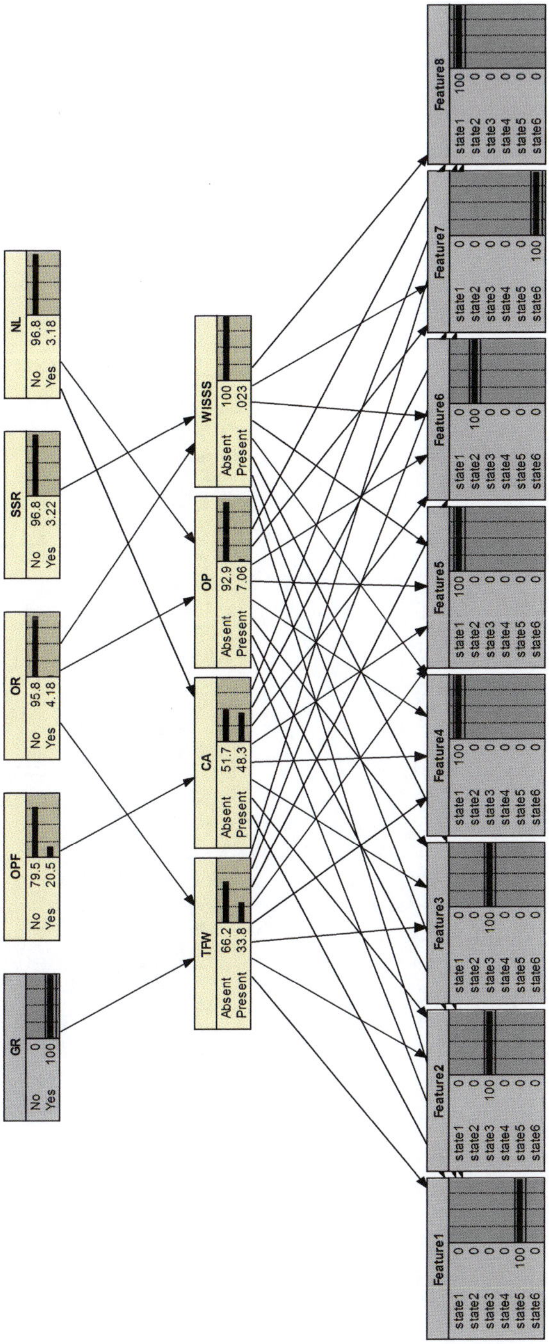

Fig. 12.8. Step 2 of the fault diagnosis for the case.

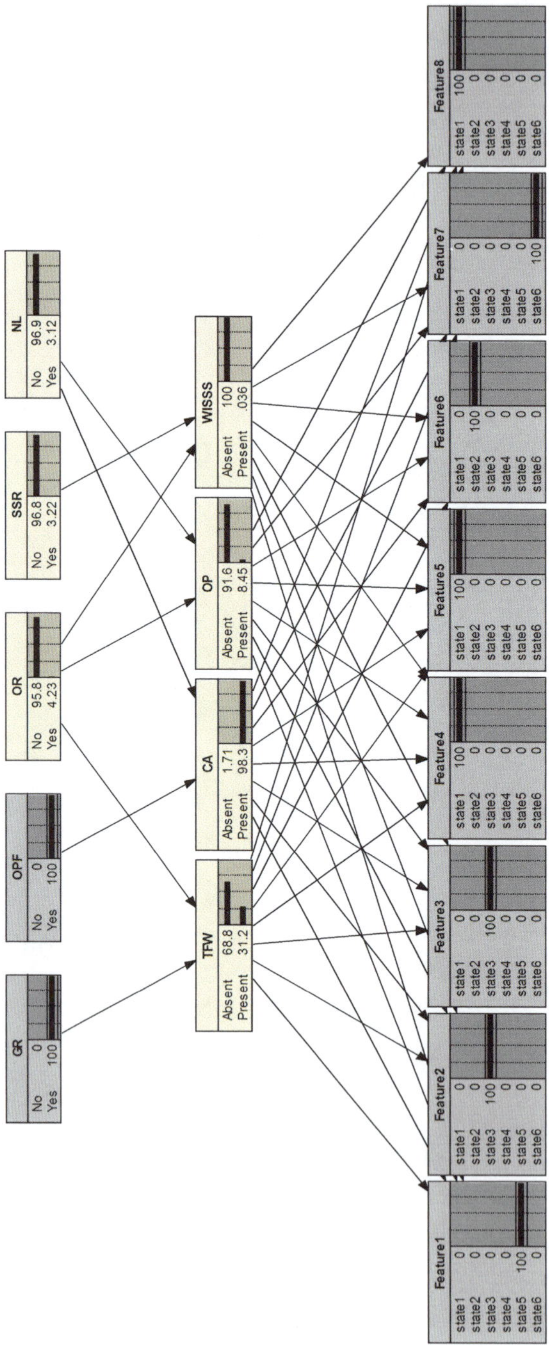

Fig. 12.9. Step 3 of the fault diagnosis for the case.

12.4.2. *Fault diagnosis using fault features and multisource information*

Take a feature set as an example, $T = \{5, 3, 3, 1, 1, 2, 6, 1\}$, and perform diagnosis using the developed Bayesian diagnostic network. The steps are shown in Figs. 12.7–12.9. At step 1 in Fig. 12.7, only fault features are applied for fault diagnosis. Probabilities of the presence of TFW, CA, OP, and WISSS are 64%, 46.5%, 3.83%, and 0.017%, respectively. So, it is hard to tell whether TFW or CA occurs without other useful information. At step 2 in Fig. 12.8, a new evidence (GR is yes) in the information layer is added. The most suspected fault is CA (48.3%). The fault probability of TFW (33.8%) is decreased under the new evidence. As shown in Fig. 12.9, at step 3 (added evidence: OPF is yes), the fault probability of CA increases from 48.3% to 98.3%. It is because the new evidence is a unique cause for CA. It demonstrates that the multisource information is helpful to fault diagnosis.

12.5. Conclusion

The main contribution of this chapter is that a methodology based on the Bayesian network and EEMD for fault diagnosis is presented. The advantages of Bayesian network and EEMD are integrated. Compared with the other conventional fault diagnosis methods, the presented methodology is able to make use of more useful information besides sensor signals. Essentially, the presented scheme based on EEMD and Bayesian network is a multisource information fusion-based methodology. With the proposed three-layered Bayesian network framework, some useful information (including naked eyes inspection, maintenance records, etc.) can be helpful to identify the fault. The proposed method has been applied to the fault diagnosis of a gear pump and it is effective and efficient based on vibration signals and other information.

The proposed methodology is applied to fault diagnosis of a gear pump. The developed diagnostic Bayesian network has three layers, namely fault feature layer, fault layer, and multisource information layer. Sensor signals and other helpful information for diagnosis could be added into the networks.

When fault features extracted from the EEMD method are only used, the developed model has better diagnostic performance than the ANN and SVM classification algorithms. It improves the average diagnosis accuracy by 4.5% and 3.5%, respectively, compared with ANN and SVM.

Sometimes, it may be hard to distinguish the faults only based on the sensor signals. A case study has demonstrated that some information from human observation or system maintenance records is very helpful for fault diagnosis.

References

[1] Y. G. Lei, Z. J. He, Y. Y. Zi, X. F. Chen, "New clustering algorithm-based fault diagnosis using compensation distance evaluation technique," *Mechanical Systems and Signal Processing*, vol. 22, pp. 419–435, 2008.
[2] F. Al-Badour, M. Sunar, L. Cheded, "Vibration analysis of rotating machinery using time-frequency analysis and wavelet techniques," *Mechanical Systems and Signal Processing*, vol. 25, pp. 2083–2101, 2011.
[3] V. Muralidharan, V. Sugumaran, "Feature extraction using wavelets and classification through decision tree algorithm for fault diagnosis of mono-block centrifugal pump," *Measurement: Journal of the International Measurement Confederation*, vol. 46, pp. 353–359, 2013.
[4] J. Jing, G. Meng,"A novel method for multi-fault diagnosis of rotor system," *Mechanism and Machine Theory*, vol. 44, pp. 697–709, 2009.
[5] F. Peng, D. Yu, J. Luo, "Sparse signal decomposition method based on multi-scale chirplet and its application to the fault diagnosis of gearboxes," *Mechanical Systems and Signal Processing*, vol. 25, pp. 549–557, 2011.
[6] J. B. Ali, N. Fnaiech, L. Saidi, B. Chebel-Morello, F. Fnaiech, "Application of empirical mode decomposition and artificial neural network for automatic bearing fault diagnosis based on vibration signals," *Applied Acoustics*, vol. 89, pp. 16–27, 2015.
[7] M. J. Ehrhardt, H. Villinger, S. Schiffler, "Evaluation of decomposition fools for sea floor pressure data: A practical comparison of modern and classical approaches," *Computers & Geosciences*, vol. 45, pp. 4–12, 2012.
[8] M. Van, H. J. Kang, K. S. Shin, "Rolling element bearing fault diagnosis based on non-local means de-noising and empirical mode decomposition," *IET Science, Measurement &Technology*, vol. 8, pp. 571–578, 2014.
[9] M. Amarnath, I. R. Praveen Krishna, "Local fault detection in helical gears via vibration and acoustic signals using EMD based statistical parameter

analysis," *Measurement: Journal of the International Measurement Confederation*, vol. 58, pp. 154–164, 2014.

[10] M. Amarnath, I. R. Praveen Krishna, "Detection and diagnosis of surface wear failure in a spur geared system using EEMD based vibration signal analysis," *Tribology International*, vol. 61, pp. 224–234, 2013.

[11] Z. H. Wu, N. E. Huang, "Ensemble empirical mode decomposition: A noise assisted data analysis method," *Advances in Adaptive Data Analysis*, vol. 1, pp. 1–41, 2009.

[12] Y. Amirat, V. Choqueuse, M. Benbouzid, "EEMD-based wind turbine bearing failure detection using the generator stator current homopolar component," *Mechanical Systems and Signal Processing*, vol. 41, pp. 667–678, 2013.

[13] Y. G. Lei, Z. J. He, Y. Y. Zi, "Application of the EEMD method to rotor fault diagnosis of rotating machinery," *Mechanical Systems and Signal Processing*, vol. 23, pp. 1327–1338, 2009.

[14] W. Caesarendra, P. B. Kosasih, A. K. Tieu, C. A. S. Moodie, B. K. Choi, "Condition monitoring of naturally damaged slow speed slewing bearing based on ensemble empirical mode decomposition," *Journal of Mechanical Science and Technology*, vol. 27, pp. 2253–2262, 2013.

[15] H. Mahgoun, R. E. Bekka, A. Felkaoui, "Gearbox fault diagnosis using ensemble empirical mode decomposition (EEMD) and residual signal," *Mechanics & Industry*, vol. 13, pp. 33–44, 2012.

[16] Y. Zhao, F. Xiao, S. W. Wang, "An intelligent chiller fault detection and diagnosis methodology using Bayesian belief network," *Energy and Buildings*, vol. 57, pp. 278–288, 2013.

[17] F. Xiao, Y. Zhao, J. Wen, S. W. Wang, "Bayesian network based FDD strategy for variable air volume terminal," *Automation in Construction*, vol. 41, pp. 106–118, 2014.

[18] B. G. Xu, "Intelligent fault inference for rotating flexible rotors using Bayesian belief network," *Expert Systems with Applications*, vol. 39, pp. 816–822, 2012.

[19] B. P. Cai *et al.*,"Multi-source information fusion based fault diagnosis of ground-source heat pump using Bayesian network," *Applied Energy*, vol. 114, pp. 1–9, 2014.

[20] L. Oukhellou, E. Come, L. Bouillaut, P. Aknin, "Combined use of sensor data and structural knowledge processed by Bayesian network: Application to a railway diagnosis aid scheme," *Transportation Research Part C: Emerging Technologies*, vol. 16, pp. 755–767, 2008.

[21] A. Onisko, M. J. Druzdzel, H. Wasyluk, "Learning Bayesian network parameters form small data sets: Application of noisy-OR gate,"*International Journal of Approximate Reasoning*, vol. 27, pp. 165–182, 2001.

[22] Y. Sun, Y. Y. Tang, S. X. Ding, S. P. Lv, Y. F. Cui, "Diagnose the mild cognitive impairment by constructing Bayesian network with missing data," *Expert Systems with Applications*, vol. 38, pp. 442–449, 2011.

[23] R. Barco, L. Diez, V. Wille, P. Lazaro, "Automatic diagnosis of mobile communication networks under imprecise parameters," *Expert Systems with Applications*, vol. 36, pp. 489–500, 2009.

[24] F. Sahin, M. C. Yavuz, Z. Arnavut, O. Uluyol, "Fault diagnosis for airplane engines using Bayesian networks and distributed particle swarm optimization," *Parallel Computing*, vol. 33, pp. 124–143, 2007.

[25] G. Kariv, V. Shani, E. Goldberg, L. Leibovici, M. Paul, "A model for diagnosis of pulmonary infections in solid-organ transplant recipients," *Computer Methods and Programs in Biomedicine*, vol. 104, pp. 135–142, 2011.

[26] P. Weber, G. Medina-Oliva, C. Simon, B. Iung, "Overview on Bayesian networks applications for dependability, risk analysis and maintenance areas," *Engineering Applications of Artificial Intelligence*, vol. 25, pp. 671–682, 2012.

[27] X. Wang, C. W. Liu, F. R. Bi, X. Y. Bi, K. Shao, "Fault diagnosis of diesel engine based on adaptive wavelet packets and EEMD-fractal dimension," *Mechanical Systems and Signal Processing*, vol. 41, pp. 581–597, 2013.

[28] N. E. Huang, Z. Shen, S. R. Long, "The empirical mode decomposition and the Hilbert spectrum for nonlinear and non-stationary time series analysis," *Proceedings of the Royal Society of London A*, vol. 454, pp. 903–995, 1998.

[29] I. Maglogiannis, E. Zafiropoulos, A. Platis, C. Lambrinoudakis, "Risk analysis of a patient monitoring system using Bayesian network modeling," *Journal of Biomedical Informatics*, vol. 39, pp. 637–647, 2006.

[30] G. Feng, J. D. Zhang, S. Y. Liao Stephen, "A novel method for combining Bayesian networks, theoretical analysis, and its applications," *Pattern Recognition*, vol. 47, pp. 2057–2069, 2014.

[31] A. Zagorecki, M. J. Druzdzel, "Knowledge engineering for Bayesian networks: How common are noisy-MAX distributions in practice?" *IEEE Transactions on Systems, Man, and Cybernetics — Part A: Systems and Humans*, vol. 43, pp. 186–195, 2013.

[32] B. A. Paya, I. I. Esat, M. N. M. Badi, "Artificial neural network based fault diagnostics of rotating machinery using wavelet transforms as a preprocessor," *Mechanical Systems and Signal Processing*, vol. 11, pp. 751–765, 1997.

[33] B. Samanta, K. R. Al-Balushi, "Artificial neural network based fault diagnostics of rolling element bearings using time-domain features," *Mechanical Systems and Signal Processing*, vol. 17, pp. 317–328, 2003.

[34] B. Samanta, "Gear fault detection using artificial neural networks and support vector machines with genetic algorithms," *Mechanical Systems and Signal Processing*, vol. 18, pp. 625–644, 2004.

Index